Building a Programmable Logic Controller with a PIC16F648A Microcontroller

CRC Press
Taylor & Francis Group
6000 Broken Sound Parkway NW, Suite 300
Boca Raton, FL 33487-2742

First issued in paperback 2019

ISBN-13: 978-1-4665-8985-8 (hbk)
ISBN-13: 978-0-367-37953-7 (pbk)

Visit the Taylor & Francis Web site at
http://www.taylorandfrancis.com

and the CRC Press Web site at
http://www.crcpress.com

Building a Programmable Logic Controller with a PIC16F648A Microcontroller

Murat Uzam

CRC Press is an imprint of the
Taylor & Francis Group, an **informa** business

To my parents and family

who love and support me

and

to my teachers and students

who enriched my knowledge

Contents

Preface

Programmable logic controllers (PLCs) have been used extensively in industry for the past five decades. PLC manufacturers offer different PLCs in terms of functions, program memories, and the number of inputs/outputs (I/Os), ranging from a few to thousands of I/Os. The design and implementation of PLCs have long been a secret of the PLC manufacturers. Recently, a serious work was reported by the author of this book to describe a microcontroller-based implementation of a PLC. With a series of 22 articles published in *Electronics World* magazine (http//www.electronicsworld.co.uk/) between the years 2008 and 2010, the design and implementation of a PIC16F648A-based PLC were described. This book is based on an improved version of the project reported in *Electronics World* magazine.

This book is written for advanced students, practicing engineers, and hobbyists who want to learn how to design and use a microcontroller-based PLC. The book assumes the reader has taken courses in digital logic design, microcontrollers, and PLCs. In addition, the reader is expected to be familiar with the PIC16F series of microcontrollers and to have been exposed to writing programs using PIC assembly language within an MPLAB integrated development environment.

The CD-ROM that accompanies this book contains all the program source files and hex files for the examples described in the book. In addition, PCB files of the CPU and I/O extension boards of the PIC16F648A-based PLC are also included on the CD-ROM.

Dr. Murat Uzam
Melikşah Üniversitesi
Mühendislik-Mimarlık Fakültesi
Elektrik-Elektronik Mühendisliği Bölümü
Talas, Kayseri
Turkey

Acknowledgments

I am grateful to Dr. Gökhan Gelen (gokhan_gelen@hotmail.com) for his great effort in drawing the printed circuit boards (PCBs) and for producing the prototypes of the CPU board and the I/O extension board. Without his help this project may have been delayed for years.

Background and Use of the Book

This project was completed during the search for an answer to the following question: How could one design and implement a programmable logic controller (PLC)? The answer to this question was partially discovered about 15 years ago by the author in a freely available PLC project called PICBIT. The file, called picbit.inc of PICBIT, contains the basic PLC macro definitions. The PIC16F648A-based PLC project has been completed by the inspiration of these macros. Of course many new features have been included within the PIC16F648A-based PLC project to make it an almost perfect PLC. The reader should be aware that this project does not include graphical interface PC software as in PICBIT or in other PLCs for developing PLC programs. Rather, PLC programs are developed by using macros as done in the Instruction List (IL) PLC programming language. An interested and skilled reader could well (and is encouraged to) develop graphical interface PC software for easy use of the PIC16F648A-based PLC.

The PIC16F648A-based PLC project was first reported in a series of 22 articles published in *Electronics World* magazine (http://www.electronicsworld.co.uk/) between the years 2008 and 2010 [1–22]. All details of this project can be viewed at http//www.meliksah.edu.tr/muzam/UZAM_PLC_with_PIC16F648A.htm [23]. This book is based on an improved version of the project reported in *Electronics World* magazine. The improvements are summarized as follows:

1. The current hardware has two boards: the CPU board and the I/O extension board. In the previous version of the hardware, the main board consisted of the CPU board and eight inputs/eight outputs, while in the current version, the CPU board excludes eight inputs/eight outputs. Thus, the CPU board is smaller than the previous main board. In addition, the current I/O extension board is also smaller than in the previous version.
2. The hardware explained in this book consists of one CPU board and two I/O extension boards. Therefore, the current version of the software supports 16 inputs and 16 outputs, while the previous one supported 8 inputs and 8 outputs.
3. Clock frequency was 4 MHz in the previous version, but is 20 MHz in the current version.
4. Some of the macros are improved compared with the previous versions.
5. Flowcharts are provided to help the understanding of all macros (functions).

In order to properly follow the topics explained in this book, it is expected that the reader will construct his or her PIC16F648A-based PLC consisting of the CPU board and two I/O extension boards using the PCB files provided within the CD-ROM attached to this book. In this book, as the PIC assembly is used as the programming language within the MPLAB integrated development environment (IDE), the reader is referred to the homepage of Microchip (http://www.microchip.com/) to obtain the latest version of MPLAB IDE. References [24] and [25] may be useful to understand some aspects of the PIC16F648A microcontroller and MPASM™ assembler, respectively.

The contents of the book's 15 chapters are explained briefly, as follows:

1. **Hardware:** In this chapter, the hardware structure of the PIC16F648A-based PLC, consisting of 16 discrete inputs and 16 discrete outputs, is explained in detail.
2. **Basic software:** This chapter explains the basic software structure of the PIC16F648A-based PLC. A PLC scan cycle includes the following: obtain the inputs, run the user program, and update the outputs. In addition, it is also necessary to define and initialize all variables used within a PLC. Necessary functions are all described as PIC assembly macros to be used in the PIC16F648A-based PLC. The macros described in this chapter can be summarized as follows: `HC165` (for handling the inputs), `HC595` (for sending the outputs), `dbncr0` and `dbncr1` (for debouncing 16 inputs), `initialize`, `get_inputs`, and `send_outputs`.
3. **Contact and relay-based macros:** The following contact and relay-based macros are described in this chapter: `ld` (load), `ld_not` (load_not), `not`, `or`, `or_not`, `nor`, `and`, `and_not`, `nand`, `xor`, `xor_not`, `xnor`, `out`, `out_not`, `in_out`, `inv_out`, `_set`, `_reset`. These macros are defined to operate on 1-bit (Boolean) variables.
4. **Flip-flop macros:** The following flip-flop–based macros are described in this chapter: `r_edge` (rising edge), `f_edge` (falling edge), `latch0`, `latch1`, `dff_r` (rising edge triggered D flip-flop), `dff_f` (falling edge triggered D flip-flop), `tff_r` (rising edge triggered T flip-flop), `tff_f` (falling edge triggered T flip-flop), `jkff_r` (rising edge triggered JK flip-flop), and `jkff_f` (falling edge triggered JK flip-flop).
5. **Timer macros:** The following timer macros are described in this chapter: `TON_8` (8-bit on-delay timer), `TOF_8` (8-bit off-delay timer), `TP_8` (8-bit pulse timer), and `TOS_8` (8-bit oscillator timer).
6. **Counter macros:** The following counter macros are described in this chapter: `CTU_8` (8-bit up counter), `CTD_8` (8-bit down counter), and `CTUD_8` (8-bit up/down counter).

7. **Comparison macros:** The comparison macros are described in this chapter. The contents of two registers (R1 and R2) are compared according to the following: `GT` (greater than, >), `GE` (greater than or equal to, ≥), `EQ` (equal to, =), `LT` (less than, <), `LE` (less than or equal to, ≤), and `NE` (not equal to, ≠). Similar comparison macros are also described for comparing the contents of an 8-bit register (R) with an 8-bit constant (K).
8. **Arithmetical macros:** The arithmetical macros are described in this chapter. The following operators are applied to the contents of two registers (R1 and R2): ADD, SUB (subtract), INC (increment), and DEC (decrement). Similar arithmetical macros are also described, to be used with the contents of an 8-bit register (R) and an 8-bit constant (K).
9. **Logical macros:** The following logical macros are described in this chapter: `inv_R`, `AND`, `NAND`, `OR`, `NOR`, `XOR`, and `XNOR`. These macros are applied to an 8-bit register (R1) with another register (R2) or an 8-bit constant (K).
10. **Shift and rotate macros:** The following shift and rotate macros are described in this chapter: `SHIFT_R` (shift right the content of register R), `SHIFT_L` (shift left the content of register R), `ROTATE_R` (rotate right the content of register R), `ROTATE_L` (rotate left the content of register R), and `SWAP` (swap the nibbles of a register).
11. **Multiplexer macros:** The following multiplexer macros are described in this chapter: `mux_2_1` (2×1 MUX), `mux_2_1_E` (2×1 MUX with enable input), `mux_4_1` (4×1 MUX), `mux_4_1_E` (4×1 MUX with enable input), `mux_8_1` (8×1 MUX), and `mux_8_1_E` (8×1 MUX with enable input).
12. **Demultiplexer macros:** The following demultiplexer macros are described in this chapter: `Dmux_1_2` (1×2 DMUX), `Dmux_1_2_E` (1×2 DMUX with enable input), `Dmux_1_4` (1×4 DMUX), `Dmux_1_4_E` (1×4 DMUX with enable input), `Dmux_1_8` (1×8 DMUX), and `Dmux_1_8_E` (1×8 DMUX with enable input).
13. **Decoder macros:** The following decoder macros are described in this chapter: `decod_1_2` (1×2 decoder), `decod_1_2_AL` (1×2 decoder with active low outputs), `decod_1_2_E` (1×2 decoder with enable input), `decod_1_2_E_AL` (1×2 decoder with enable input and active low outputs), `decod_2_4` (2×4 decoder), `decod_2_4_AL` (2×4 decoder with active low outputs), `decod_2_4_E` (2×4 decoder with enable input), `decod_2_4_E_AL` (2×4 decoder with enable input and active low outputs), `decod_3_8` (3×8 decoder), `decod_3_8_AL` (3×8 decoder with active low outputs), `decod_3_8_E` (3×8 decoder with enable input), and

decod_3_8_E_AL (3×8 decoder with enable input and active low outputs).

14. **Priority encoder macros:** The following priority encoder macros are described in this chapter: encod_4_2_p (4×2 priority encoder), encod_4_2_p_E (4×2 priority encoder with enable input), encod_8_3_p (8×3 priority encoder), encod_8_3_p_E (8×3 priority encoder with enable input), encod_dec_bcd_p (decimal to binary coded decimal [BCD] priority encoder), and encod_dec_bcd_p_E (decimal to BCD priority encoder with enable input).
15. **Application example:** This chapter describes an example remotely controlled model gate system and makes use of the PIC16F648A-based PLC to control it for different control scenarios.

Table 1 shows the general characteristics of the PIC16F648A-based PLC.

TABLE 1

General Characteristics of the PIC16F648A-Based PLC

Inputs/Outputs/Functions	Byte Addresses/ Related Bytes	Bit Addresses or Function Numbers
16 discrete inputs (external inputs: 5 or 24 V DC)	I0 I1	I0.0, I0.1, …, I0.7 I1.0, I1.1, …, I1.7
16 discrete outputs (relay type outputs)	Q0 Q1	Q0.0, Q0.1, …, Q0.7 Q1.0, Q1.1, …, Q1.7
32 internal relays (memory bits)	M0 M1 M2 M3	M0.0, M0.1, …, M0.7 M1.0, M1.1, …, M1.7 M2.0, M2.1, …, M2.7 M3.0, M3.1, …, M3.7
8 rising edge detectors	RED	r_edge (0, 1, …, 7)
8 falling edge detectors	FED	f_edge (0, 1, …, 7)
8 rising edge triggered D flip-flop	DFF_RED	dff_r (0, 1, …, 7), regi,biti, rego,bito
8 falling edge triggered D flip-flop	DFF_FED	dff_f (0, 1, …, 7), regi,biti, rego,bito
8 rising edge triggered T flip-flop	TFF_RED	tff_r (0, 1, …, 7), regi,biti, rego,bito
8 falling edge triggered T flip-flop	TFF_FED	tff_f (0, 1, …, 7), regi,biti, rego,bito
8 rising edge triggered JK flip-flop	JKFF_RED	jkff_r (0, 1, …, 7), regi,biti, rego,bito
8 falling edge triggered JK flip-flop	JKFF_FED	jkff_f (0, 1, …, 7), regi,biti, rego,bito

TABLE 1 (CONTINUED)

General Characteristics of the PIC16F648A-Based PLC

Inputs/Outputs/Functions	Byte Addresses/ Related Bytes	Bit Addresses or Function Numbers
8 on-delay timers	TON8, TON8+1, ..., TON8+7 TON8_Q TON8_RED	TON8_Q0 TON8_Q1, ... TON8_Q7
8 off-delay timers	TOF8, TOF8+1, ..., TOF8+7, TOF8_Q TOF8_RED	TOF8_Q0 TOF8_Q1, ... TOF8_Q7
8 pulse timers	TP8, TP8+1, ..., TP8+7, TP8_Q TP8_RED1 TP8_RED2	TP8_Q0 TP8_Q1, ... TP8_Q7
8 oscillator timers	TOS8, TOS8+1, ..., TOS8+7 TOS8_Q TOS8_RED	TOS8_Q0 TOS8_Q1, ... TOS8_Q7
8 counters CTU: up counter CTD: down counter CTUD: up/down counter	CV8, CV8+1, ..., CV8+7 CTU8_Q CTU8_RED CTD8_Q CTD8_RED CTUD8_Q CTUD8_RED	CTU8_Q0 CTU8_Q1, ... CTU8_Q7 or CTD8_Q0 CTD8_Q1, ... CTD8_Q7 or CTUD8_Q0 CTUD8_Q1, ... CTUD8_Q7

Note: `regi`, `biti`, input bit; `rego`, `bito`, output bit.
At any time, a total of eight different counters can be used.

About the Author

Murat Uzam was borned in Söke, Turkey, in 1968. He received his BSc and MSc degrees from the Electrical Engineering Department of Yıldız Technical University, İstanbul, Turkey, in 1989 and 1991, respectively. He received his PhD degree from the University of Salford, Salford, UK, in 1998. He is currently a professor in the Department of Electrical and Electronics Engineering at Melikşah University in Kayseri, Turkey.

Dr. Uzam's research interests include the design and implementation of discrete event control systems modeled by Petri nets (PN) and, in particular, deadlock prevention/liveness enforcement in flexible manufacturing systems, Programmable Logic Controllers (PLCs), microcontrollers (especially PIC microcontrollers), and the design of microcontroller-based PLCs. The details of his studies are accessible from his web page: http://www.meliksah.edu.tr/muzam.

1

Hardware of the PIC16F648A-Based PLC

The hardware of the PIC16F648A-based programmable logic controller (PLC) consists of two parts: the *CPU board* and the *I/O extension board*. The schematic diagram and the photograph of the PIC16F648A-based PLC CPU board are shown in Figures 1.1 and 1.2, respectively. The CPU board contains mainly three sections: power, programming, and CPU (central processor unit).

The *power section* accepts 12 V AC input and produces two DC outputs: 12 V DC, to be used as the operating voltage of relays, and 5 V DC, to be used for ICs, inputs, etc. The *programming section* deals with the programming of the PIC16F648A microcontroller. For programming the PIC16F648A in circuit, it is necessary to use PIC programmer hardware and software with In Circuit Serial Programming (ICSP) capability. For related hardware and software to be used for programming the PIC16F648A-based PLC, please visit the following web page: http://www.meliksah.edu.tr/muzam/. For other types of USB, serial, or parallel port PIC programmers the reader is expected to make necessary arrangements. The ICSP connector takes the lines VPP(MCLR), VDD, VSS(GND), DATA (RB7), and CLOCK (RB6) from the PIC programmer hardware through a properly prepared cable, and it connects them to a four-pole double-throw (4PDT) switch. There are two positions of the 4PDT switch. As seen from Figure 1.1, in one position of the 4PDT switch, PIC16F648A is ready to be programmed, and in the other position the loaded program is run. For properly programming the PIC16F648A by means of a PIC programmer and the 4PDT switch, it is also a necessity to *switch off* the power switch. The *CPU section* consists of the PIC16F648A microcontroller. In the project reported in this book, the PLC is fixed to run at 20 MHz with an external oscillator. This frequency is fixed because time delays are calculated based on this speed. By means of two switches, SW1 and SW2, it is also possible to use another internal or external oscillator with different crystal frequencies. When doing so, time delay functions must be calculated accordingly. SW3 connects the RA5 pin either to one pole of the 4PDT switch or to the future extension connector. When programming PIC16F648A, RA5 should be connected to the 4PDT switch. RB0, RB6, and RB7 pins are all reserved to be used for 8-bit parallel-to-serial converter register 74HC/LS165. Through these three pins and with added 74HC/LS165 registers, we can describe as many inputs as necessary. RB0, RB6, and RB7 are the `data in`, `clock in`, and `shift/load` pins, respectively. Similarly, RB3, RB4, and RB5 pins are all reserved to be used for 8-bit serial-to-parallel converter register/driver TPIC6B595. Through these three pins and with added TPIC6B595 registers,

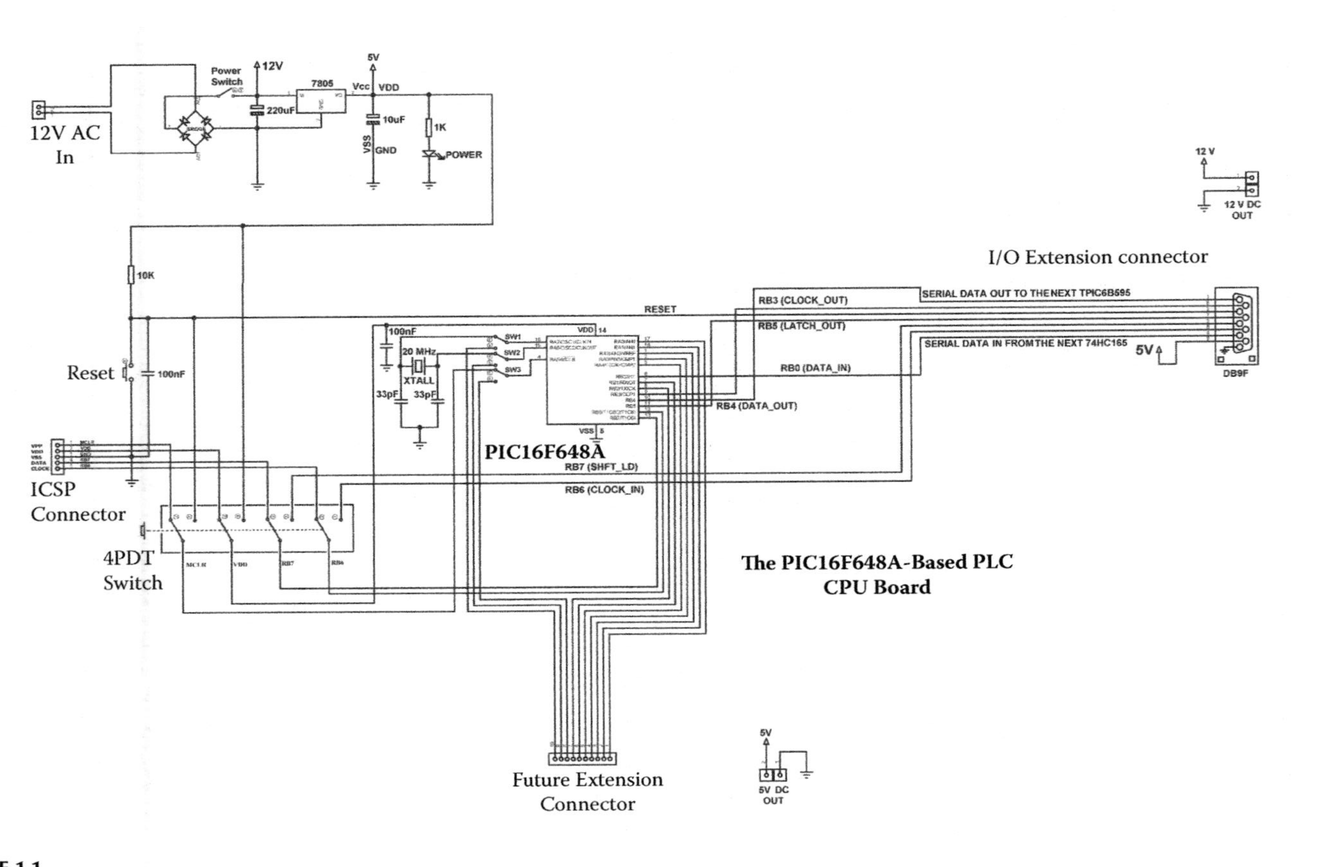

FIGURE 1.1
Schematic diagram of the CPU board.

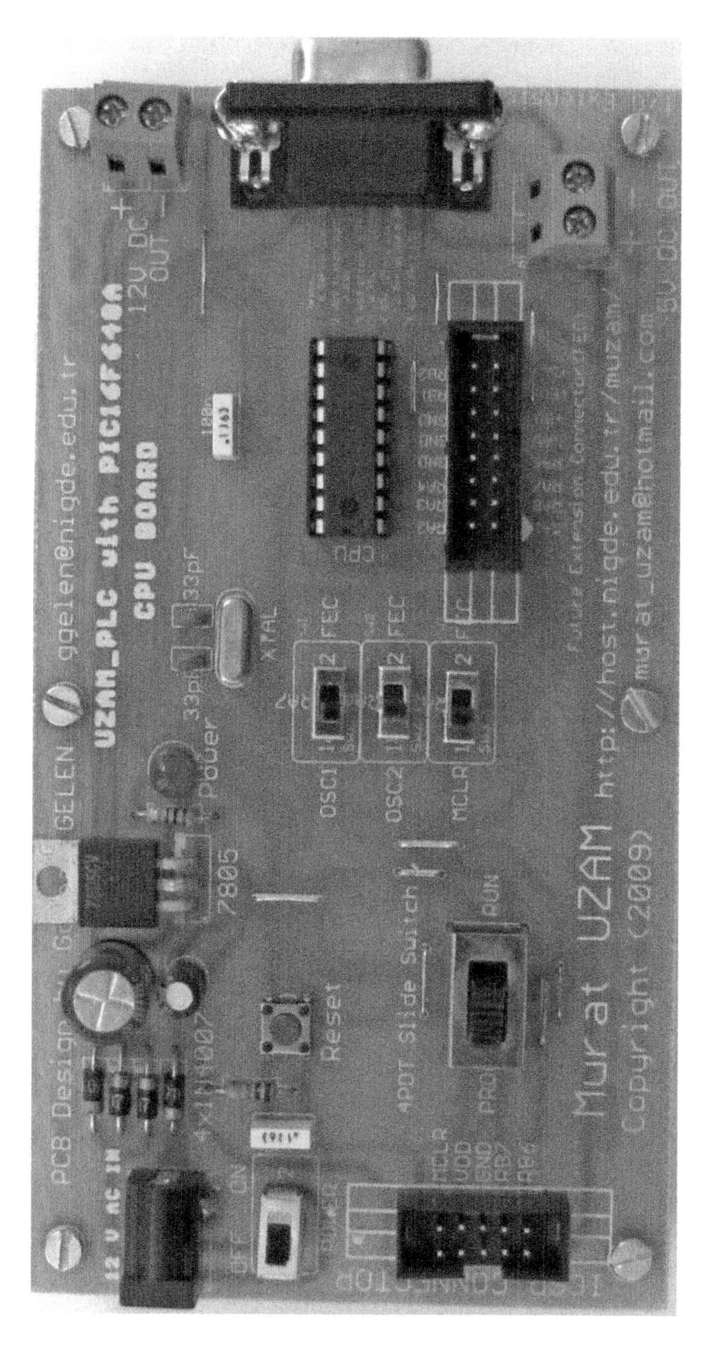

FIGURE 1.2
Photograph of the CPU board.

we can describe as many outputs as necessary. RB3, RB4, and RB5 are the `clock out`, `data out`, and `latch out` pins, respectively. The remaining unused pins of the PIC16F648A are connected to the future extension connector. PIC16F648A provides the following: flash program memory (words), 4096; RAM data memory (bytes), 256; and EEPROM data memory (bytes), 256. The PIC16F648A-based PLC macros make use of registers defined in RAM data memory. Note that it may be possible to use PIC16F628A as the CPU, but one has to bear in mind that PIC16F628A provides the following: flash program memory (words), 2048; RAM data memory (bytes), 224; and EEPROM data memory (bytes), 128. In that case, it is necessary to take care of the usage of RAM data memory.

Figures 1.3 and 1.4 show the schematic diagram and photograph of the I/O extension board, respectively. The I/O extension board contains mainly two sections: eight discrete inputs and eight discrete outputs. The I/O extension connector DB9M seen on the left connects the I/O extension board to the CPU board or to a previous I/O extension board. Similarly, the I/O extension

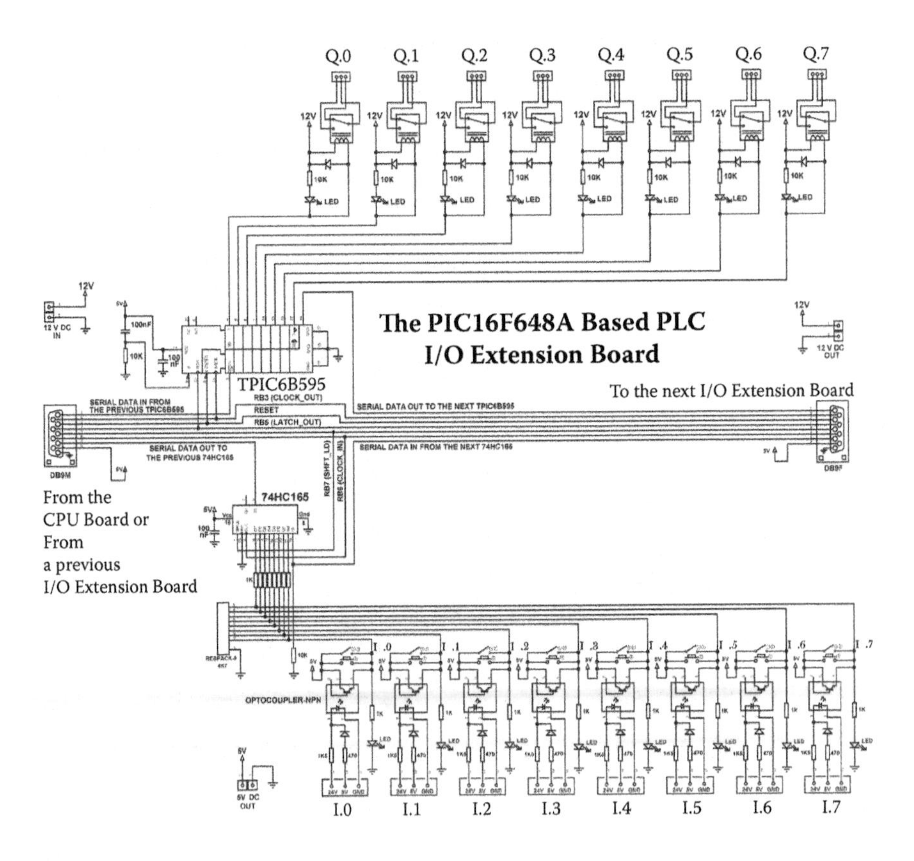

FIGURE 1.3
Schematic diagram of the I/O extension board.

FIGURE 1.4
Photograph of the I/O extension board.

connector DB9F seen on the right connects the I/O extension board to a next I/O extension board. In this way we can connect as many I/O extension boards as necessary. Five-volt DC and 12 V DC are taken from the CPU board or from a previous I/O extension board, and they are passed to the next I/O extension boards. All I/O data are sent to and taken from all the connected extension I/O boards by means of I/O extension connectors DB9M and DB9F.

The *inputs section* introduces eight discrete inputs for the PIC16F648A-based PLC (called I0.0, I0.1, …, I0.7 for the first I/O extension board). Five-volt DC or 24 V DC input signals can be accepted by each input. These external input signals are isolated from the other parts of the hardware by using NPN type opto-couplers (e.g., 4N25). For simulating input signals, one can use onboard push buttons as temporary inputs and slide switches as permanent inputs. In the beginning of each PLC scan cycle (`get_inputs`) the 74HC/LS165 is loaded (RB7 (`shift/load`) = 0) with the level of eight inputs and then these

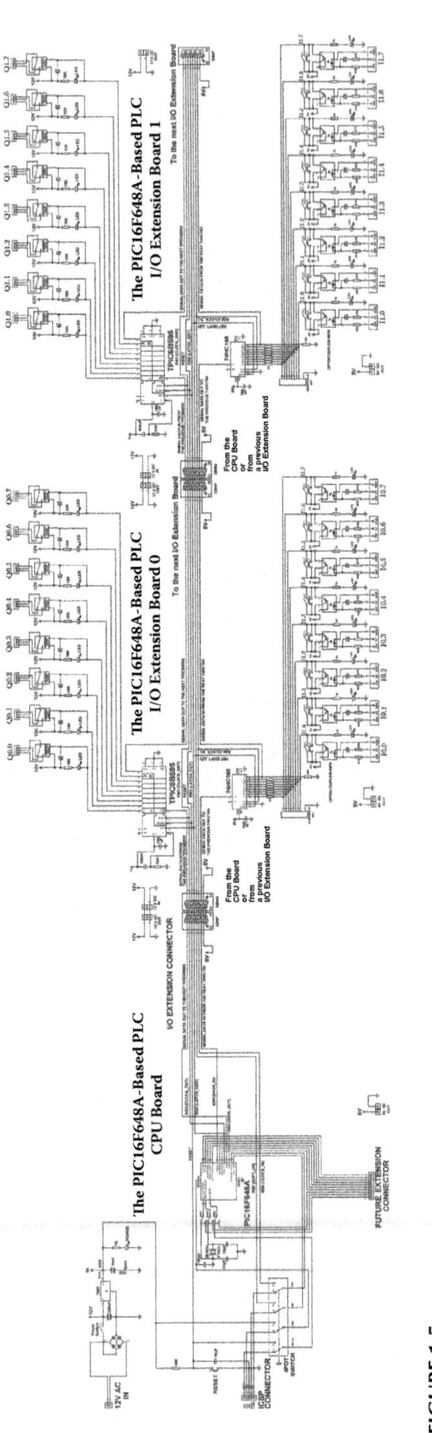

FIGURE 1.5
Schematic diagram of the CPU board plus two I/O extension boards.

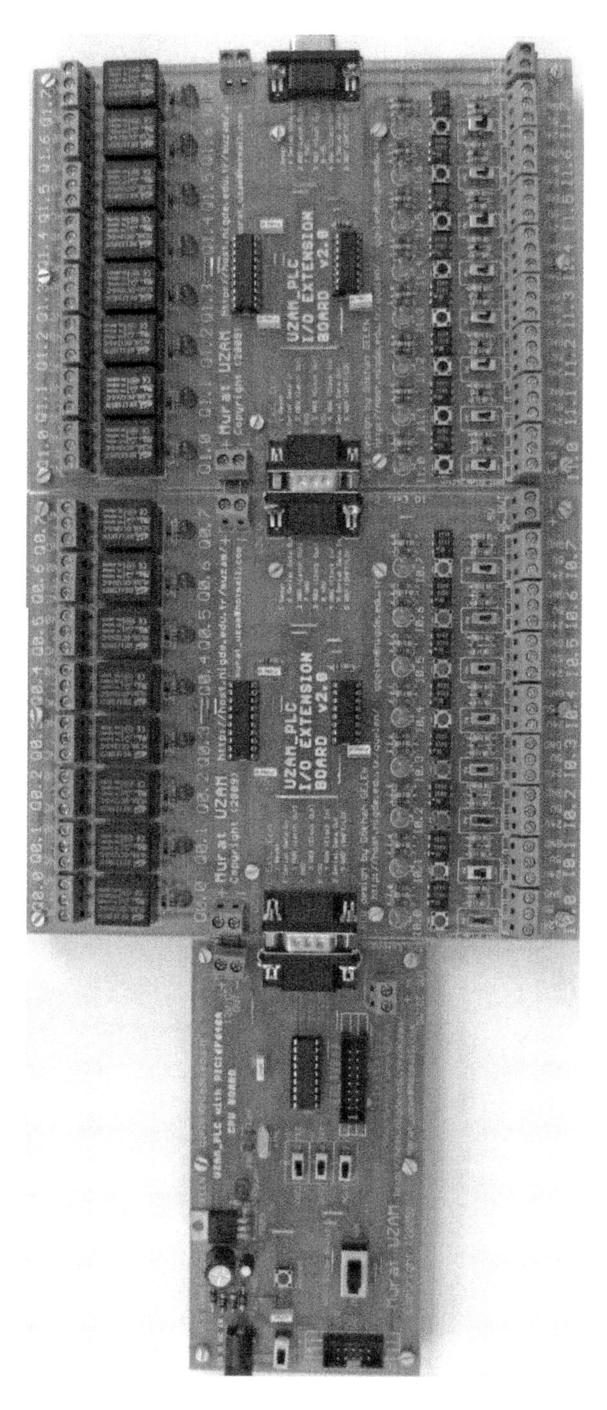

FIGURE 1.6
Photograph of the CPU board plus two I/O extension boards.

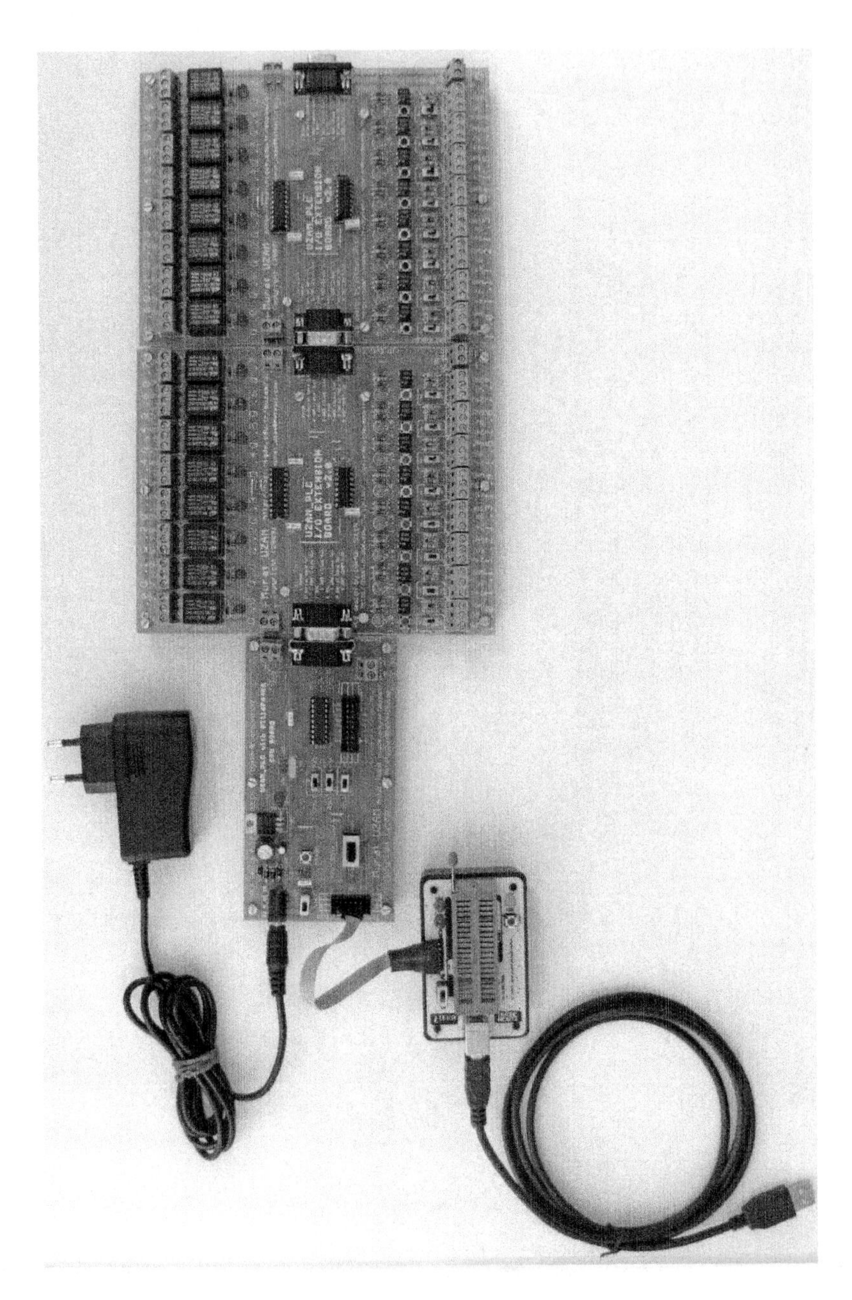

FIGURE 1.7
Photograph of the CPU board plus two I/O extension boards and a USB PIC programmer.

data are serially clocked in (when RB7 = 1; through RB0 `data in` and RB6 `clock in` pins). If there is only one I/O extension board used, then eight `clock_in` signals are enough to get the eight input signals. For each additional I/O extension board, eight more `clock_in` signals are necessary. The serial data coming from the I/O extension board(s) are taken from the SI input of the 74HC/LS165.

The *outputs section* introduces eight discrete relay outputs for the PIC16F648A-based PLC (called Q0.0, Q0.1, ..., Q0.7 for the first I/O extension board). Each relay operates with 12 V DC and is driven by an 8-bit serial-to-parallel converter register/driver TPIC6B595. Relays have single-pole double-throw (SPDT) contacts with C (common), NC (normally closed), and NO (normally open) terminals. At the end of each PLC scan cycle (`send_outputs`) the output data are serially clocked out (through RB3 `clock out` and RB4 `data out` pins) and finally latched within the TPIC6B595. If there is only one I/O extension board used, then eight `clock_out` signals are enough to send the eight output signals. For each additional I/O extension board, eight more `clock_out` signals are necessary. The serial data going to the I/O extension board(s) are sent out from the SER OUT (pin 18) of the TPIC6B595.

The PCB design files of both the CPU board and the I/O extension board can be obtained from the CD-ROM attached to this book. Note that in the PCB design of the CPU board and the I/O extension board, some lines of I/O extension connectors DB9M and DB9F are different from the ones shown in Figures 1.1 and 1.3.

The project reported in this book makes use of a CPU board and two I/O extension boards, as can be seen from the schematic diagram and photograph depicted in Figures 1.5 and 1.6, respectively. Thus, in total there are 16 inputs and 16 outputs. Figure 1.7 shows the PIC16F648A-based PLC consisting of a CPU board, I/O extension boards, 12 V DC adapter, and USB PIC programmer.

2

Basic Software

In this chapter, the basic software of the PIC16F648A-based PLC is explained. A PLC scan cycle includes the following: obtain the inputs, run the user program, and update the outputs. It is also necessary to define and initialize all variables used within a PLC. Necessary functions are all described as PIC assembly macros to be used in the PIC16F648A-based PLC. The macros described in this chapter could be summarized as follows: `HC165` (for handling the inputs), `HC595` (for sending the outputs), `dbncr0` and `dbncr1` (for debouncing the inputs), `initialize`, `get_inputs`, and `send_outputs`. In addition, the concept of *contact bouncing* and how it is solved in the PIC16F648A-based PLC is explained in detail.

2.1 Basic Software Structure

The basic software of the PIC16F648A-based PLC makes use of general purpose 8-bit registers of static random-access memory (SRAM) data memory of the PIC16F648A microcontroller. For the sake of simplicity, we restrict ourselves to use only BANK 0; i.e., all macros, including the basic definitions explained here, are defined by means of 8-bit SRAM registers of BANK 0. The file definitions.inc, included within the CD-ROM attached to this book, contains all basic macros and definitions necessary for the PIC16F648A-based PLC. In this chapter, we will explain the contents of this file. First, let us look at the file called UZAM_plc_16i16o_ex1.asm, the view of which is shown in Figure 2.1. As is well known, a PLC scan cycle includes the following: obtain the inputs, run the user program, and update the outputs. This cycle is repeated as long as the PLC runs. Before getting into these endless PLC scan cycles, the initial conditions of the PLC are set up in the initialization stage. These main steps can be seen from Figure 2.1, where `initialize` is a macro for setting up the initial conditions, `get_inputs` is a macro for getting and handling the inputs, and `send_outputs` is a macro for updating the outputs. The user PLC program must be placed between `get_inputs` and `send_outputs`. The endless PLC scan cycles are obtained by means of the label "scan" and the instruction "goto scan."

The PIC16F648A-based PLC is fixed to run at 20 MHz with an external oscillator. The watchdog timer is used to prevent user program lockups. As

```
;------------------------------------------------------------;
;Filename:  UZAM_plc_16i16o_ex1.asm                          ;
;Date:      27 September 2011                                ;
;Author:    Prof.Dr. Murat UZAM                              ;
;Company:   Melikşah Üniversitesi                            ;
;           Mühendislik-Mimarlık Fakültesi                   ;
;           Elektrik-Elektronik Mühendisliği Bölümü          ;
;           Talas, 38280, Kayseri, TURKEY                    ;
;           http://www.meliksah.edu.tr/muzam/                ;
;           murat_uzam@meliksah.edu.tr                       ;
;           murat_uzam@hotmail.com                           ;
;           Tel: ++ 90 352 207 73 00 / 7351                  ;
;           Fax: ++ 90 352 207 73 49                         ;
;------------------------------------------------------------;
;Notes:     This is the basic program                        ;
;           for PIC16F648A microcontroller                   ;
;           based UZAM PLC with                              ;
;           16 Inputs and 16 Outputs                         ;
;           and 32 Memory Bits (Internal Relays)             ;
;------------------------------------------------------------;
      list      p=16F648A           ;list directive to define processor
      #include <p16F648A.inc>       ;processor specific variable definitions
      #include <definitions.inc>    ;basic PLC definitions, macros, etc.
      __CONFIG   _CP_OFF & DATA_CP_OFF & _LVP_OFF & _BOREN_OFF & _MCLRE_ON
      & _WDT_OFF & _PWRTE_ON & _HS_OSC
      org    0x00                   ;Reset Vector
main
      initialize
scan
      get_inputs
;--------------- user program starts here -----------------------


;--------------- user program ends here -------------------------
      send_outputs
      goto   scan
      end                           ;directive 'end of program'
```

FIGURE 2.1
View of the file UZAM_plc_16i16o_ex1.asm.

will be explained later, the hardware timer TMR0 is utilized to obtain free-running reference timing signals.

2.1.1 Variable Definitions

Next, let us now consider the inside of the file definitions.inc. The definitions of 8-bit variables to be used for the basic software and their allocation in BANK 0 of SRAM data memory are shown in Figure 2.2(a) and (b), respectively. Although we can define as many inputs and outputs as we want, in this book we restrict ourselves to BANK 0 and define two 8-bit input registers and two 8-bit output registers (Q0 and Q1).

It is well known that inputs taken from contacts always suffer from contact bouncing. To circumvent this problem we define a debouncing mechanism for the inputs; this will be explained later. In the `get_inputs` stage of the PLC scan cycle, the input signals are serially taken from the related 74HC/LS165 registers and stored in the SRAM registers. As a result, bI0 and bI1 will hold these bouncing input signals. After applying the debouncing mechanism to the bouncing input signals of bI0 and bI1 we obtain debounced input signals, and they are stored in SRAM registers I0 and I1, respectively.

In the `send_outputs` stage of the PLC scan cycle, the output information stored in the 8-bit SRAM registers Q0 and Q1 is serially sent out to and stored in the related TPIC6B595 registers. This means that Q0 and Q1 registers will hold output information, and they will be copied into the TPIC6B595 registers at the end of each PLC scan cycle. Four 8-bit registers, namely, M0, M1, M2, and M3, are defined for obtaining 32 memory bits (internal relays, in PLC jargon). To be used for the debouncer macros `dbncr0` and `dbncr1`, we define sixteen 8-bit registers (DBNCR0, DBNCR0+1, …, DBNCR0+7) and (DBNCR1, DBNCR1+1, …, DBNCR1+7). In addition, the registers DBNCRRED0 and DBNCRRED1 are also defined to be used for the debouncer macros `dbncr0` and `dbncr1`, respectively. Temp_1 is a general temporary register declared to be used in the macros. Temp_2 is declared to be used especially for obtaining special memory bits, as will be explained later. Timer_2 is defined for storing the high byte of the free-running timing signals. The low byte of the free-running timing signals is stored in TMR0 (recalled as Timer_1).

For accessing the SRAM data memory easily, BANK macros are defined as shown in Figure 2.3.

```
;--------------- VARIABLE DEFINITIONS ----------------
        CBLOCK 0x20             ;
        bI0,bI1                 ;
        endc                    ;
        CBLOCK 0x22             ;
        I0,I1                   ;
        endc                    ;
        CBLOCK 0x24             ;
        Q0,Q1                   ;
        endc                    ;
        CBLOCK 0x26
        M0,M1,M2,M3             ;4x8=32 Memory bits(Internal Relays)
        endc
        CBLOCK 0x2A             ;
        DBNCR0                  ;DBNCR0, DBNCR0+1, ..., DBNCR0+7
        endc                    ;
        CBLOCK 0x32             ;
        DBNCR1                  ;DBNCR1, DBNCR1+1, ..., DBNCR1+7
        endc                    ;
        CBLOCK 0x3A
        Temp_1,Temp_2,Timer_2,DBNCRRED0,DBNCRRED1
        endc
;-----------------------------------------------------
```

(a)

FIGURE 2.2
(a) The definition of 8-bit variables to be used in the basic software. (*Continued*)

Address	BANK 0
20h	**bI0**
21h	**bI1**
22h	**I0**
23h	**I1**
24h	**Q0**
25h	**Q1**
26h	**M0**
27h	**M1**
28h	**M2**
29h	**M3**
2Ah	**DBNCR0**
2Bh	**DBNCR0+1**
2Ch	**DBNCR0+2**
2Dh	**DBNCR0+3**
2Eh	**DBNCR0+4**
2Fh	**DBNCR0+5**
30h	**DBNCR0+6**
31h	**DBNCR0+7**
32h	**DBNCR1**
33h	**DBNCR1+1**
34h	**DBNCR1+2**
35h	**DBNCR1+3**
36h	**DBNCR1+4**
37h	**DBNCR1+5**
38h	**DBNCR1+6**
39h	**DBNCR1+7**
3Ah	**Temp_1**
3Bh	**Temp_2**
3Ch	**Timer_2**
3Dh	**DBNCRRED0**
3Eh	**DBNCRRED1**

BANK 0

(b)

FIGURE 2.2 (*Continued*)
(b) Their allocation in BANK 0 of SRAM data memory.

The definitions of 1-bit (Boolean) variables are depicted in Figure 2.4. The following definitions are self-explanatory: 74HC165, TPIC6B595, 16 INPUTS, 16 OUTPUTS, and 32 memory bits.

The individual bits (1-bit variables) of 8-bit SRAM registers bI0, bI1, I0, I1, Q0, Q1, M0, M1, M2, and M3 are shown below:

bI0 is an 8-bit register:

bI0

The individual bits of bI0 are as follows:

bI0.7	bI0.6	bI0.5	bI0.4	bI0.3	bI0.2	bI0.1	bI0.0

```
;------------------ BANK macros --------------
BANK0 macro
            bcf STATUS,RP0
            bcf STATUS,RP1
            endm
BANK1 macro
            bsf STATUS,RP0
            bcf STATUS,RP1
            endm
BANK2 macro
            bcf STATUS,RP0
            bsf STATUS,RP1
            endm
BANK3 macro
            bsf STATUS,RP0
            bsf STATUS,RP1
            endm
;-------------------------------------------
```

FIGURE 2.3
BANK macros.

bI1 is an 8-bit register:

bI1

The individual bits of bI1 are as follows:

bI1.7	bI1.6	bI1.5	bI1.4	bI1.3	bI1.2	bI1.1	bI1.0

I0 is an 8-bit register:

I0

The individual bits of I0 are as follows:

I0.7	I0.6	I0.5	I0.4	I0.3	I0.2	I0.1	I0.0

I1 is an 8-bit register:

I1

The individual bits of I1 are as follows:

I1.7	I1.6	I1.5	I1.4	I1.3	I1.2	I1.1	I1.0

Q0 is an 8-bit register:

Q0

```
;----------------- 16 INPUTS --------------------
#define bI0.0  bI0,0    ;b:bouncing
#define bI0.1  bI0,1
#define bI0.2  bI0,2
#define bI0.3  bI0,3
#define bI0.4  bI0,4
#define bI0.5  bI0,5
#define bI0.6  bI0,6
#define bI0.7  bI0,7

#define I0.0  I0,0      ;I0 = debounced bI0
#define I0.1  I0,1
#define I0.2  I0,2
#define I0.3  I0,3
#define I0.4  I0,4
#define I0.5  I0,5
#define I0.6  I0,6
#define I0.7  I0,7

#define bI1.0  bI1,0    ;b:bouncing
#define bI1.1  bI1,1
#define bI1.2  bI1,2
#define bI1.3  bI1,3
#define bI1.4  bI1,4
#define bI1.5  bI1,5
#define bI1.6  bI1,6
#define bI1.7  bI1,7

#define I1.0  I1,0      ;I1 = debounced bI1
#define I1.1  I1,1
#define I1.2  I1,2
#define I1.3  I1,3
#define I1.4  I1,4
#define I1.5  I1,5
#define I1.6  I1,6
#define I1.7  I1,7
;---------------------------------------------
```

(a)

```
;----------------- 16 OUTPUTS ------------------
#define Q0.0  Q0,0
#define Q0.1  Q0,1
#define Q0.2  Q0,2
#define Q0.3  Q0,3
#define Q0.4  Q0,4
#define Q0.5  Q0,5
#define Q0.6  Q0,6
#define Q0.7  Q0,7

#define Q1.0  Q1,0
#define Q1.1  Q1,1
#define Q1.2  Q1,2
#define Q1.3  Q1,3
#define Q1.4  Q1,4
#define Q1.5  Q1,5
#define Q1.6  Q1,6
#define Q1.7  Q1,7
;---------------------------------------------
```

(b)

FIGURE 2.4
Definitions of 1-bit (Boolean) variables: (a) 16 inputs, (b) 16 outputs. (*Continued*)

```
;--------- LOGIC VALUES -----------
#define LOGIC0    Temp_2,0
#define LOGIC1    Temp_2,1
;----------------------------------

;--------- SPECIAL BITS -----------
#define FRSTSCN   Temp_2,2
#define SCNOSC    Temp_2,3
;----------------------------------
```

(c)

```
;--------- Definitions for 74HC165 -------------
#define data_in   PORTB,0
#define clock_in  PORTB,6
#define shft_ld   PORTB,7
;-----------------------------------------------

;--------- Definitions for TPIC6B595 -----------
#define data_out  PORTB,4
#define clock_out PORTB,3
#define latch_out PORTB,5
;-----------------------------------------------
```

(d)

FIGURE 2.4 (*Continued*)
Definitions of 1-bit (Boolean) variables: (c) logic values and special bits, (d) definitions for 74HC165 and TPIC6B595. (*Continued*)

The individual bits of Q0 are as follows:

Q0.7	Q0.6	Q0.5	Q0.4	Q0.3	Q0.2	Q0.1	Q0.0

Q1 is an 8-bit register:

Q1

The individual bits of Q1 are as follows:

Q1.7	Q1.6	Q1.5	Q1.4	Q1.3	Q1.2	Q1.1	Q1.0

M0 is an 8-bit SRAM register:

M0

The individual bits of M0 are as follows:

M0.7	M0.6	M0.5	M0.4	M0.3	M0.2	M0.1	M0.0

M1 is an 8-bit SRAM register:

M1

```
;--- 32 Memory Bits(Internal Relays) ----------
#define M0.0  M0,0
#define M0.1  M0,1
#define M0.2  M0,2
#define M0.3  M0,3
#define M0.4  M0,4
#define M0.5  M0,5
#define M0.6  M0,6
#define M0.7  M0,7

#define M1.0  M1,0
#define M1.1  M1,1
#define M1.2  M1,2
#define M1.3  M1,3
#define M1.4  M1,4
#define M1.5  M1,5
#define M1.6  M1,6
#define M1.7  M1,7

#define M2.0  M2,0
#define M2.1  M2,1
#define M2.2  M2,2
#define M2.3  M2,3
#define M2.4  M2,4
#define M2.5  M2,5
#define M2.6  M2,6
#define M2.7  M2,7

#define M3.0  M3,0
#define M3.1  M3,1
#define M3.2  M3,2
#define M3.3  M3,3
#define M3.4  M3,4
#define M3.5  M3,5
#define M3.6  M3,6
#define M3.7  M3,7
;---------------------------------------------
```

(e)

FIGURE 2.4 (*Continued*)
Definitions of 1-bit (Boolean) variables: (e) 32 memory bits (internal relays). (*Continued*)

The individual bits of M1 are as follows:

M1.7	M1.6	M1.5	M1.4	M1.3	M1.2	M1.1	M1.0

M2 is an 8-bit SRAM register:

M2

The individual bits of M2 are as follows:

M2.7	M2.6	M2.5	M2.4	M2.3	M2.2	M2.1	M2.0

```
;---------- REFERENCE TIMING SIGNALS -----------
#define Timer_1   TMR0  ;at 20 MHz clock frequency:
#define T0.0 Timer_1,0  ;Timer clock :    0.1024 ms
#define T0.1 Timer_1,1  ;Timer clock :    0.2048 ms
#define T0.2 Timer_1,2  ;Timer clock :    0.4096 ms
#define T0.3 Timer_1,3  ;Timer clock :    0.8192 ms
#define T0.4 Timer_1,4  ;Timer clock :    1.6384 ms
#define T0.5 Timer_1,5  ;Timer clock :    3.2768 ms
#define T0.6 Timer_1,6  ;Timer clock :    6.5536 ms
#define T0.7 Timer_1,7  ;Timer clock :   13.1072 ms
#define T1.0 Timer_2,0  ;Timer clock :   26.2144 ms
#define T1.1 Timer_2,1  ;Timer clock :   52.4288 ms
#define T1.2 Timer_2,2  ;Timer clock :  104.8576 ms
#define T1.3 Timer_2,3  ;Timer clock :  209.7152 ms
#define T1.4 Timer_2,4  ;Timer clock :  419.4304 ms
#define T1.5 Timer_2,5  ;Timer clock :  838.8608 ms
#define T1.6 Timer_2,6  ;Timer clock : 1677.7216 ms = 1.6777216 s.
#define T1.7 Timer_2,7  ;Timer clock : 3355.4432 ms = 3.3554432 s.
;-----------------------------------------------
```

(f)

FIGURE 2.4 (*Continued*)
Definitions of 1-bit (Boolean) variables: (f) 16 reference timing signals.

M3 is an 8-bit SRAM register:

M3

The individual bits of M3 are as follows:

M3.7	M3.6	M3.5	M3.4	M3.3	M3.2	M3.1	M3.0

Register Temp_2 has the following individual bits:

7	6	5	4	3	2	1	0
				SCNOSC	FRSTSCN	LOGIC1	LOGIC0

LOGIC0: Set to 0 after the first scan.

LOGIC1: Set to 1 after the first scan.

FRSTSCN: Set to 1 during the first scan and set to 0 after the first scan.

SCNOSC: Toggled between 0 and 1 at each scan.

The variable LOGIC0 is defined to hold a logic 0 value throughout the PLC operation. At the initialization stage it is deposited with this value. Similarly, the variable LOGIC1 is defined to hold a logic 1 value throughout the PLC operation. At the initialization stage it is deposited with this value. The special memory bit FRSTSCN is arranged to hold the value of 1 at the first PLC scan cycle only. In the other PLC scan cycles following the first one it is reset. The special memory bit SCNOSC is arranged to work as a *scan oscillator*. This means that in one PLC scan cycle this special bit will hold the value of 0, in the next one the value of 1, in the next one the value of 0, and so on. This will keep on going for every PLC scan cycle.

Timer_1 (TMR0) is an 8-bit register:

Timer_1 (TMR0)

The individual bits of Timer_1 are as follows:

T0.7	T0.6	T0.5	T0.4	T0.3	T0.2	T0.1	T0.0

Timer_2 is an 8-bit register:

Timer_2

The individual bits of Timer_2 are as follows:

T1.7	T1.6	T1.5	T1.4	T1.3	T1.2	T1.1	T1.0

Let us now consider the 16 reference timing signals. As will be explained later, TMR0 of PIC16F648A is set up to count the ¼ of 20 MHz oscillator signal, i.e., 5 MHz with a prescaler arranged to divide the signal to 256. As a result, by means of TMR0 bits (also called Timer_1), we obtain eight free-running reference timing signals with the T timing periods starting from 0.1024 ms to 13.1072 ms. As will be explained later, the register Timer_2 is incremented on Timer_1 overflow. This also gives us (by means of Timer_2 bits) eight more free-running reference timing signals with the T timing periods starting from 26.2144 ms to 3355.4432 ms. The timing diagram of the free-running reference timing signals is depicted in Figure 2.5. Note that the evaluation of TMR0 (Timer_1) is independent from the PLC scan cycles, but Timer_2 is incremented within the `get_inputs` stage of the PLC scan cycle on Timer_1 overflow. This is justified as long as the PLC scan cycle takes less than 13.1072 ms.

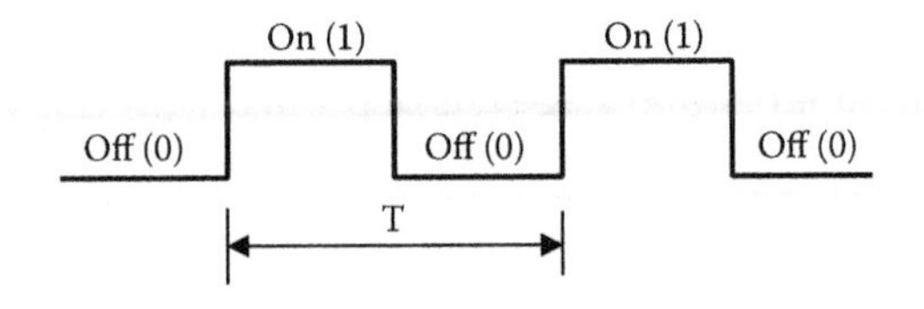

FIGURE 2.5
Timing diagram of the free-running reference timing signals (T = 0.1024, 0.2048, ..., 3355.4432 ms).

```
;------------------- Macro HC165 ------------------------------------------
HC165       macro num,var0          ;This macro can be used for 74HC/HCT/LS165
            local i=0,j=0           ;parallel to serial shift register ICs
            bcf   shft_ld           ;latch
            nop                     ;the inputs
            bsf   shft_ld           ;of all 74HC165s
            while j < num           ;carry on while j < num
                 while i < 8        ;for each 74HC165, 8 times do the following
                 rlf   var0+j,f     ;rotate the register "var0+j" one position left
                 btfss data_in      ;if the data_in is set then skip
                 bcf   var0+j,0     ;if the data_in is reset then reset "var0+j,0"
                 btfsc data_in      ;if the data_in is reset then skip
                 bsf   var0+j,0     ;if the data_in is set then set "var0+j,0"
                 bcf   clock_in     ;generate
                 nop                ;a clock_in
                 bsf   clock_in     ;pulse
i += 1                              ;increment "i"
                 endw               ;after 8 iterations end the while loop for "i"
i=0                                 ;i=0            : get ready
j += 1                              ;increment "j"  : for a new 74HC165
            endw                    ;after 'num' iterations end the while loop for "j"
            endm                    ;end macro HC165
;--------------------------------------------------------------------------
```

FIGURE 2.6
The macro HC165.

2.1.2 Macro HC165

The macro HC165 is shown in Figure 2.6. The input signals are serially taken from the related 74HC/LS165 registers and stored in the SRAM registers bI0 and bI1 by means of this macro. The num defines the number of 74HC/LS165 registers to be considered. This means that with this macro we can obtain inputs from as many 74HC/LS165 registers as we wish. However, as explained before, in this book we restrict this number to be 2, because we have 16 discrete inputs. var0 is the beginning of the registers to which the state of inputs taken from 74HC/LS165 registers will be stored. This implies that there should be enough SRAM locations reserved after var0, and also there should be enough 74HC/LS165 registers to get the inputs from. There are some explanations within the macro to describe how it works. As can be seen, this macro makes use of previously defined data_in, clock_in, and sfht_ld bits to obtain the input signals from 74HC/LS165 registers.

2.1.3 Macro HC595

The macro HC595 is shown in Figure 2.7. The output signals are stored in the 8-bit SRAM registers Q0 and Q1 and serially sent out to and stored in the related TPIC6B595 registers by means of this macro. The num defines the number of TPIC6B595 registers to be used. This means that with this macro we can send output data serially to as many TPIC6B595 registers as we wish. However as explained before, in this book we restrict this number to 2, because we have 16 discrete outputs. var0 is the beginning of the 8-bit registers, such as Q0 in SRAM from which the state of outputs are taken and serially sent out to TPIC6B595 registers. This implies that there should

```
;------------------- Macro HC595 -------------------------------------------------
HC595       macro num,var0            ;This macro can be used for 74HC/HCT/LS595
            local i=0,j=num-1         ;or TPIC6B595 serial to parallel shift register ICs
            while j >= 0              ;carry on while j >= 0
                 while i < 8          ;for each TPIC6B595, 8 times do the following:
                 rlf   var0+j,f       ;rotate the register "var0+j" one position left
                 btfss STATUS,C       ;if the Carry flag is set then skip
                 bcf   data_out       ;if the Carry flag is reset then reset data_out
                 btfsc STATUS,C       ;if the Carry flag is reset then skip
                 bsf   data_out       ;if the Carry flag is set then set data_out
                 bsf   clock_out      ;generate
                 nop                  ;a clock_out
                 bcf   clock_out      ;pulse
i += 1                                ;increment "i"
                 endw                 ;after 8 iterations end the while loop for "i"
            rlf   var0+j,f            ;rotate the register "var0+j" one position left
i=0                                   ;i=0          : get ready
j-= 1                                 ;decrement "j" : for a new TPIC6B595
            endw                      ;after 'num' iterations end the while loop for "j"
            bsf   latch_out           ;Latch the serially shifted out data
            nop                       ;on all
            bcf   latch_out           ;TPIC6B595's
            endm                      ;end macro HC595
;---------------------------------------------------------------------------------
```

FIGURE 2.7
The macro HC595.

be enough SRAM locations reserved after var0, and also there should be enough TPIC6B595 registers to hold the outputs. There are some explanations within the macro to describe how it works. As can be seen, this macro makes use of previously defined data_out, clock_out, and latch_out bits to send the output signals serially to TPIC6B595 registers.

2.2 Elimination of Contact Bouncing Problem in the PIC16F648A-Based PLC

2.2.1 Contact Bouncing Problem

When a mechanical contact, such as a push-button switch, examples of which are shown in Figure 2.8, user interface button, limit switch, relay, or contactor contact, is opened or closed, the contact seldom demonstrates a clean transition from one state to another. There are two types of contacts: normally open (NO) and normally closed (NC). When a contact is closed or opened, it will close and open (technically speaking, make and break) many times before finally settling in a stable state due to mechanical vibration. As can be seen from Figure 2.9, this behavior of a contact is interpreted as multiple false input signals, and a digital circuit will respond to each of these on-off or off-on transitions. This problem is well known as *contact bounce* and has always been a very important problem when interfacing switches, relays, etc., to a digital control system.

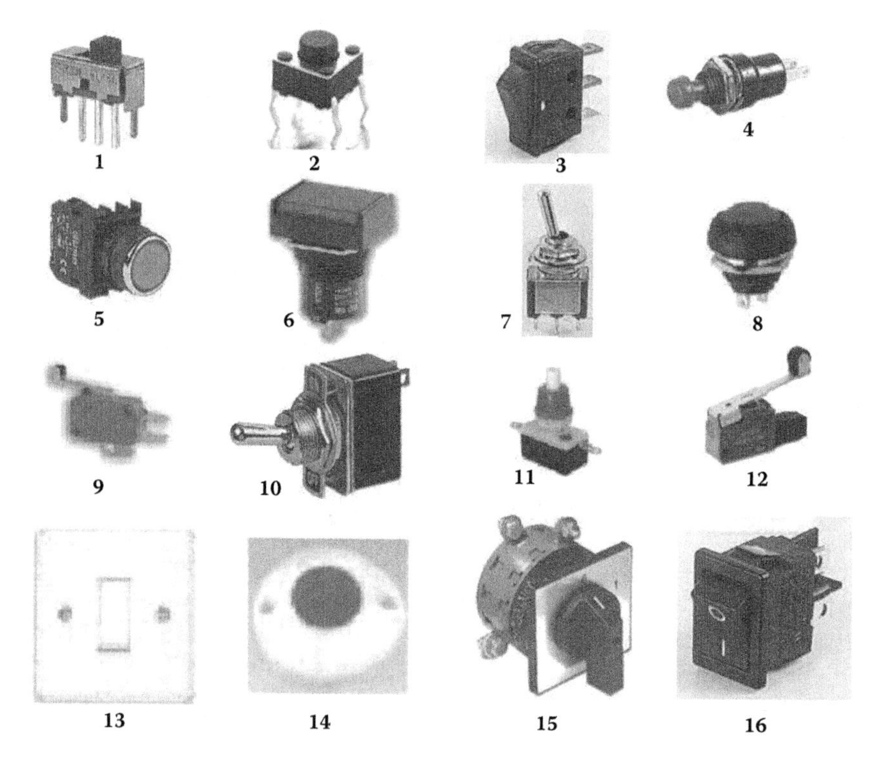

FIGURE 2.8
Different types and makes of switches and buttons.

In some industrial applications *debouncing* is required to eliminate both mechanical and electrical effects. Most switches seem to exhibit bounce duration under 10 ms, and therefore it is reasonable to pick a debounce period in the 20 to 50 ms range. On the other hand, when dealing with relay contacts, the debounce period should be large enough, i.e., within the 20 to 200 ms range. Nevertheless, a reasonable switch will not bounce longer than 500 ms. Both closing and opening contacts suffer from the bouncing problem, and therefore in general, both rising and falling edges of an input signal should be debounced, as seen from the timing diagram of Figure 2.10.

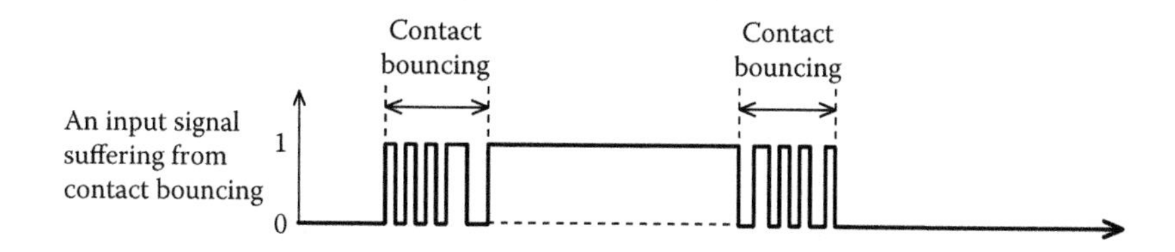

FIGURE 2.9
Contact bouncing problem, causing an input signal to bounce between 0 and 1.

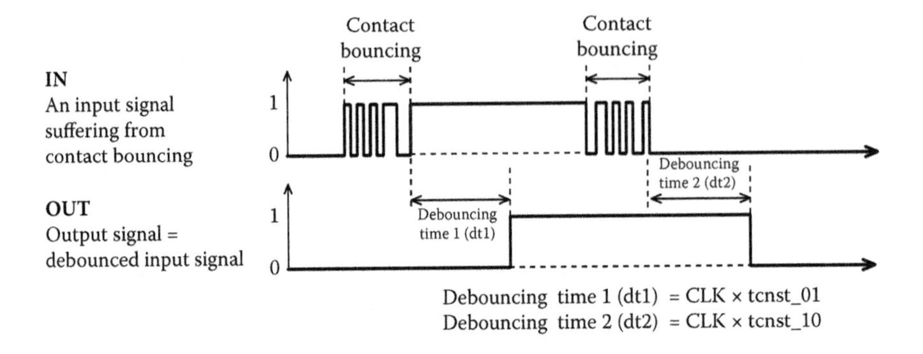

FIGURE 2.10
The timing diagram of a single I/O debouncer (also the timing diagram of each channel of the independent 8-bit I/O contact debouncers, dbncr0 and dbncr1).

2.2.2 Understanding a Generic Single I/O Contact Debouncer

In order to understand how a debouncer works, let us now consider a generic single I/O debouncer. We can think of the generic single I/O debouncer as being a single INput/single OUTput system, whose state transition diagram is shown in Figure 2.11. In the state transition diagram there are four states,

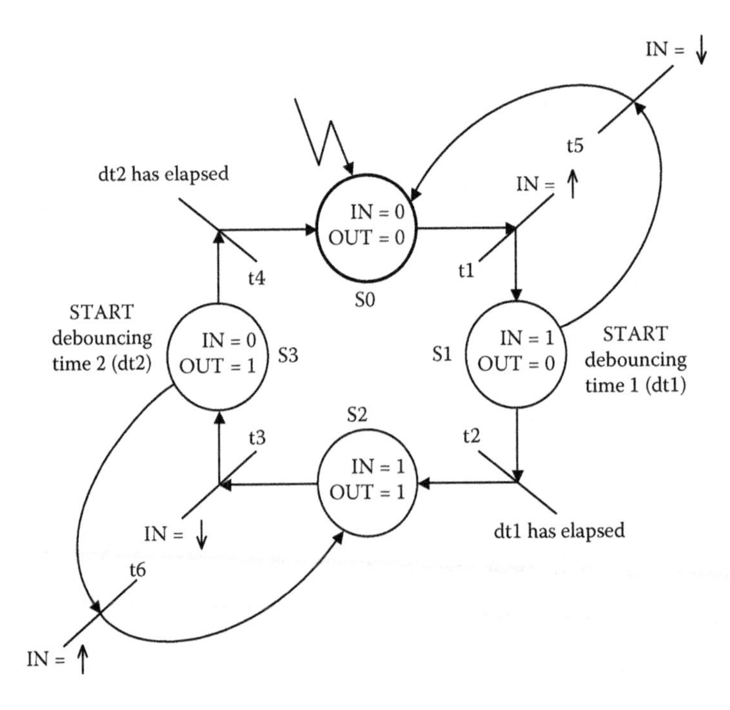

FIGURE 2.11
State transition diagram of a generic single I/O debouncer.

S0, S1, S2, and S3, drawn as circles, and six transitions, t1, t2, ..., t6, drawn as bars. States and transitions are connected by directed arcs. The following explains the behavior of the generic single I/O debouncer (also each channel of the independent 8-bit I/O contact debouncers, `dbncr0` and `dbncr1`) based on the state transition diagram shown in Figure 2.11:

1. Initially, it is assumed that the input signal IN and the output signal OUT are both LOW (state S0).
2. When the system is in S0 (the IN is LOW and the OUT is LOW), if the rising edge (↑) of IN is detected (transition t1), then the system moves from S0 to S1 and the debouncer starts a time delay, called debouncing time 1 (`dt1`).
3. While the system is in S1 (the IN is HIGH and the OUT is LOW), before the `dt1` ms time delay ends, if the falling edge (↓) of IN is detected (transition t5), then the system goes back to S0 from S1, and the time delay `dt1` is canceled and the OUT remains LOW (no state change is issued).
4. When the system is in S1 (the IN is HIGH and the OUT is LOW), if the input signal is still HIGH and the time delay `dt1` has elapsed (transition t2), then the system moves from S1 to S2. In this case, the state change is issued, i.e., the OUT is set to HIGH.
5. When the system is in S2 (the IN is HIGH and the OUT is HIGH), if the falling edge (↓) of IN is detected (transition t3), then the system moves from S2 to S3 and the debouncer starts a time delay, called debouncing time 2 (`dt2`).
6. While the system is in S3 (the IN is LOW and the OUT is HIGH), before the `dt2` ms time delay ends, if the rising edge (↑) of IN is detected (transition t6), then the system goes back to S2 from S3, and the time delay `dt2` is canceled and the OUT remains HIGH (no state change is issued).
7. When the system is in S3 (the IN is LOW and the OUT is HIGH), if the input signal is still LOW and the time delay `dt2` has elapsed (transition t4), then the system moves from S3 to S0. In this case, the state change is issued, i.e., the OUT is set to LOW.

2.2.3 Debouncer Macros `dbncr0` and `dbncr1`

The macro `dbncr0` and its flowchart are shown in Figures 2.12 and 2.13, respectively. Table 2.1 shows the schematic symbol of the macro `dbncr0`. The detailed timing diagram of one channel of this debouncer is provided in Figure 2.14. It can be used for debouncing eight independent buttons, switches, relay or contactor contacts, etc. It is seen that the output changes its state only after the input becomes stable and waits in the stable state for the

```
;------------------- macro: debouncer0 ----------------
dbncr0 macro num,regi,biti,t_reg,t_bit,tcnst_01,tcnst_10,rego,bito
        local   L1,L2,L3,L4
        btfsc   rego,bito
        goto    L4
        btfsc   regi,biti
        goto    L2
        clrf    DBNCR0+num
        goto    L1
L4      btfss   regi,biti
        goto    L3
        clrf    DBNCR0+num
        goto    L1
L3      btfss   t_reg,t_bit
        bsf     DBNCRRED0,num
        btfss   t_reg,t_bit
        goto    L1
        btfss   DBNCRRED0,num
        goto    L1
        bcf     DBNCRRED0,num
        incf    DBNCR0+num,f
        movf    DBNCR0+num,w
        xorlw   tcnst_10
        skpnz
        bcf     rego,bito
        goto    L1
L2      btfss   t_reg,t_bit
        bsf     DBNCRRED0,num
        btfss   t_reg,t_bit
        goto    L1
        btfss   DBNCRRED0,num
        goto    L1
        bcf     DBNCRRED0,num
        incf    DBNCR0+num,f
        movf    DBNCR0+num,w
        xorlw   tcnst_01
        skpnz
        bsf     rego,bito
L1
        endm
;-----------------------------------------------------
```

FIGURE 2.12
The macro `dbncr0`.

predefined debouncing time `dt1` or `dt2`. The debouncing is applied to both rising and falling edges of the input signal. In this macro, each channel is intended for a *normally open contact* connected to the PIC by means of a pull-down resistor, as this is the case with the PIC16F648A-based PLC. It can also be used without any problem for a *normally closed contact* connected to the PIC by means of a pull-up resistor. The debouncing times, such as 20, 50, or 100 ms, can be selected as required depending on the application. It is possible to pick up different debouncing times for each channel. It is also possible to choose different debouncing times for rising and falling edges of the same input signal if necessary. This gives a good deal of flexibility. This is simply

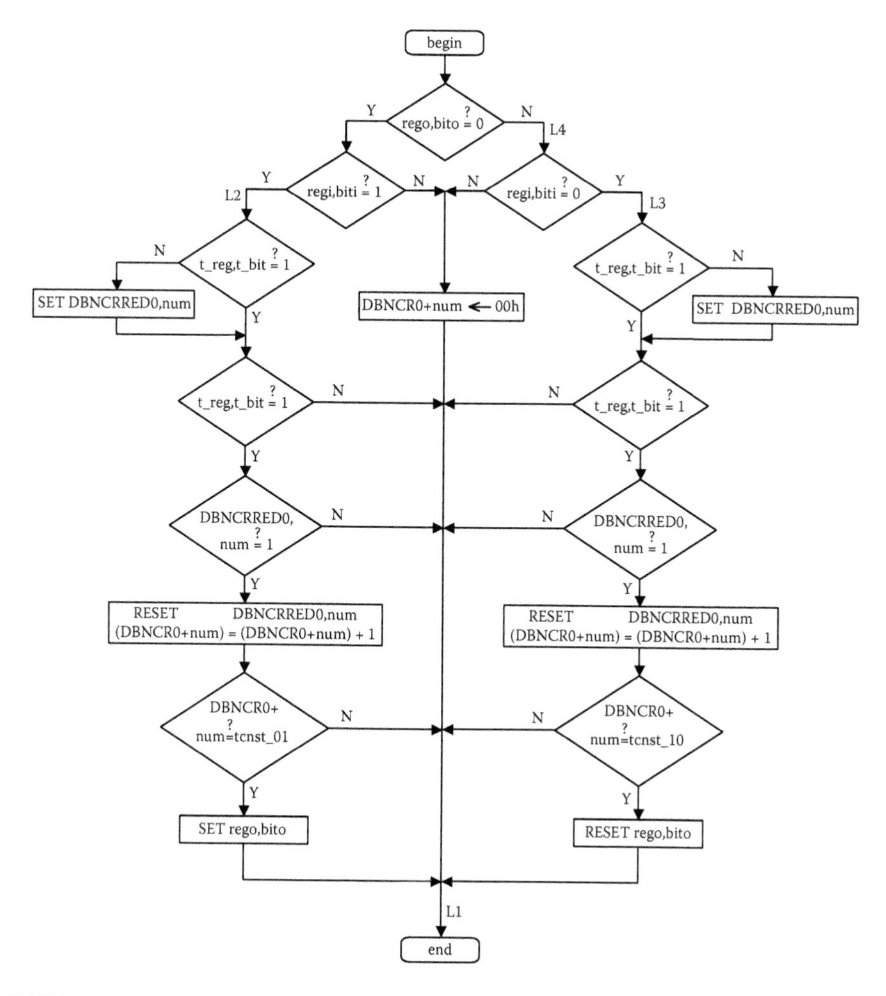

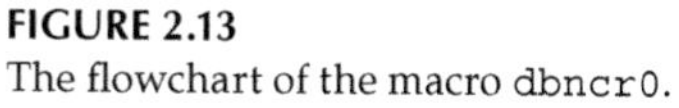
FIGURE 2.13
The flowchart of the macro `dbncr0`.

done by changing the related time constant `tcnst_01` or `tcnst_10` defining the debouncing time delay for each channel and for both edges within the assembly program. Note that if the state change of the contact is shorter than the predefined debouncing time, this will also be regarded as bouncing, and it will not be taken into account. Therefore, no state change will be issued in this case. Each of the eight input channels of the debouncer may be used independently from other channels. The activity of one channel does not affect that of the other channels.

Let us now briefly consider how the macro `dbncr0` works. First, one of the previously defined reference timing signals is chosen as `t_reg,t_bit`, to be used within this macro. Then, we can set up both debouncing times `dt1` and `dt2` by means of time constants `tcnst_01` and `tcnst_10`, as

TABLE 2.1
Schematic Symbol of the Macro `dbncr0`

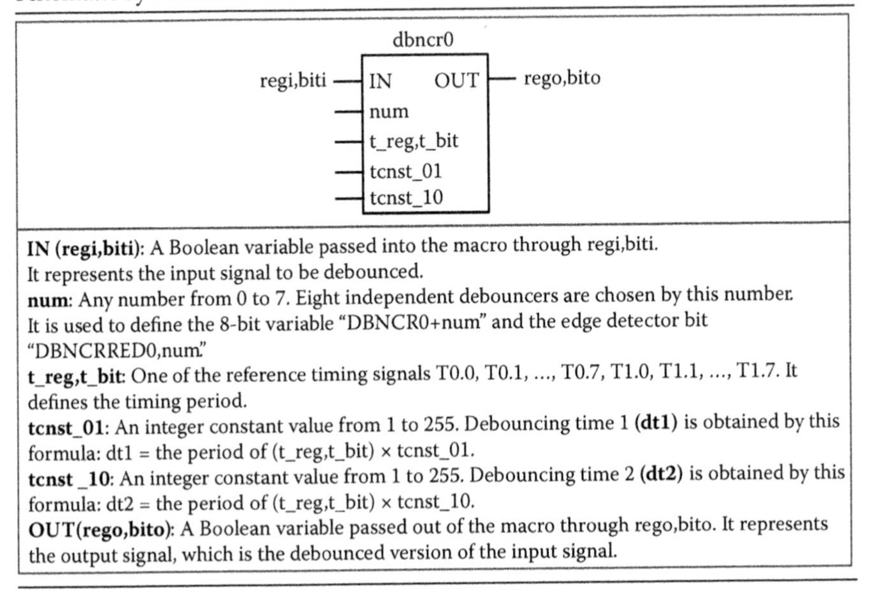

IN (regi,biti): A Boolean variable passed into the macro through regi,biti. It represents the input signal to be debounced.
num: Any number from 0 to 7. Eight independent debouncers are chosen by this number. It is used to define the 8-bit variable "DBNCR0+num" and the edge detector bit "DBNCRRED0,num."
t_reg,t_bit: One of the reference timing signals T0.0, T0.1, ..., T0.7, T1.0, T1.1, ..., T1.7. It defines the timing period.
tcnst_01: An integer constant value from 1 to 255. Debouncing time 1 **(dt1)** is obtained by this formula: dt1 = the period of (t_reg,t_bit) × tcnst_01.
tcnst _10: An integer constant value from 1 to 255. Debouncing time 2 **(dt2)** is obtained by this formula: dt2 = the period of (t_reg,t_bit) × tcnst_10.
OUT(rego,bito): A Boolean variable passed out of the macro through rego,bito. It represents the output signal, which is the debounced version of the input signal.

`dt1` = the period of (`t_reg,t_bit`) × `tcnst_01` and `dt2` = the period of (`t_reg,t_bit`) × `tcnst_10`, respectively. If the input signal (`regi,biti`) = 0 and the output signal (`rego,bito`) = 0 or the input signal (`regi,biti`) = 1 and the output signal (`rego,bito`) = 1, then the related counter DBNCR0+num is loaded with 00h and no state change is issued. If the output signal (`rego,bito`) = 0 and the input signal (`regi,biti`) = 1, then with each rising edge of the reference timing signal `t_reg,t_bit` the related counter DBNCR0+num is incremented by one. In this case, when the count value of DBNCR0+num is equal to the number `tcnst_01`, this means that the input signal is debounced properly and then state change from 0 to 1 is issued for the output signal (`rego,bito`). Similarly, if the output signal (`rego,bito`) = 1 and the input signal (`regi,biti`) = 0, then with each rising edge of the reference timing signal `t_reg,t_bit` the related counter DBNCR0+num is incremented by one. In this case, when the count value of DBNCR0+num is equal to the number `tcnst_10`, this means that the input signal is debounced properly and then state change from 1 to 0 is issued for the output signal (`rego,bito`). For this macro it is necessary to define the following 8-bit variables in SRAM: Temp_1 and DBNCRRED0. In addition, it is also necessary to define eight 8-bit variables in successive SRAM locations, the first of which is to be defined as DBNCR0. It is not necessary to name the other seven variables. Each bit of the variable DBNCRRED0 is used to detect the rising edge of the reference timing signal `t_reg,t_bit` for the related channel.

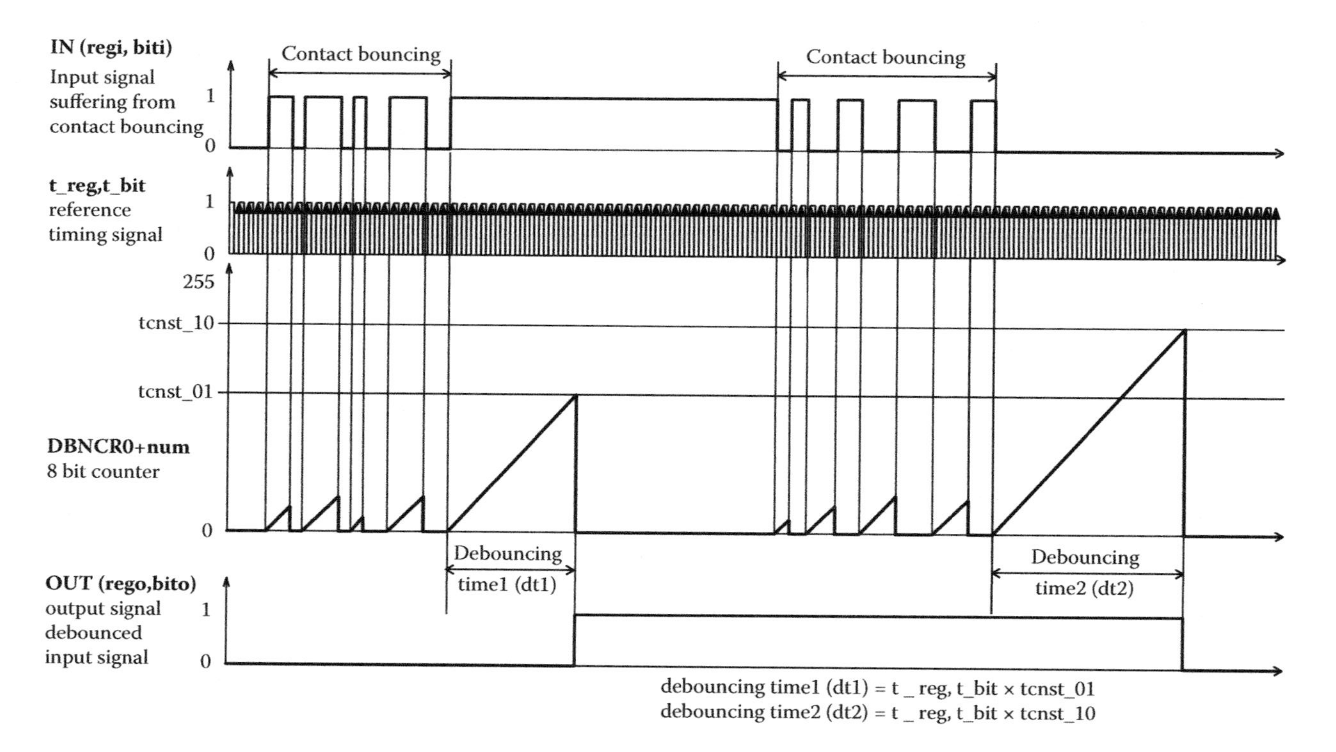

FIGURE 2.14
Detailed timing diagram of one of the channels of the macro `dbncr0`.

With the use of the macro dbncr0 it is possible to debounce 8 input signals; as we commit to have 16 discrete inputs in the PIC16F648A-based PLC project, there are 8 more input signals to be debounced. To solve this problem the macro dbncr1 is introduced. It works in the same manner as the macro dbncr0. The macro dbncr1 is shown in Figure 2.15. Table 2.2 shows the schematic symbol of the macro dbncr1. For this macro it is necessary to define the following 8-bit variables in SRAM: Temp_1 and DBNCRRED1. Each bit of the variable DBNCRRED1 is used to detect the rising edge of the reference timing signal t_reg,t_bit for the related channel. In addition, it

```
;------------------- macro: debouncer1 -----------------
dbncr1 macro num,regi,biti,t_reg,t_bit,tcnst_01,tcnst_10,rego,bito
       local   L1,L2,L3,L4
       btfsc   rego,bito
       goto    L4
       btfsc   regi,biti
       goto    L2
       clrf    DBNCR1+num
       goto    L1
L4     btfss   regi,biti
       goto    L3
       clrf    DBNCR1+num
       goto    L1
L3     btfss   t_reg,t_bit
       bsf     DBNCRRED1,num
       btfss   t_reg,t_bit
       goto    L1
       btfss   DBNCRRED1,num
       goto    L1
       bcf     DBNCRRED1,num
       incf    DBNCR1+num,f
       movf    DBNCR1+num,w
       xorlw   tcnst_10
       skpnz
       bcf     rego,bito
       goto    L1
L2     btfss   t_reg,t_bit
       bsf     DBNCRRED1,num
       btfss   t_reg,t_bit
       goto    L1
       btfss   DBNCRRED1,num
       goto    L1
       bcf     DBNCRRED1,num
       incf    DBNCR1+num,f
       movf    DBNCR1+num,w
       xorlw   tcnst_01
       skpnz
       bsf     rego,bito
L1
       endm
;---------------------------------------------------
```

FIGURE 2.15
The macro dbncr1.

TABLE 2.2
Schematic Symbol of the Macro `dbncr1`

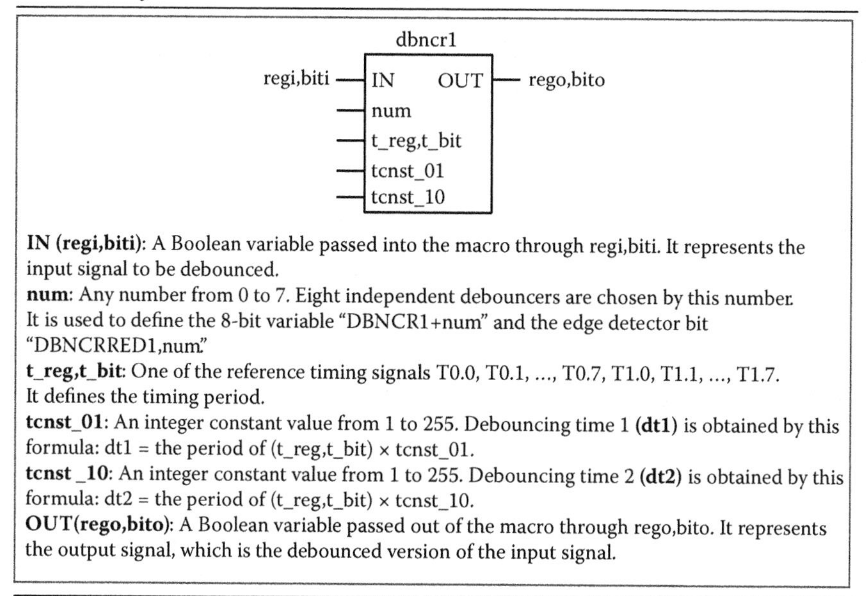

IN (regi,biti): A Boolean variable passed into the macro through regi,biti. It represents the input signal to be debounced.
num: Any number from 0 to 7. Eight independent debouncers are chosen by this number. It is used to define the 8-bit variable "DBNCR1+num" and the edge detector bit "DBNCRRED1,num."
t_reg,t_bit: One of the reference timing signals T0.0, T0.1, ..., T0.7, T1.0, T1.1, ..., T1.7. It defines the timing period.
tcnst_01: An integer constant value from 1 to 255. Debouncing time 1 (**dt1**) is obtained by this formula: dt1 = the period of (t_reg,t_bit) × tcnst_01.
tcnst _10: An integer constant value from 1 to 255. Debouncing time 2 (**dt2**) is obtained by this formula: dt2 = the period of (t_reg,t_bit) × tcnst_10.
OUT(rego,bito): A Boolean variable passed out of the macro through rego,bito. It represents the output signal, which is the debounced version of the input signal.

is also necessary to define eight 8-bit variables in successive SRAM locations, the first of which is to be defined as DBNCR1.

2.3 Basic Macros of the PIC16F648A-Based PLC

In this section the following basic three macros are considered: `initialize`, `get_inputs`, and `send_outputs`.

2.3.1 Macro `initialize`

The macro `initialize` is shown in Figure 2.16. There are mainly two tasks carried out within this macro. In the former, first, TMR0 is set up as a free-running hardware timer with the ¼ of 20 MHz oscillator signal, i.e., 5 MHz, and with a prescaler arranged to divide the signal to 256. In addition, PORTB is initialized to make RB0 `(data_in)` as input, and the following as outputs: RB3 `(clock_out)`, RB4 `(data_out)`, RB5 `(latch_out)`, RB6 `(clock_in)`, and RB7 `(shift/load)`. In the latter, all utilized SRAM registers are loaded with initial "safe values." In other words, all utilized SRAM registers are cleared (loaded with 00h) except for Temp_2, which is loaded with 06h.

```
;------------------ macro: initialize ---------------
initialize macro
        local   L1
        BANK1                       ;goto BANK1
        movlw   b'00000111'         ;W <-- b'00000111' : Fosc/4 --> TMR0, PS=256
        movwf   OPTION_REG          ;pull-up on PORTB, OPTION_REG <-- W
        movlw   b'00000001'         ;PORTB is both input and output port
        movwf   TRISB               ;TRISB <-- b'00000001'
        BANK0                       ;goto BANK0
        clrf    PORTA               ;Clear PortA
        clrf    PORTB               ;Clear PortB
        clrf    TMR0                ;Clear TMR0
        movlw   h'20'               ;initialize the pointer
        movwf   FSR                 ;to RAM
L1      clrf    INDF                ;clear INDF register
        incf    FSR,f               ;increment pointer
        btfss   FSR,7               ;all done?
        goto    L1                  ;if not goto L1
                                    ;if yes carry on
        movlw   06h                 ;W <--- 06h
        movwf   Temp_2              ;Temp_2 <--- W(06h)
        endm
;-------------------------------------------------------
```

FIGURE 2.16
The macro `initialize`.

As explained before, Temp_2 holds some special memory bits; therefore, the initial values of these special memory bits are put into Temp_2 within this macro. As a result, these special memory bits are loaded with the following initial values: LOGIC0 (Temp_2,0) = 0, LOGIC1 (Temp_2,1) = 1, FRSTSCN (Temp_2,2) = 1, SCNOSC (Temp_2,3) = 0.

2.3.2 Macro `get_inputs`

The macro `get_inputs` is shown in Figure 2.17. There are mainly three tasks carried out within this macro. In the first one, the macro `HC165` is called with the parameters .2 and bI0. This means that we will use the CPU board and two I/O extension boards; therefore, the macro `HC165` is called with the parameter .2. As explained before, the input information taken from the macro is rated as bouncing information, and therefore these 16-bit data are stored in bI0 and bI1 registers. For example, if we decide to use the CPU board connected to four I/O extension boards, then we must call the macro `HC165` as follows: `HC165 .4,bI0`. Then, this will take four 8-bit bouncing input data from the 74HC/LS165 ICs and put them to the four successive registers starting with the register bI0. In the second task within this macro, each bit of bI0,i (i = 0, 1, …, 7) is debounced by the macro `dbncr0`, and each debounced input signal is stored in the related bit I0,i (i = 0, 1, …, 7). Likewise, each bit of bI1,i (i = 0, 1, …, 7) is debounced by the macro `dbncr1`, and each debounced input signal is stored in the related bit I1,i (i = 0, 1, …, 7). In general, a 10 ms time delay is enough for debouncing both rising and falling edges of an input signal. Therefore, to achieve these time delays, the

```
;------------------- macro: get_inputs ---------------
get_inputs  macro
        local   Nzero
        HC165   .2,bI0                          ;obtain the 16 inputs from
        dbncr0  0,bI0.0,T0.2,.25,.25,I0.0       ;2 input registers (74HC165)
        dbncr0  1,bI0.1,T0.2,.25,.25,I0.1       ;and put them into bI0 and bI1
        dbncr0  2,bI0.2,T0.2,.25,.25,I0.2       ;registers within PIC16F648A.
        dbncr0  3,bI0.3,T0.2,.25,.25,I0.3       ;Then debounce all bits of
        dbncr0  4,bI0.4,T0.2,.25,.25,I0.4       ;bI0.
        dbncr0  5,bI0.5,T0.2,.25,.25,I0.5       ;The debounced input signals
        dbncr0  6,bI0.6,T0.2,.25,.25,I0.6       ;are stored in the register
        dbncr0  7,bI0.7,T0.2,.25,.25,I0.7       ;I0
        ;dt1=dt2=0.4096 ms x 25 = 10,24 ms      ;
        dbncr1  0,bI1.0,T0.2,.25,.25,I1.0       ;Likewise debounce all bits of
        dbncr1  1,bI1.1,T0.2,.25,.25,I1.1       ;
        dbncr1  2,bI1.2,T0.2,.25,.25,I1.2       ;bI1.
        dbncr1  3,bI1.3,T0.2,.25,.25,I1.3       ;
        dbncr1  4,bI1.4,T0.2,.25,.25,I1.4       ;The debounced input signals
        dbncr1  5,bI1.5,T0.2,.25,.25,I1.5       ;
        dbncr1  6,bI1.6,T0.2,.25,.25,I1.6       ;are stored in the register
        dbncr1  7,bI1.7,T0.2,.25,.25,I1.7       ;I1

        btfsc   Timer_1,7               ;
        bsf     Temp_2,4                ;Increment Timer_2 on Timer_1 overflow
        btfsc   Timer_1,7               ;
        goto    Nzero                   ;
        btfss   Temp_2,4                ;
        goto    Nzero                   ;
        incf    Timer_2,f               ;
        bcf     Temp_2,4                ;
Nzero
        endm
;-----------------------------------------------------
```

FIGURE 2.17
The macro `get_inputs`.

reference timing signal, obtained from Timer_1, is chosen as T0.2 (0.4096 ms period), and both `tcnst_01` and `tcnst_10` are chosen to be 25. Then we obtain the following: `dt1` = T0.2 × `tcnst_01` = (0.4096 ms) × 25 = 10.24 ms, `dt2` = T0.2 × `tcnst_01` = (0.4096 ms) × 25 = 10.24 ms. The last task is about incrementing the Timer_2 on overflow of Timer_1. In this task, Timer_2 is incremented by one when the falling edge of the bit Timer_1,7 is detected. In order to detect the falling edge of the bit Timer_1,7, Temp_2,4 bit is utilized.

2.3.3 Macro `send_outputs`

The macro `send_outputs` is shown in Figure 2.18. There are mainly four tasks carried out within this macro. In the first one, the macro `HC595` is called with the parameters .2 and Q0. This means that we will use the CPU board and two I/O extension boards; therefore, the macro `HC595` is called with the parameter .2. As explained before, 16-bit output data are taken from the registers Q0 and Q1, and this macro sends the bits of Q0 and Q1 serially to TPIC6B595 registers. For example, if we decide to use the CPU board connected to four I/O extension boards, then we must call the macro `HC595`

```
;------------------- macro: send_outputs -------------
send_outputs        macro
        local   L1,L2
        HC595   .2,Q0               ;take the registers Q0 and Q1 from PIC16F648A
                                    ;and put them into output registers Q0 and Q1(TPIC6B595)

        clrwdt                      ;clear the watchdog timer

        bcf     FRSTSCN             ;reset the FRSTSCN bit

        btfss   SCNOSC              ;toggle
        goto    L2                  ;the SCNOSC bit
        bcf     SCNOSC              ;after a program
        goto    L1                  ;scan
L2      bsf     SCNOSC              ;
L1
        endm
;------------------------------------------------------
```

FIGURE 2.18
The macro send_outputs.

as follows: HC595.4,Q0. Then, the macro HC595 will take four 8-bit output data stored in Q3, Q2, Q1, and Q0 and send them serially to the four TPIC6B595 register ICs, respectively. In the second task within this macro, the watchdog timer is cleared. In the third task, the FRSTSCN special memory bit is reset. As the final task, within this macro the SCNOSC special memory bit is toggled after a program scan; i.e., when it is 1 it is reset, and when it is 0 it is set.

2.4 Example Program

Up to now we have seen the hardware and basic software necessary for the PIC16F648A-based PLC. It is now time to consider a simple example. Before you can run the simple example considered here, you are expected to construct your own PIC16F648A-based PLC hardware by using the necessary PCB files, and producing your PCBs, with their components. The user program of the example UZAM_plc_16i16o_ex2.asm is shown in Figure 2.19. The file UZAM_plc_16i16o_ex2.asm is included within the CD-ROM attached to this book. Please open it by MPLAB integrated development environment

```
.
;--------------- user program starts here -----------------------
        movfw I0
        movwf Q0
        movfw Timer_2
        movwf Q1
;--------------- user program ends here -------------------------
.
```

FIGURE 2.19
The user program of UZAM_plc_16i16o_ex2.asm.

(IDE) and compile it. After that, by using the PIC programmer software, take the compiled file UZAM_plc_16i16o_ex2.hex, and by your PIC programmer hardware send it to the program memory of PIC16F648A microcontroller within the PIC16F648A-based PLC. To do this, switch the 4PDT in PROG position and the power switch in OFF position. After loading the UZAM_plc_16i16o_ex2.hex file, switch the 4PDT in RUN position and the power switch in ON position. Now, you are ready to test the first example program. There are mainly two different operations done. In the first part, eight inputs, namely, bits I0.0, I0.1, …, I0.7, are transferred to the respective eight outputs, namely, bits Q0.0, Q0.1, …, Q0.7. That is, if I0.0 = 0, then Q0.0 = 0, and similarly, if I0.0 = 1, then Q0.0 = 1. This applies to all eight inputs I0 – eight outputs Q0. In the second part, the contents of the Timer_2 register, namely, T1.0, T1.1, …, T1.7, are transferred to eight outputs Q1, namely, Q1.0, Q1.1, …, Q1.7, respectively.

3

Contact and Relay-Based Macros

In this chapter, the following contact and relay-based macros are described:

`ld` (load)
`ld_not` (load not)
`not`
`or`
`or_not`
`nor`
`and`
`and_not`
`nand`
`xor`
`xor_not`
`xnor`
`out`
`out_not`
`in_out`
`inv_out`
`_set`
`_reset`

The file definitions.inc, included within the CD-ROM attached to this book, contains all macros defined for the PIC16F648A-based PLC. The contact and relay-based macros are defined to operate on Boolean (1-bit) variables. The working register W is utilized to transfer the information to or from the contact and relay-based macros, except for macros `in_out` and `inv_out`. Let us now briefly consider these macros.

TABLE 3.1

Truth Table and Symbols of the Macro `ld`

Truth Table		Ladder Diagram Symbol	Schematic Symbol
IN	OUT	reg,bit — W (NO contact)	reg,bit —— W
reg,bit	W		
0	0		
1	1		

3.1 Macro `ld` (load)

The truth table and symbols of the macro `ld` are depicted in Table 3.1. Figure 3.1 shows the macro `ld` and its flowchart. This macro has a Boolean input variable passed into it as `reg,bit` and a Boolean output variable passed out through W. In ladder logic, this macro is represented by a normally open (NO) contact. When the input variable is 0 (respectively 1), the output (W) is forced to 0 (respectively to 1). Operands for the instruction `ld` are shown in Table 3.2.

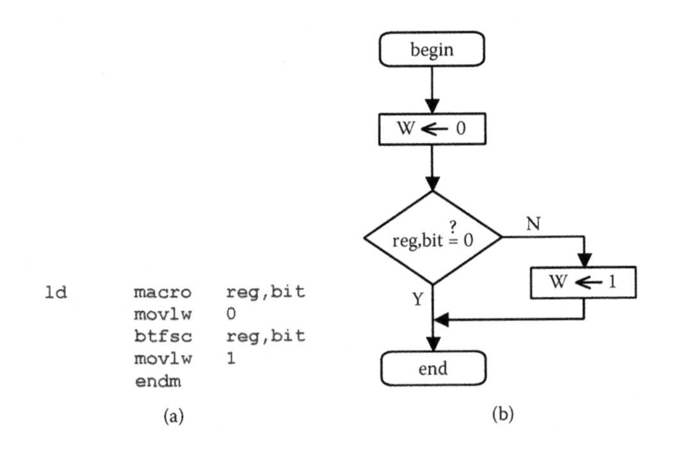

FIGURE 3.1
(a) The macro `ld` and (b) its flowchart.

TABLE 3.2

Operands for the Instruction `ld`

Input (reg,bit)	Data Type	Operands
Bit	BOOL	I, Q, M, TON8_Q, TOF8_Q, TP8_Q, TOS8_Q, CTU8_Q, CTD8_Q, CTUD8_Q, LOGIC1, LOGIC0, FRSTSCN, SCNOSC

TABLE 3.3

Truth Table and Symbols of the Macro `ld_not`

Truth Table		Ladder Diagram Symbol	Schematic Symbol
IN	OUT	reg,bit — W (NC contact)	reg,bit → (NOT) → W
reg,bit	W		
0	1		
1	0		

3.2 Macro `ld_not` (load not)

The truth table and symbols of the macro `ld_not` are depicted in Table 3.3. Figure 3.2 shows the macro `ld_not` and its flowchart. This macro has a Boolean input variable passed into it as `reg,bit`, and a Boolean output variable passed out through W. In ladder logic, this macro is represented by a normally closed (NC) contact. When the input variable is 0 (respectively 1), the output (W) is forced to 1 (respectively to 0). Operands for the instruction `ld_not` are shown in Table 3.4.

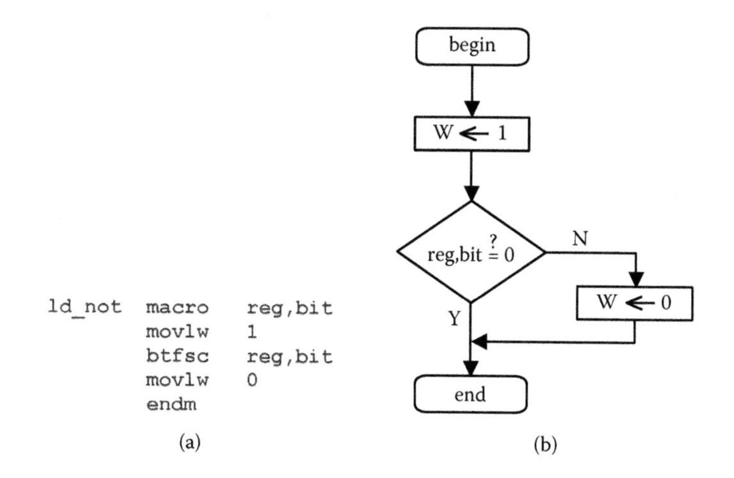

FIGURE 3.2
(a) The macro `ld_not` and (b) its flowchart.

TABLE 3.4

Operands for the Instruction `ld_not`

Input (reg,bit)	Data Type	Operands
Bit	BOOL	I, Q, M, TON8_Q, TOF8_Q, TP8_Q, TOS8_Q, CTU8_Q, CTD8_Q, CTUD8_Q, LOGIC1, LOGIC0, FRSTSCN, SCNOSC

TABLE 3.5

Truth Table and Symbols of the Macro `not`

Truth Table		Ladder Diagram symbol	Schematic Symbol
IN	OUT	W —|NOT|— W	W —▷o— W
W	W		
0	1		
1	0		

3.3 Macro `not`

The truth table and symbols of the macro `not` are depicted in Table 3.5. Figure 3.3 shows the macro `not` and its flowchart. This macro is used as a logical NOT gate. The input is taken from W, and the output is send out by W. When the input variable is 0 (respectively 1), the output (W) is forced to 1 (respectively to 0).

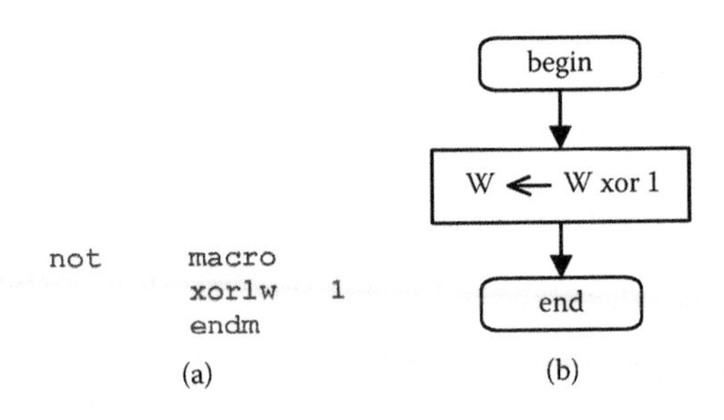

FIGURE 3.3
(a) The macro `not` and (b) its flowchart.

TABLE 3.6

Truth Table and Symbols of the Macro `or`

Truth Table			Ladder diagram symbol	Schematic symbol
IN1	IN2	OUT		
W	reg,bit	W	W, reg,bit → W	W, reg,bit → W
0	0	0		
0	1	1		
1	0	1		
1	1	1		

3.4 Macro `or`

The truth table and symbols of the macro `or` are depicted in Table 3.6. Figure 3.4 shows the macro `or` and its flowchart. This macro is used as a two-input logical OR gate. One input is taken from W, and the other one is taken from `reg,bit`. The result is passed out of the macro through W. Operands for the instruction `or` are shown in Table 3.7.

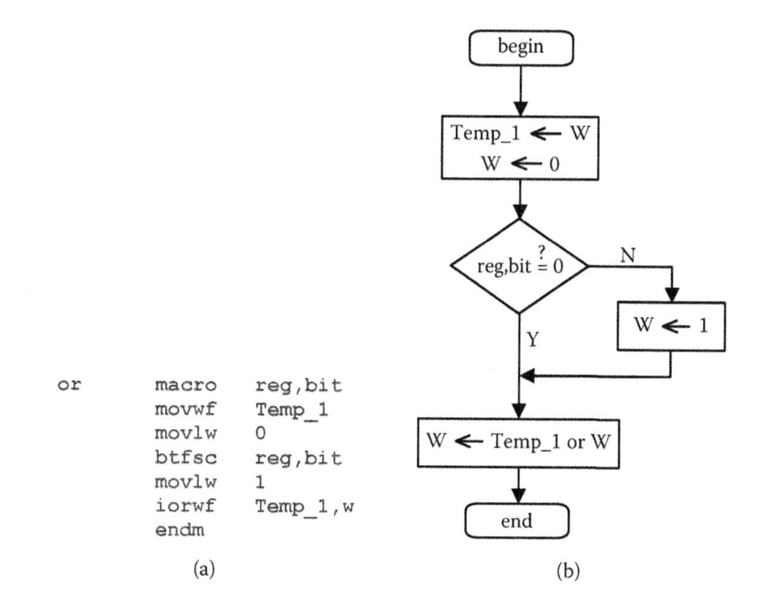

```
or      macro   reg,bit
        movwf   Temp_1
        movlw   0
        btfsc   reg,bit
        movlw   1
        iorwf   Temp_1,w
        endm
```


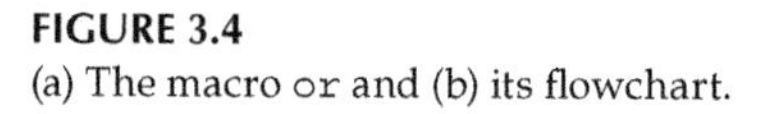

FIGURE 3.4
(a) The macro `or` and (b) its flowchart.

TABLE 3.7

Operands for the Instruction `or`

Input (reg,bit)	Data Type	Operands
Bit	BOOL	I, Q, M, TON8_Q, TOF8_Q, TP8_Q, TOS8_Q, CTU8_Q, CTD8_Q, CTUD8_Q, LOGIC1, LOGIC0, FRSTSCN, SCNOSC

3.5 Macro `or_not`

The truth table and symbols of the macro `or_not` are depicted in Table 3.8. Figure 3.5 shows the macro `or_not` and its flowchart. This macro is also used as a two-input logical OR gate, but this time one of the inputs is inverted. One input is taken from W, and the inverted input is taken from `reg,bit`. The result is passed out of the macro through W. Operands for the instruction `or_not` are shown in Table 3.9.

3.6 Macro `nor`

The truth table and symbols of the macro `nor` are depicted in Table 3.10. Figure 3.6 shows the macro `nor` and its flowchart. This macro is used as a two-input logical NOR gate. One input is taken from W, and the other input is taken from `reg,bit`. The result is passed out of the macro through W. Operands for the instruction `nor` are shown in Table 3.11.

TABLE 3.8

Truth Table and Symbols of the Macro `or_not`

Truth Table			Ladder Diagram Symbol	Schematic Symbol
IN1	IN2	OUT	W, reg,bit → W	W, reg,bit → W
W	reg,bit	W		
0	0	1		
0	1	0		
1	0	1		
1	1	1		

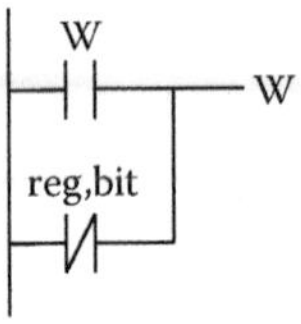

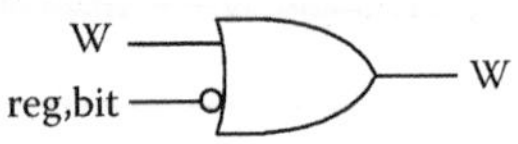

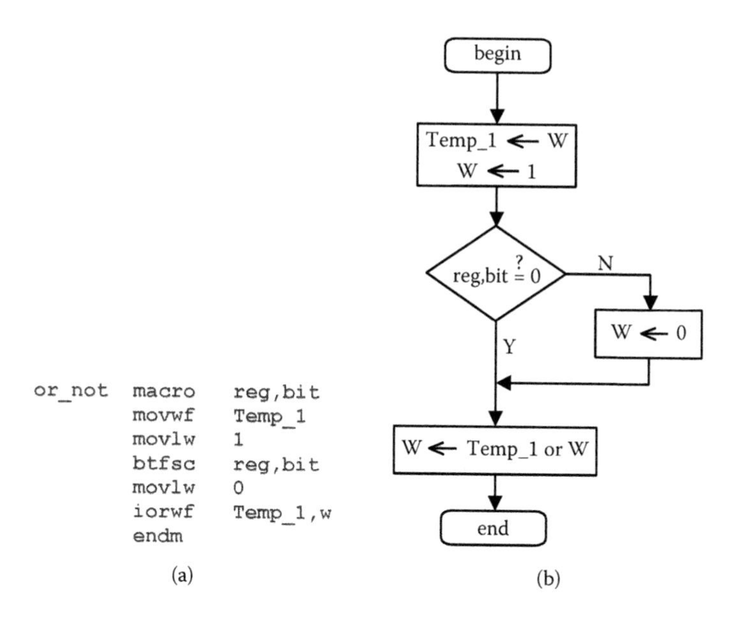

FIGURE 3.5
(a) The macro `or_not` and (b) its flowchart.

TABLE 3.9

Operands for the Instruction `or_not`

Input (reg,bit)	Data Type	Operands
Bit	BOOL	I, Q, M, TON8_Q, TOF8_Q, TP8_Q, TOS8_Q, CTU8_Q, CTD8_Q, CTUD8_Q, LOGIC1, LOGIC0, FRSTSCN, SCNOSC

TABLE 3.10

Truth Table and Symbols of the Macro `nor`

Truth Table			Ladder Diagram Symbol	Schematic Symbol
IN1	IN2	OUT	W, reg,bit, NOT, W	W, reg,bit, W
W	reg,bit	W		
0	0	1		
0	1	0		
1	0	0		
1	1	0		

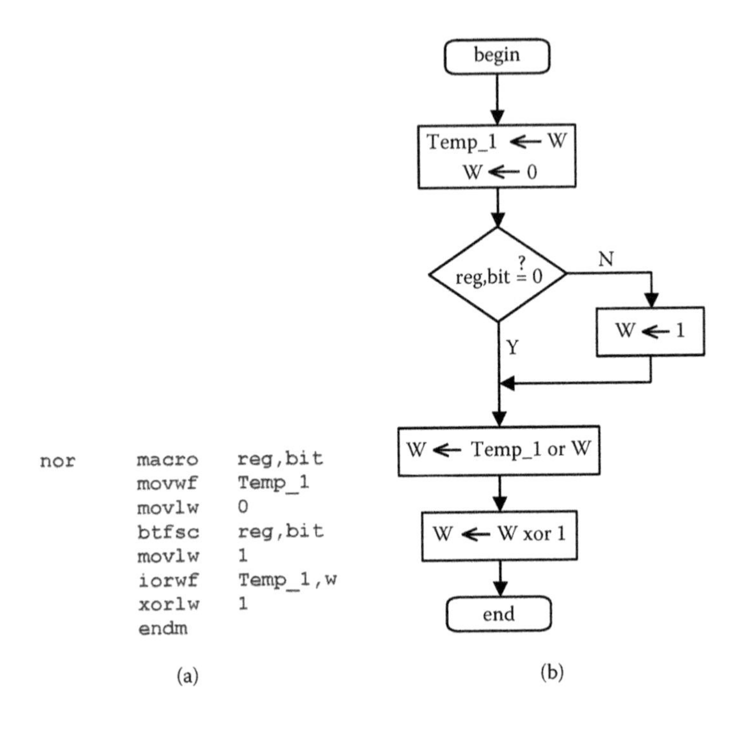


```
nor     macro   reg,bit
        movwf   Temp_1
        movlw   0
        btfsc   reg,bit
        movlw   1
        iorwf   Temp_1,w
        xorlw   1
        endm
```

(a)

FIGURE 3.6
(a) The macro nor and (b) its flowchart.

TABLE 3.11

Operands for the Instruction nor

Input (reg,bit)	Data Type	Operands
Bit	BOOL	I, Q, M, TON8_Q, TOF8_Q, TP8_Q, TOS8_Q, CTU8_Q, CTD8_Q, CTUD8_Q, LOGIC1, LOGIC0, FRSTSCN, SCNOSC

3.7 Macro and

The truth table and symbols of the macro and are depicted in Table 3.12. Figure 3.7 shows the macro and and its flowchart. This macro is used as a two-input logical AND gate. One input is taken from W, and the other one is taken from `reg,bit`. The result is passed out of the macro through W. Operands for the instruction and are shown in Table 3.13.

TABLE 3.12

Truth Table and Symbols of the Macro and

Truth Table:

IN1	IN2	OUT
W	reg,bit	W
0	0	0
0	1	0
1	0	0
1	1	1

Ladder Diagram Symbol: W, reg,bit, W

Schematic Symbol: W, reg,bit, W

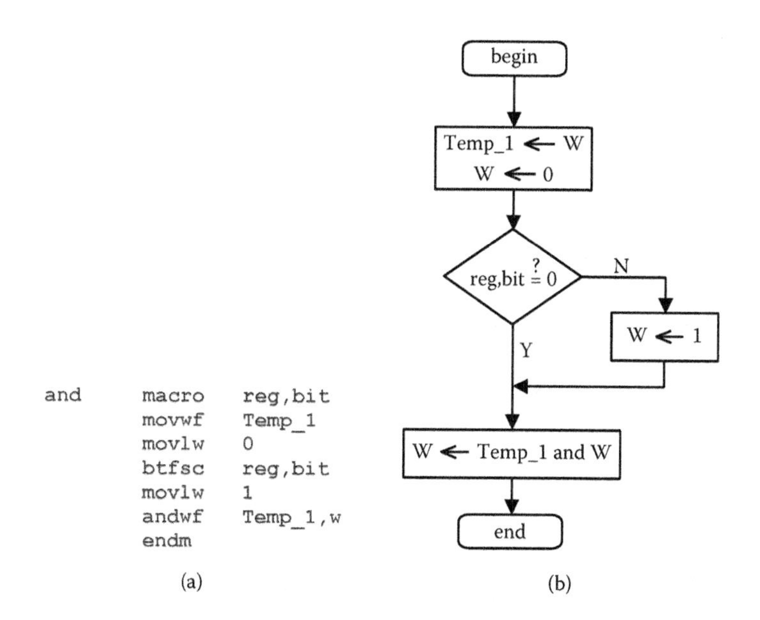

FIGURE 3.7
(a) The macro and and (b) its flowchart.

TABLE 3.13

Operands for the Instruction and

Input (reg,bit)	Data Type	Operands
Bit	BOOL	I, Q, M, TON8_Q, TOF8_Q, TP8_Q, TOS8_Q, CTU8_Q, CTD8_Q, CTUD8_Q, LOGIC1, LOGIC0, FRSTSCN, SCNOSC

TABLE 3.14

Truth Table and Symbols of the Macro `and_not`

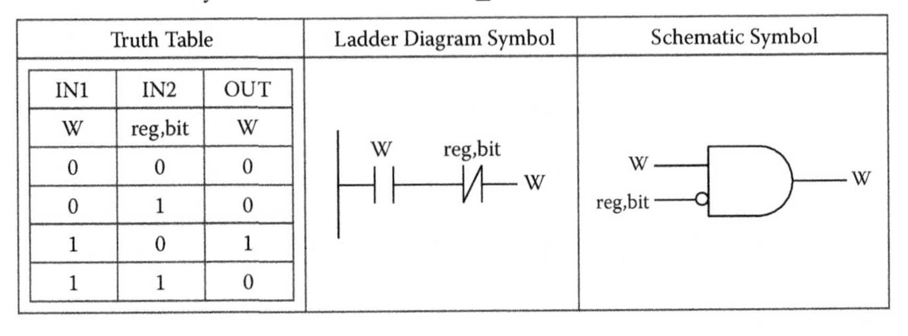

Truth Table			Ladder Diagram Symbol	Schematic Symbol
IN1	IN2	OUT		
W	reg,bit	W		
0	0	0		
0	1	0		
1	0	1		
1	1	0		

3.8 Macro `and_not`

The truth table and symbols of the macro `and_not` are depicted in Table 3.14. Figure 3.8 shows the macro `and_not` and its flowchart. This macro is also used as a two-input logical AND gate, but this time one of the inputs is inverted. One input is taken from W, and the inverted input is taken from `reg,bit`. The result is passed out of the macro through W. Operands for the instruction `and_not` are shown in Table 3.15.

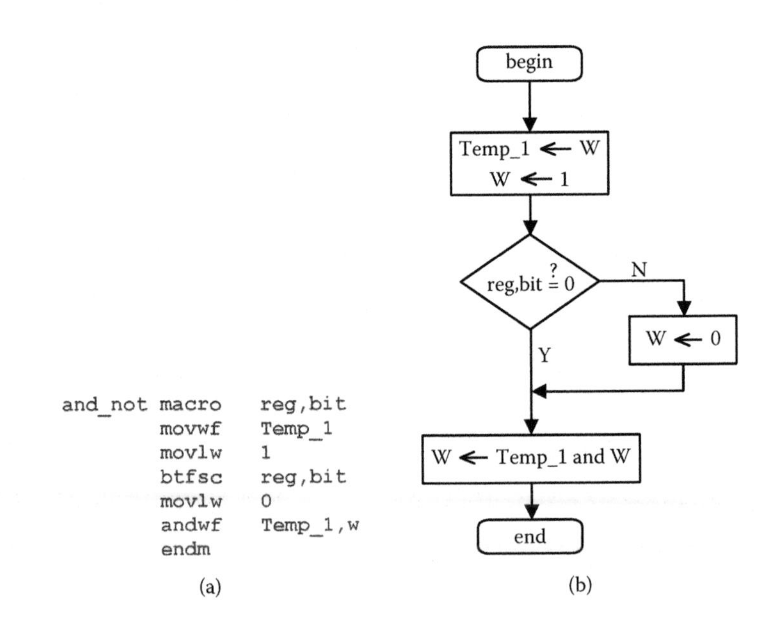

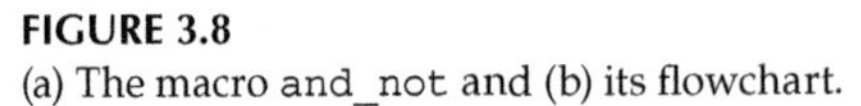

FIGURE 3.8
(a) The macro `and_not` and (b) its flowchart.

TABLE 3.15

Operands for the Instruction `and_not`

Input (reg,bit)	Data Type	Operands
Bit	BOOL	I, Q, M, TON8_Q, TOF8_Q, TP8_Q, TOS8_Q, CTU8_Q, CTD8_Q, CTUD8_Q, LOGIC1, LOGIC0, FRSTSCN, SCNOSC

3.9 Macro nand

The truth table and symbols of the macro `nand` are depicted in Table 3.16. Figure 3.9 shows the macro `nand` and its flowchart. This macro is used as a two-input logical NAND gate. One input is taken from W, and the other one is taken from `reg,bit`. The result is passed out of the macro through W. Operands for the instruction `nand` are shown in Table 3.17.

3.10 Macro xor

The truth table and symbols of the macro `xor` are depicted in Table 3.18. Figure 3.10 shows the macro `xor` and its flowchart. This macro is used as a two-input logical EXOR gate. One input is taken from W, and the other one is taken from `reg,bit`. The result is passed out of the macro through W. Operands for the instruction `xor` are shown in Table 3.19.

TABLE 3.16

Truth Table and Symbols of the Macro `nand`

Truth Table			Ladder Diagram Symbol	Schematic Symbol
IN1	IN2	OUT	W reg,bit NOT W	W reg,bit W
W	reg,bit	W		
0	0	1		
0	1	1		
1	0	1		
1	1	0		

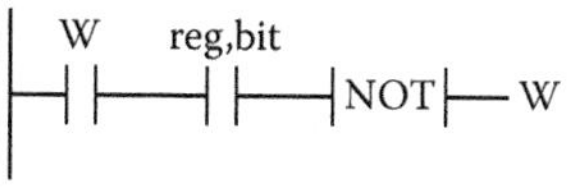

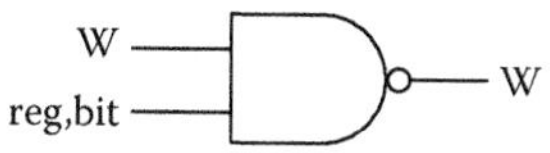

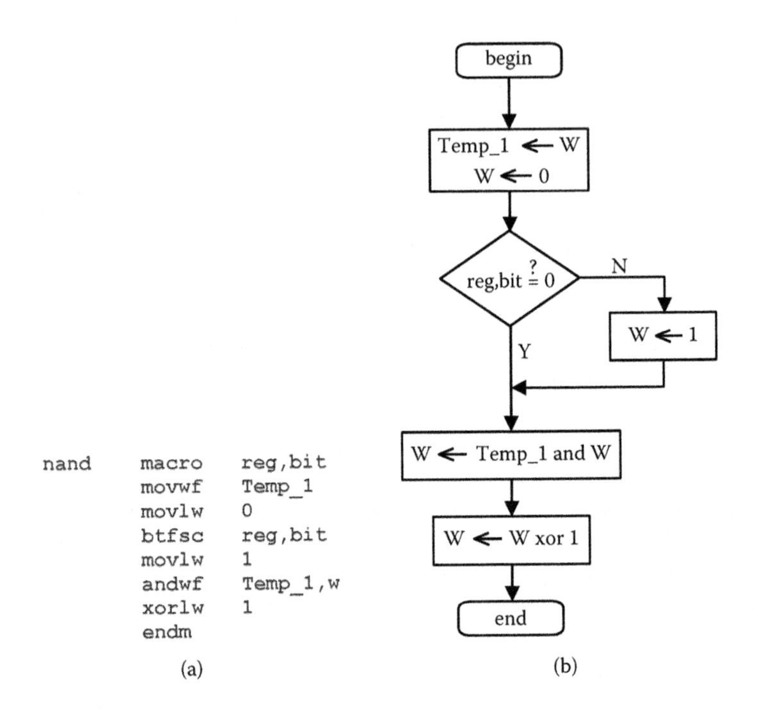

FIGURE 3.9
(a) The macro nand and (b) its flowchart.

TABLE 3.17
Operands for the Instruction nand

Input (reg,bit)	Data Type	Operands
Bit	BOOL	I, Q, M, TON8_Q, TOF8_Q, TP8_Q, TOS8_Q, CTU8_Q, CTD8_Q, CTUD8_Q, LOGIC1, LOGIC0, FRSTSCN, SCNOSC

TABLE 3.18
Truth Table and Symbols of the Macro xor

Truth Table			Ladder Diagram Symbol	Schematic Symbol
IN1	IN2	OUT	W reg,bit W W reg,bit	W reg,bit W
W	reg,bit	W		
0	0	0		
0	1	1		
1	0	1		
1	1	0		

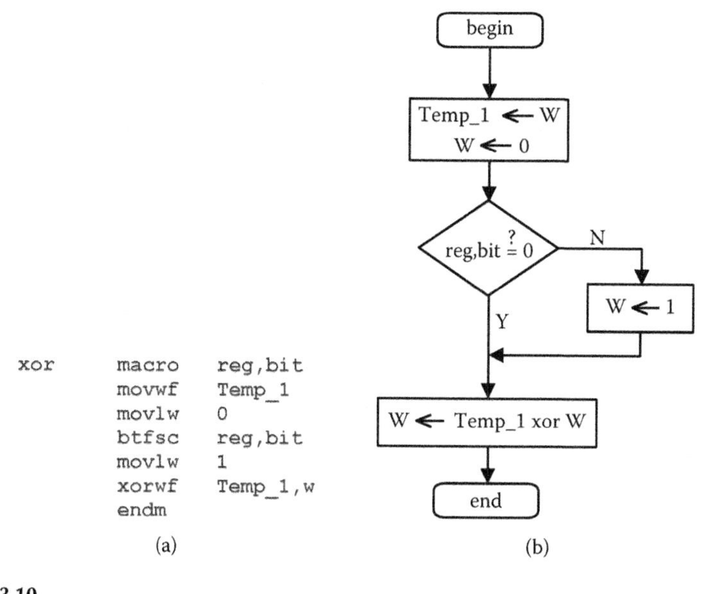

FIGURE 3.10
(a) The macro xor and (b) its flowchart.

3.11 Macro xor_not

The truth table and symbols of the macro xor_not are depicted in Table 3.20. Figure 3.11 shows the macro xor_not and its flowchart. This macro is also used as a two-input logical EXOR gate, but this time one of the inputs is inverted. One input is taken from W, and the inverted input is taken from reg,bit. The result is passed out of the macro through W. Operands for the instruction xor_not are shown in Table 3.21.

3.12 Macro xnor

The truth table and symbols of the macro xnor are depicted in Table 3.22. Figure 3.12 shows the macro xnor and its flowchart. This macro is used as a two-input logical EXNOR gate. One input is taken from W, and the other

TABLE 3.19

Operands for the Instruction xor

Input (reg,bit)	Data Type	Operands
Bit	BOOL	I, Q, M, TON8_Q, TOF8_Q, TP8_Q, TOS8_Q, CTU8_Q, CTD8_Q, CTUD8_Q, LOGIC1, LOGIC0, FRSTSCN, SCNOSC

TABLE 3.20

Truth Table and Symbols of the Macro `xor_not`

Truth Table	Ladder Diagram Symbol	Schematic Symbol
(see below)	W, reg,bit (normally open contacts in series) parallel with W, reg,bit (normally closed contacts in series) → W	W, reg,bit (inverted input) → XOR gate → W

IN1	IN2	OUT
W	reg,bit	W
0	0	1
0	1	0
1	0	0
1	1	1

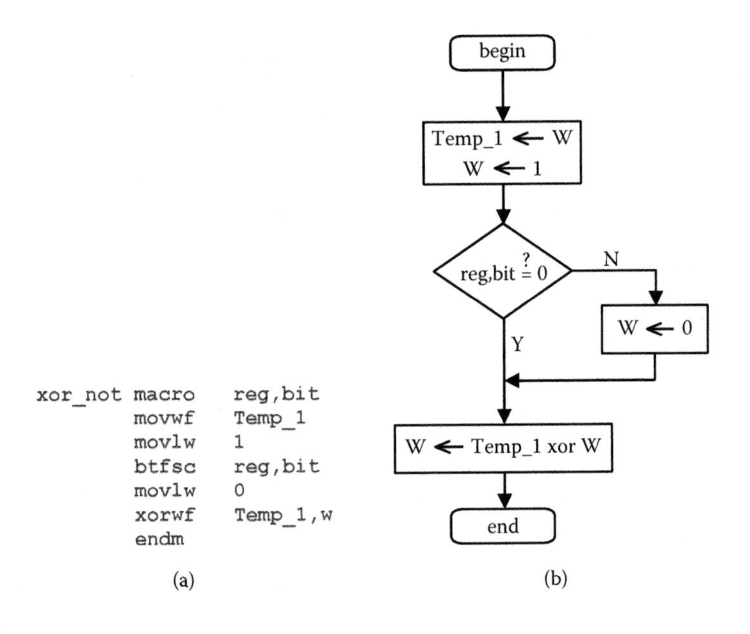

FIGURE 3.11
(a) The macro `xor_not` and (b) its flowchart.

TABLE 3.21

Operands for the Instruction `xor_not`

Input (reg,bit)	Data Type	Operands
Bit	BOOL	I, Q, M, TON8_Q, TOF8_Q, TP8_Q, TOS8_Q, CTU8_Q, CTD8_Q, CTUD8_Q, LOGIC1, LOGIC0, FRSTSCN, SCNOSC

TABLE 3.22

Truth Table and Symbols of the Macro xnor

Truth Table			Ladder Diagram Symbol	Schematic Symbol
IN1	IN2	OUT	W reg,bit NOT W / W reg,bit	W, reg,bit → W
W	reg,bit	W		
0	0	1		
0	1	0		
1	0	0		
1	1	1		

one is taken from `reg,bit`. The result is passed out of the macro through W. Operands for the instruction xnor are shown in Table 3.23.

3.13 Macro out

The truth table and symbols of the macro out are depicted in Table 3.24. Figure 3.13 shows the macro out and its flowchart. This macro has a Boolean input variable passed into it by W and a Boolean output variable passed out

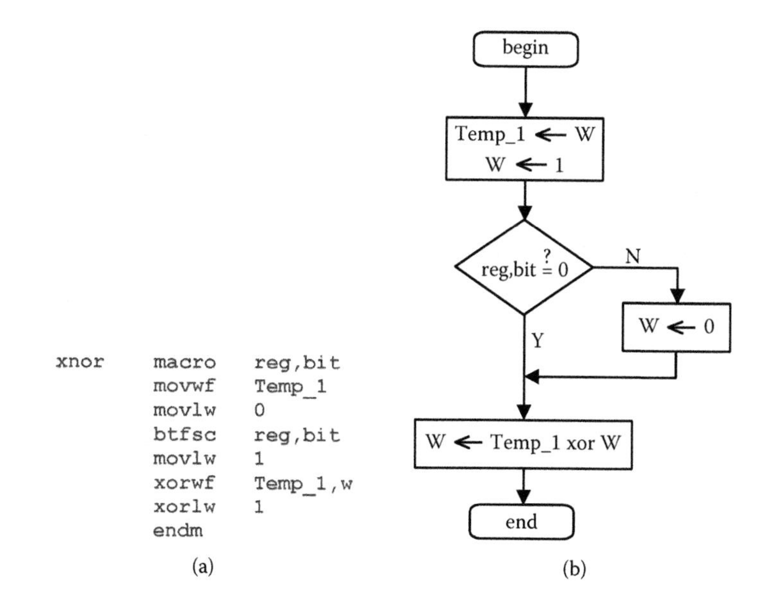

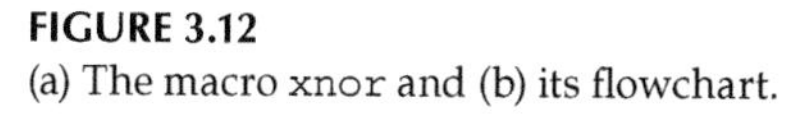

FIGURE 3.12
(a) The macro xnor and (b) its flowchart.

TABLE 3.23

Operands for the Instruction `xnor`

Input (reg,bit)	Data Type	Operands
Bit	BOOL	I, Q, M, TON8_Q, TOF8_Q, TP8_Q, TOS8_Q, CTU8_Q, CTD8_Q, CTUD8_Q, LOGIC1, LOGIC0, FRSTSCN, SCNOSC

TABLE 3.24

Truth Table and Symbols of the Macro `out`

Truth Table		Ladder diagram symbol	Schematic symbol
IN	OUT	W —( )— reg,bit	W —[reg,bit>
W	reg,bit		
0	0		
1	1		

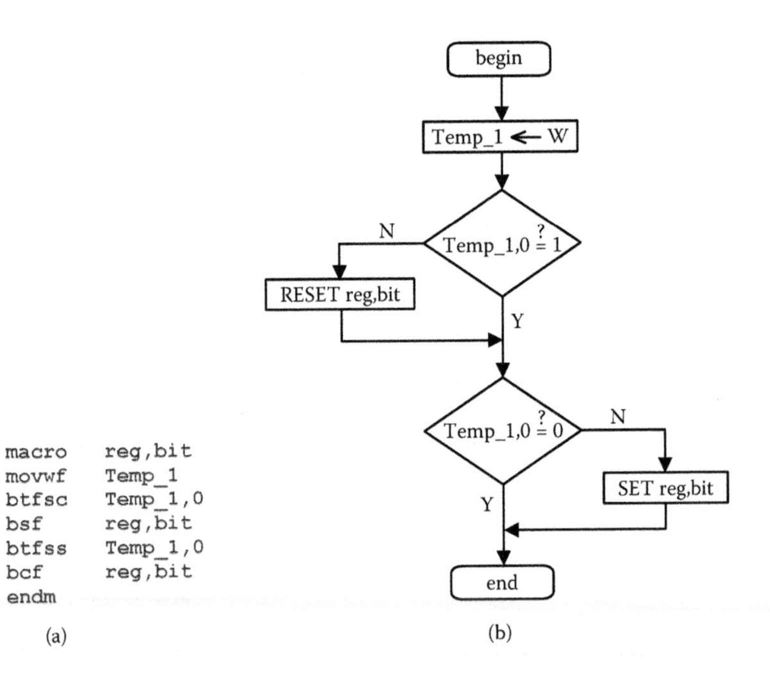

FIGURE 3.13
(a) The macro `out` and (b) its flowchart.

TABLE 3.25

Operands for the Instruction out

Output (reg,bit)	Data Type	Operands
Bit	BOOL	Q, M, TON8_Q, TOF8_Q, TP8_Q, TOS8_Q, CTU8_Q, CTD8_Q, CTUD8_Q

through reg,bit. In ladder logic, this macro is represented by an output relay (internal or external relay). When the input variable is 0 (respectively 1), the output (W) is forced to 0 (respectively to 1). Operands for the instruction out are shown in Table 3.25.

3.14 Macro out_not

The truth table and symbols of the macro out_not are depicted in Table 3.26. Figure 3.14 shows the macro out_not and its flowchart. This macro has a Boolean input variable passed into it by W and a Boolean output variable passed out through reg,bit. In ladder logic, this macro is represented by an inverted output relay (internal or external relay). When the input variable is 0 (respectively 1), the output (W) is forced to 1 (respectively to 0). Operands for the instruction out_not are shown in Table 3.27.

TABLE 3.26

The Truth Table and Symbols of the Macro out_not

Truth Table		Ladder Diagram Symbol	Schematic Symbol
IN	OUT	W —(/)— reg,bit	W —▷o— reg,bit
W	reg,bit		
0	1		
1	0		

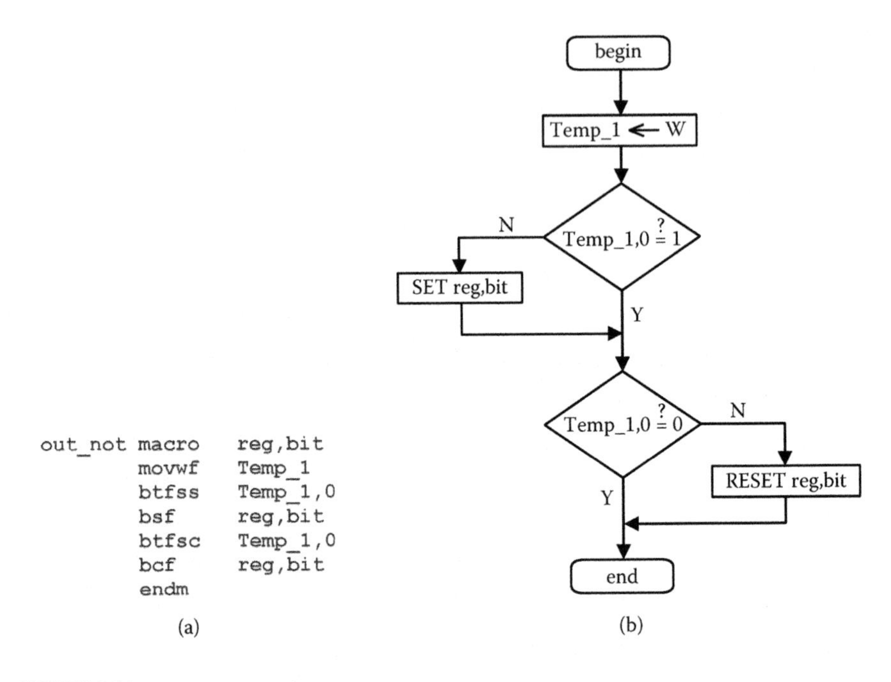

FIGURE 3.14
(a) The macro `out_not` and (b) its flowchart.

3.15 Macro `in_out`

The truth table and symbols of the macro `in_out` are depicted in Table 3.28. Figure 3.15 shows the macro `in_out` and its flowchart. This macro has a Boolean input variable passed into it by `regi,biti` and a Boolean output variable passed out through `rego,bito`. When the input variable `regi,biti` is 0 (respectively 1), the output variable `rego,bito` is forced to 0 (respectively to 1). Operands for the instruction `in_out` are shown in Table 3.29.

TABLE 3.27

Operands for the Instruction `out_not`

Output (reg,bit)	Data Type	Operands
Bit	BOOL	Q, M, TON8_Q, TOF8_Q, TP8_Q, TOS8_Q, CTU8_Q, CTD8_Q, CTUD8_Q

TABLE 3.28

Truth Table and Symbols of the Macro `in_out`

Truth Table		Ladder Diagram Symbol	Schematic Symbol
IN	OUT	regi,biti —| |— rego,bito —( )—	regi,biti ▷ — rego,bito ▷
regi,biti	rego,bito		
0	0		
1	1		

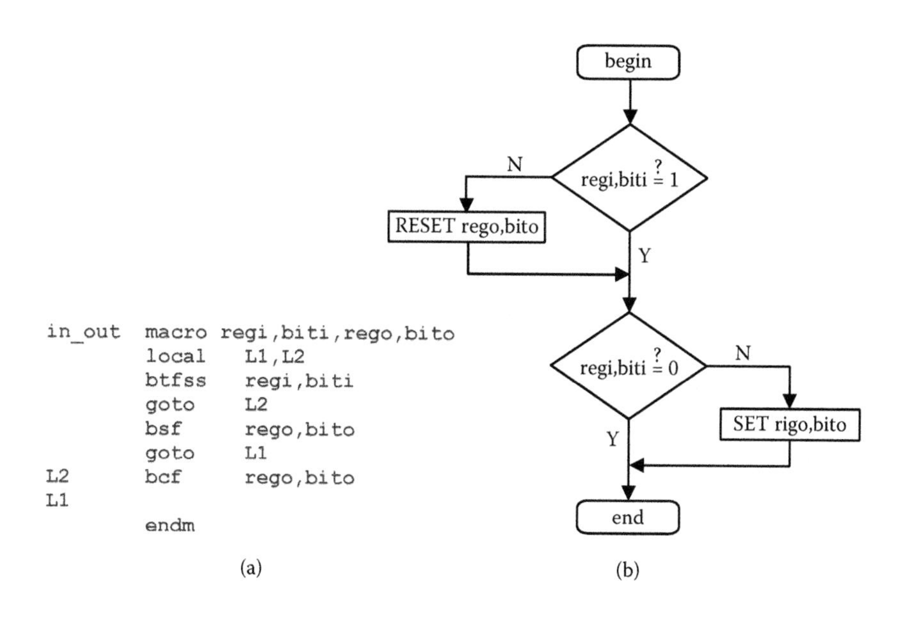

FIGURE 3.15
(a) The macro `in_out` and (b) its flowchart.

TABLE 3.29

Operands for the Instruction `in_out`

Input/Output	Data Type	Operands
Input (regi,biti) Bit	BOOL	I, Q, M, TON8_Q, TOF8_Q, TP8_Q, TOS8_Q, CTU8_Q, CTD8_Q, CTUD8_Q, LOGIC1, LOGIC0, FRSTSCN, SCNOSC
Output (rego,bito) Bit	BOOL	Q, M, TON8_Q, TOF8_Q, TP8_Q, TOS8_Q, CTU8_Q, CTD8_Q, CTUD8_Q

TABLE 3.30

Truth Table and Symbols of the Macro `inv_out`

Truth Table	Ladder Diagram Symbol	Schematic Symbol
IN: regi,biti — OUT: rego,bito; 0 → 1; 1 → 0	regi,biti (normally closed contact) — rego,bito (coil) or regi,biti (normally open contact) — rego,bito (negated coil)	regi,biti → NOT → rego,bito

IN	OUT
regi,biti	rego,bito
0	1
1	0

3.16 Macro `inv_out`

The truth table and symbols of the macro `inv_out` are depicted in Table 3.30. Figure 3.16 shows the macro `inv_out` and its flowchart. This macro has a Boolean input variable passed into it by `regi,biti` and a Boolean output variable passed out through `rego,bito`. When the input variable `regi,biti`

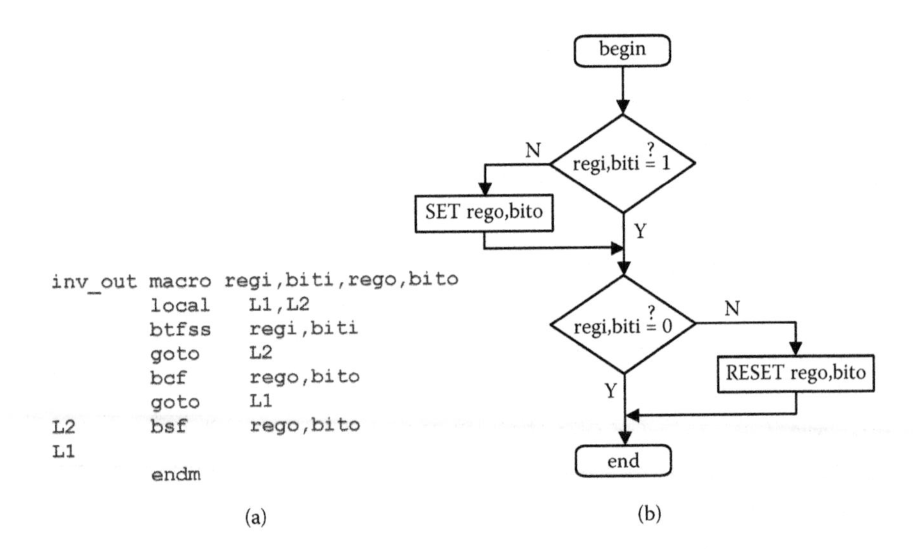

FIGURE 3.16
(a) The macro `inv_out` and (b) its flowchart.

TABLE 3.31

Operands for the Instruction `inv_out`

Input/Output	Data Type	Operands
Input (regi,biti) Bit	BOOL	I, Q, M, TON8_Q, TOF8_Q, TP8_Q, TOS8_Q, CTU8_Q, CTD8_Q, CTUD8_Q, LOGIC1, LOGIC0, FRSTSCN, SCNOSC
Output (rego,bito) Bit	BOOL	Q, M, TON8_Q, TOF8_Q, TP8_Q, TOS8_Q, CTU8_Q, CTD8_Q, CTUD8_Q

is 0 (respectively 1), the output variable `rego,bito` is forced to 1 (respectively to 0). Operands for the instruction `inv_out` are shown in Table 3.31.

3.17 Macro `_set`

The truth table and symbols of the macro `_set` are depicted in Table 3.32. Figure 3.17 shows the macro `_set` and its flowchart. This macro has a Boolean input variable passed into it by W and a Boolean output variable passed out through `reg,bit`. When the input variable is 0, no action is taken, but when

TABLE 3.32

Truth Table and Symbols of the Macro `_set`

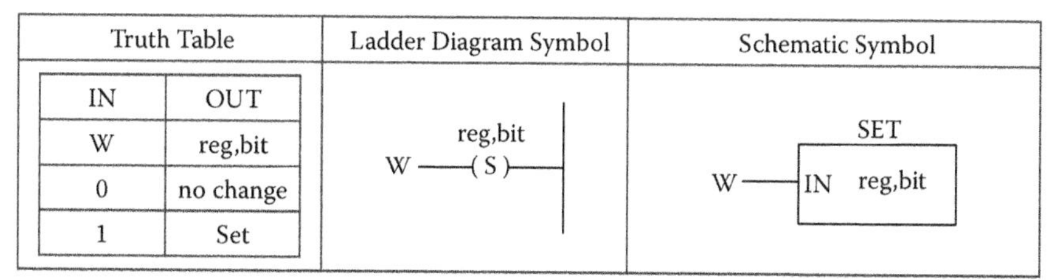

Truth Table		Ladder Diagram Symbol	Schematic Symbol
IN	OUT	W —(S)— reg,bit	SET: W — IN reg,bit
W	reg,bit		
0	no change		
1	Set		

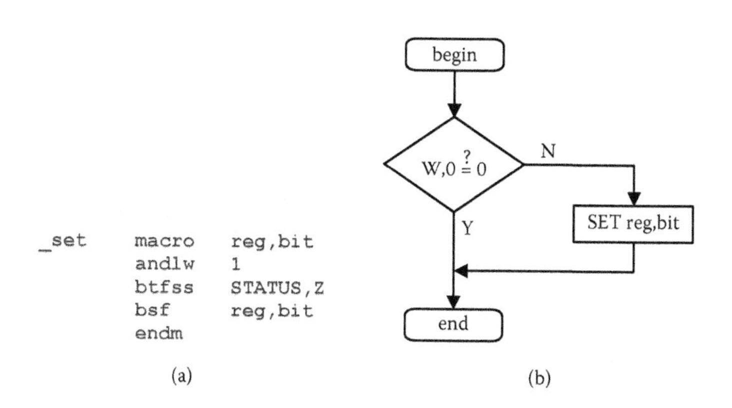

FIGURE 3.17
(a) The macro `_set` and (b) its flowchart.

TABLE 3.33

Operands for the Instruction `_set`

Output (reg,bit)	Data Type	Operands
Bit	BOOL	Q, M, TON8_Q, TOF8_Q, TP8_Q, TOS8_Q, CTU8_Q, CTD8_Q, CTUD8_Q

the input variable is 1, the output variable `reg,bit` is set to 1. Operands for the instruction `_set` are shown in Table 3.33.

3.18 Macro `_reset`

The truth table and symbols of the macro `_reset` are depicted in Table 3.34. Figure 3.18 shows the macro `_reset` and its flowchart. This macro has a Boolean input variable passed into it by W and a Boolean output variable

TABLE 3.34

Truth Table and Symbols of the Macro `_reset`

Truth Table	Ladder Diagram Symbol	Schematic Symbol
IN: W / OUT: reg,bit 0 → no change 1 → Reset	reg,bit W —(R)—	RESET W — IN reg,bit

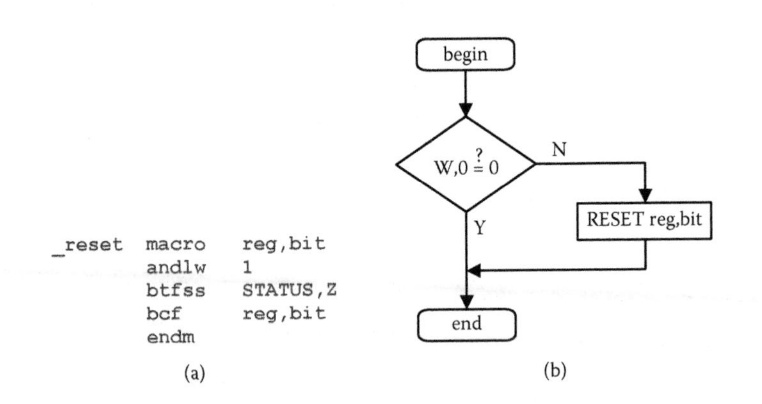

FIGURE 3.18
(a) The macro `_reset` and (b) its flowchart.

TABLE 3.35

Operands for the Instruction `_reset`

Output (reg,bit)	Data Type	Operands
Bit	BOOL	Q, M, TON8_Q, TOF8_Q, TP8_Q, TOS8_Q, CTU8_Q, CTD8_Q, CTUD8_Q

passed out through `reg,bit`. When the input variable is 0, no action is taken, but when the input variable is 1, the output variable `reg,bit` is reset. Operands for the instruction `_reset` are shown in Table 3.35.

3.19 Examples for Contact and Relay-Based Macros

In this section, we will consider two examples, UZAM_plc_16i16o_ex3.asm and UZAM_plc_16i16o_ex4.asm, to show the usage of contact and relay-based macros. In order to test the respective example, please take the files from the CD-ROM attached to this book and then open the respective program by MPLAB IDE and compile it. After that, by using the PIC programmer software, take the compiled file UZAM_plc_16i16o_ex3.hex or UZAM_plc_16i16o_ex4.hex, and by your PIC programmer hardware send it to the program memory of the PIC16F648A microcontroller within the PIC16F648A-based PLC. To do this, switch the 4PDT in the PROG position and the power switch in the OFF position. After loading the UZAM_plc_16i16o_ex3.hex or UZAM_plc_16i16o_ex4.hex, switch the 4PDT in RUN and the power switch in the ON position. Please check each program's accuracy by cross-referencing it with the related macros.

Let us now consider these two example programs: The first example program, UZAM_plc_16i16o_ex3.asm, is shown in Figure 3.19. It shows the usage of the following contact and relay-based macros: `ld`, `ld_not`, `not`, `out`, `out_not`, `in_out`, `inv_out`, `or`, `or_not`, and `nor`. The schematic and ladder diagrams of the user program of UZAM_plc_16i16o_ex3.asm, shown in Figure 3.19, are depicted in Figure 3.20(a) and (b), respectively.

```
;-------------- user program starts here -
        ld          I0.0            ;rung 1
        out         Q0.0

        ld_not      I0.1            ;rung 2
        out         Q0.1

        ld          I0.2            ;rung 3
        out         M2.7

        ld          M2.7            ;rung 4
        out_not     Q0.2

        ld          I0.3            ;rung 5
        not
        out         Q0.3

        in_out      I0.4,Q0.4       ;rung 6

        inv_out     I0.5,Q0.5       ;rung 7

        in_out      LOGIC1,Q0.6 ;rung 8

        in_out      T1.5,Q0.7       ;rung 9

        ld          I1.0            ;rung 10
        or          I1.1
        out         Q1.0

        ld          I1.0            ;rung 11
        or          I1.1
        or          I1.2
        out         Q1.1

        ld          I1.0            ;rung 12
        or_not      I1.4
        out         Q1.2

        ld          I1.2            ;rung 13
        or          I1.3
        or_not      I1.4
        out         Q1.3

        ld          I1.4            ;rung 14
        nor         I1.5
        out         Q1.4

        ld          I1.4            ;rung 15
        nor         I1.5
        nor         I1.6
        out         Q1.5

        ld          I1.4            ;rung 16
        or          I1.5
        or_not      I1.6
        nor         I1.7
        out         Q1.6
;-------------- user program ends here ---
```

FIGURE 3.19
The user program of UZAM_plc_16i16o_ex3.asm.

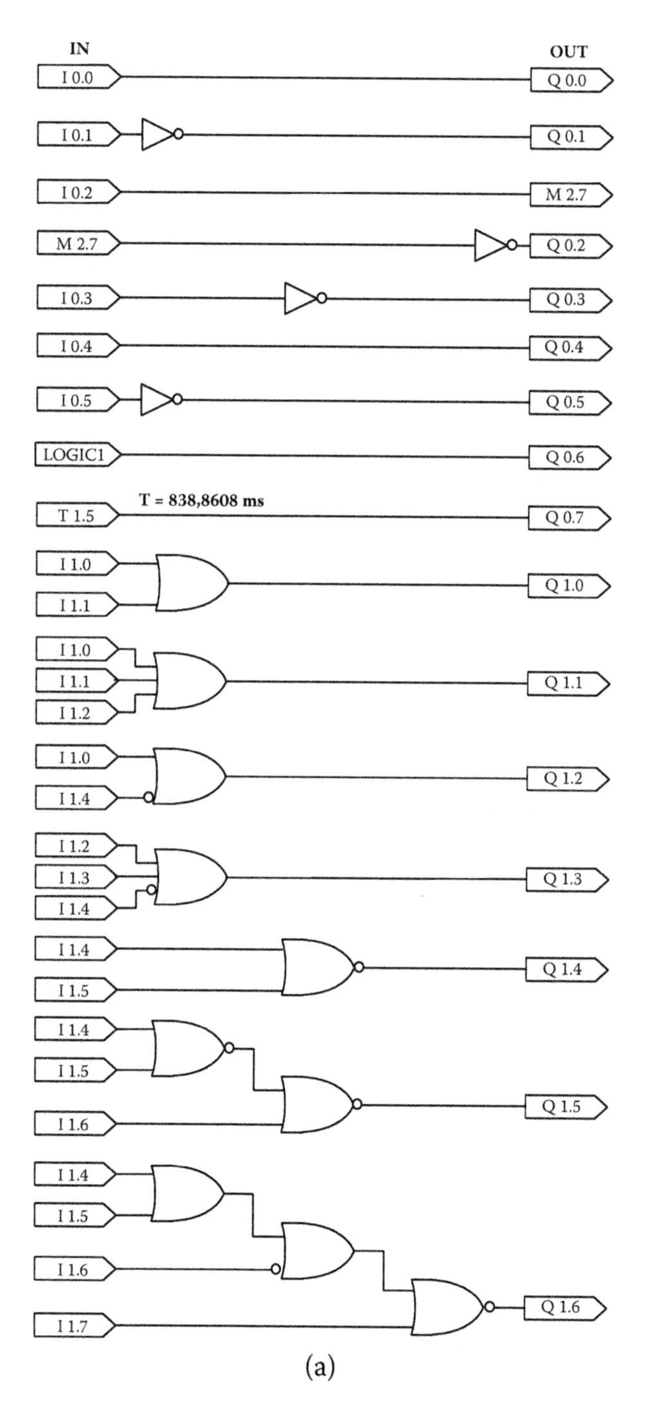

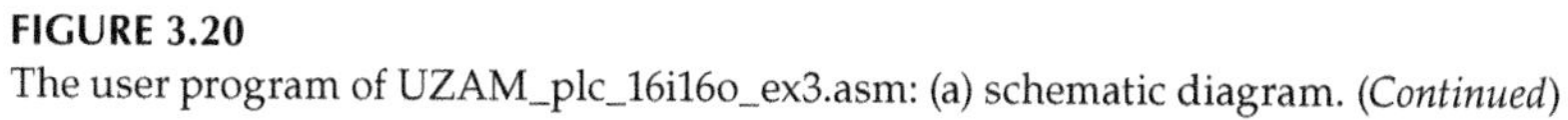

FIGURE 3.20
The user program of UZAM_plc_16i16o_ex3.asm: (a) schematic diagram. (*Continued*)

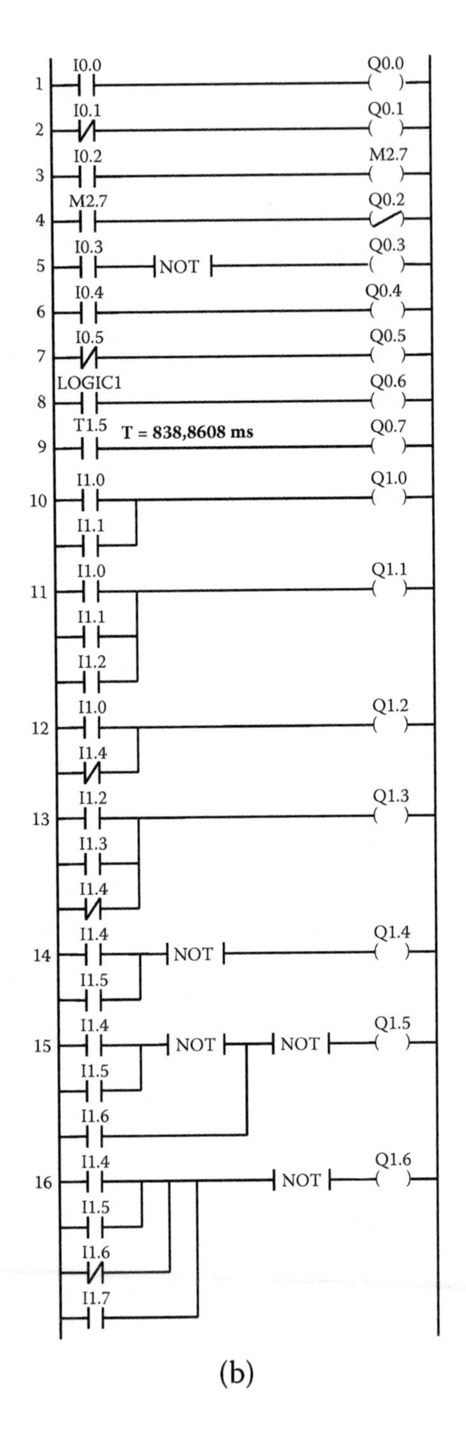

(b)

FIGURE 3.20 *(Continued)*
The user program of UZAM_plc_16i16o_ex3.asm: (b) ladder diagram.

```
;--------------- user program starts here --
      ld          I0.0         ;rung 1
      and         I0.1
      out         Q0.0

      ld          I0.0         ;rung 2
      and         I0.1
      and         I0.2
      out         Q0.1

      ld          I0.0         ;rung 3
      and_not     I0.4
      out         Q0.2

      ld          I0.2         ;rung 4
      and         I0.3
      and_not     I0.4
      out         Q0.3

      ld          I0.4         ;rung 5
      nand        I0.5
      out         Q0.4

      ld          I0.4         ;rung 6
      nand        I0.5
      nand        I0.6
      out         Q0.5

      ld          I0.4         ;rung 7
      and         I0.5
      and_not     I0.6
      nand        I0.7
      out         Q0.6

      ld          I1.0         ;rung 8
      xor         I1.1
      out         Q1.0

      ld          I1.2         ;rung 9
      xor_not     I1.3
      out         Q1.2

      ld          I1.4         ;rung 10
      xnor        I1.5
      out         Q1.4

      ld          I1.6         ;rung 11
      _set        Q1.7

      ld          I1.7         ;rung 12
      _reset      Q1.7
;--------------- user program ends here ----
```

FIGURE 3.21
The user program of UZAM_plc_16i16o_ex4.asm.

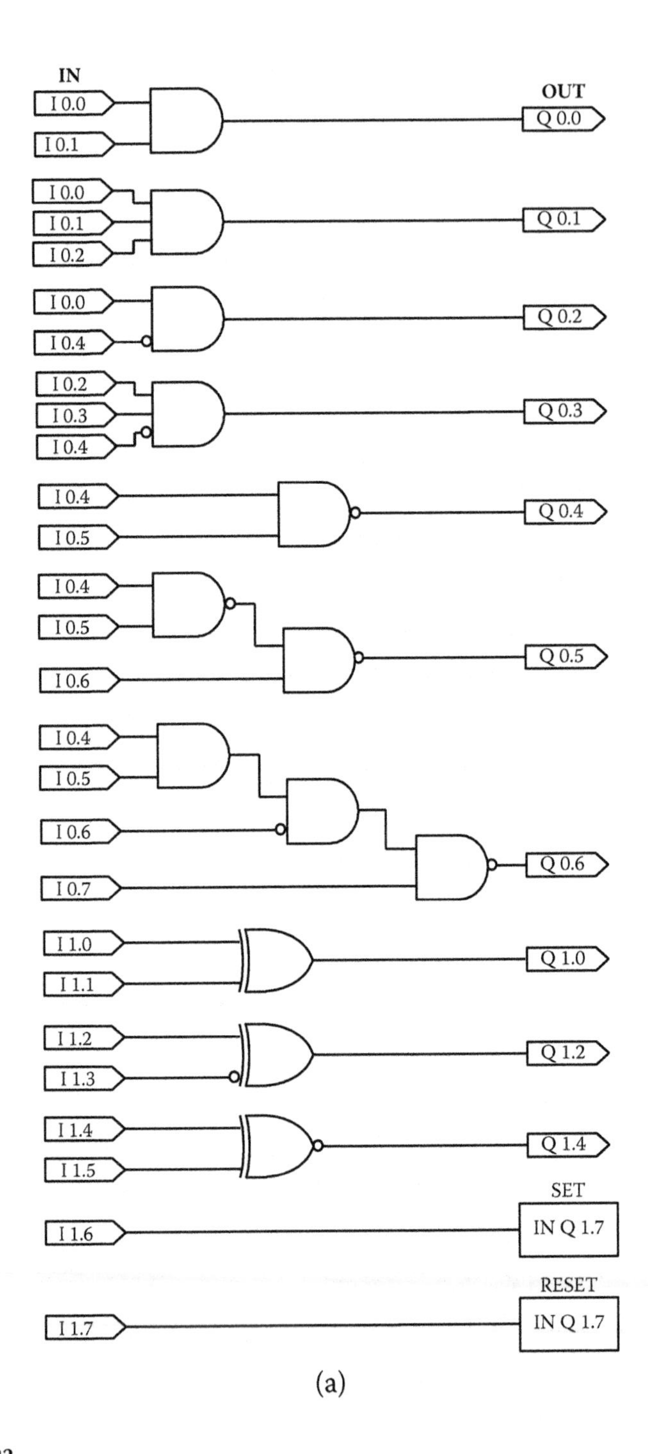

(a)

FIGURE 3.22
The user program of UZAM_plc_16i16o_ex4.asm: (a) schematic diagram. (*Continued*)

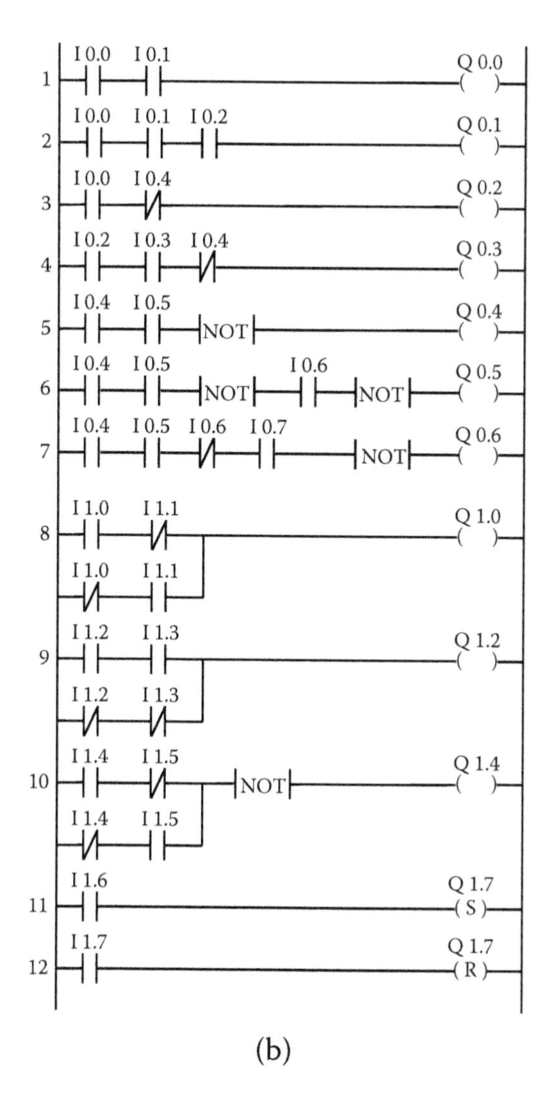

FIGURE 3.22 *(Continued)*
The user program of UZAM_plc_16i16o_ex4.asm: (b) ladder diagram.

The second example program, UZAM_plc_16i16o_ex4.asm, is shown in Figure 3.21. It shows the usage of the following contact and relay-based macros: `ld`, `and`, `and_not`, `nand`, `xor`, `xor_not`, `xnor`, `_set`, and `_reset`. The schematic and ladder diagrams of the user program of UZAM_plc_16i16o_ex4.asm, shown in Figure 3.21, are depicted in Figure 3.22(a) and (b), respectively.

4

Flip-Flop Macros

In this chapter, the following flip-flop macros are described:

r_edge (rising edge detector)
f_edge (falling edge detector)
latch1 (D latch with active high enable)
latch0 (D latch with active low enable)
dff_r (rising edge triggered D flip-flop)
dff_f (falling edge triggered D flip-flop)
tff_r (rising edge triggered T flip-flop)
tff_f (falling edge triggered T flip-flop)
jkff_r (rising edge triggered JK flip-flop)
jkff_f (falling edge triggered JK flip-flop)

Each macro defined here requires an edge detection mechanism except for latch0 and latch1. The following 8-bit variables are used for this purpose:

RED: Rising edge detector
FED: Falling edge detector
DFF_RED: Rising edge detector for D flip-flop
DFF_FED: Falling edge detector for D flip-flop
TFF_RED: Rising edge detector for T flip-flop
TFF_FED: Falling edge detector for T flip-flop
JKFF_RED: Rising edge detector for JK flip-flop
JKFF_FED: Falling edge detector for JK flip-flop

They are declared within the SRAM data memory as shown in Figure 4.1. Each 8-bit variable enables us to declare and use eight different functions defined by the related macro. The macros latch0 and latch1 are an exception to this, which means that we can use as many latches of latch0 or latch1 as we wish. The file definitions.inc, included within the CD-ROM attached to this book, contains all flip-flop macros defined for the PIC16F648A-based PLC.

Let us now briefly consider these macros.

```
;--------------- VARIABLE DEFINITIONS ---
     CBLOCK 0x3F
     RED,FED,DFF_RED,DFF_FED,TFF_RED,TFF_FED,JKFF_RED,JKFF_FED
     endc
;----------------------------------------
```

(a)

3Fh	**RED**
40h	**FED**
41h	**DFF_RED**
42h	**DFF_FED**
43h	**TFF_RED**
44h	**TFF_FED**
45h	**JKFF_RED**
46h	**JKFF_FED**
	BANK 0

(b)

FIGURE 4.1
(a) The definition of 8-bit variables to be used for the flip-flop-based macros. (b) Their allocation in BANK 0 of SRAM data memory.

4.1 Macro `r_edge` (Rising Edge Detector)

The symbols and the timing diagram of the macro `r_edge` are depicted in Table 4.1. Figure 4.2 shows the macro `r_edge` and its flowchart. The macro `r_edge` defines eight rising edge detector functions (or contacts) selected with the num = 0, 1, …, 7. It has a Boolean input variable, namely, IN, passed

TABLE 4.1

Symbols and Timing Diagram of the Macro `r_edge`

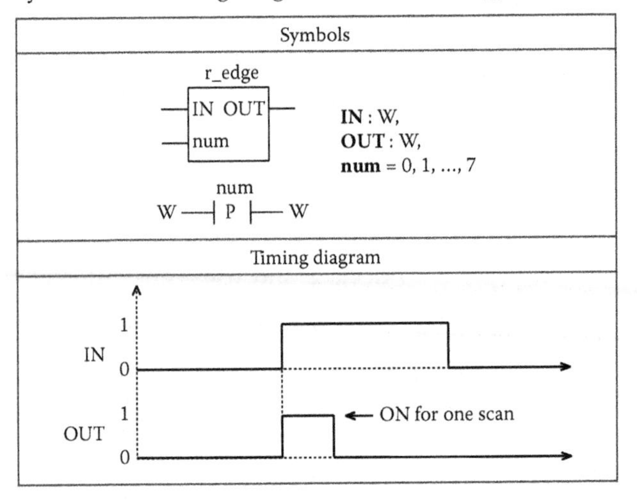

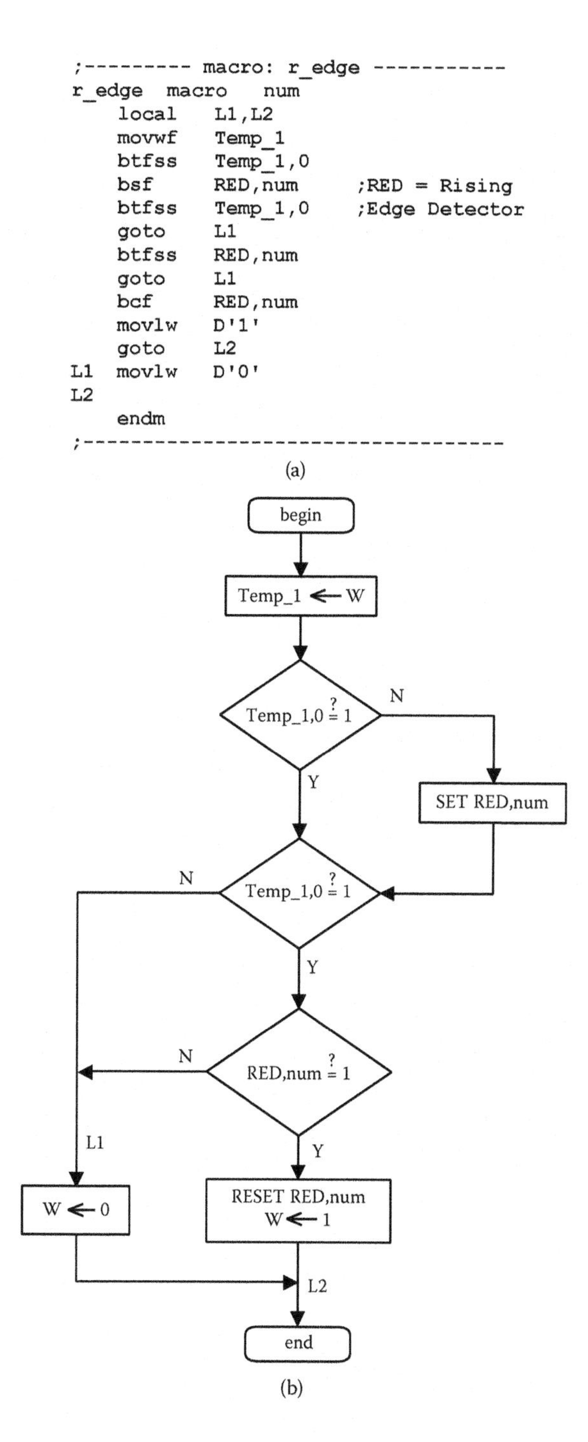

```
;--------- macro: r_edge ----------
r_edge  macro   num
    local   L1,L2
    movwf   Temp_1
    btfss   Temp_1,0
    bsf     RED,num      ;RED = Rising
    btfss   Temp_1,0     ;Edge Detector
    goto    L1
    btfss   RED,num
    goto    L1
    bcf     RED,num
    movlw   D'1'
    goto    L2
L1  movlw   D'0'
L2
    endm
;---------------------------------
```

(a)

(b)

FIGURE 4.2
(a) The macro `r_edge` and (b) its flowchart.

into the macro through W, and a Boolean output variable, namely, OUT, passed out of the macro through W. This means that the input signal IN should be loaded into W before this macro is run, and the output signal OUT will be provided within the W at the end of the macro. In ladder logic, this macro is represented by a normally open (NO) contact with the identifier P, meaning positive transition-sensing contact. As can be seen from the timing diagram, the OUT is ON (1) for only one scan time when the IN changes its state from OFF (0) to ON (1). In the other instances, the OUT remains OFF (0).

4.2 Macro `f_edge` (Falling Edge Detector)

The symbols and the timing diagram of the macro `f_edge` are depicted in Table 4.2. Figure 4.3 shows the macro `f_edge` and its flowchart. The macro `f_edge` defines eight falling edge detector functions (or contacts) selected with the num = 0, 1, …, 7. It has a Boolean input variable, namely, IN, passed into the macro through W, and a Boolean output variable, namely, OUT, passed out of the macro through W. This means that the input signal IN should be loaded into W before this macro is run, and the output signal OUT will be provided within the W at the end of the macro. In ladder logic, this macro is represented by a normally open (NO) contact with the identifier N, meaning negative transition-sensing contact. As can be seen from the timing diagram, the OUT is ON (1) for only one scan time when the IN changes its state from ON (1) to OFF (0). In the other instances, the OUT remains OFF (0).

TABLE 4.2

Symbols and Timing Diagram of the Macro `f_edge`

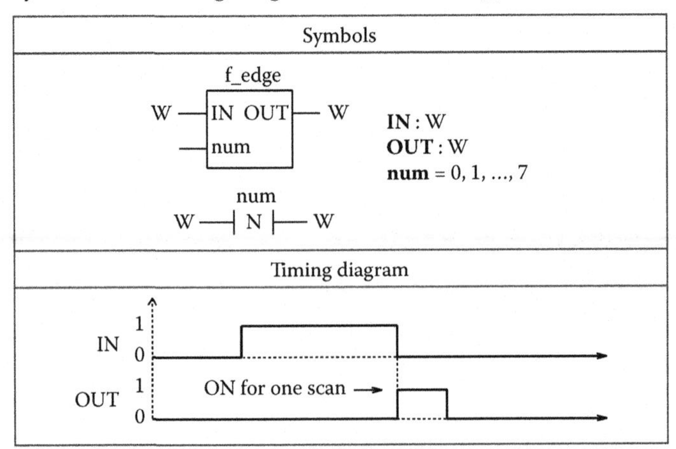

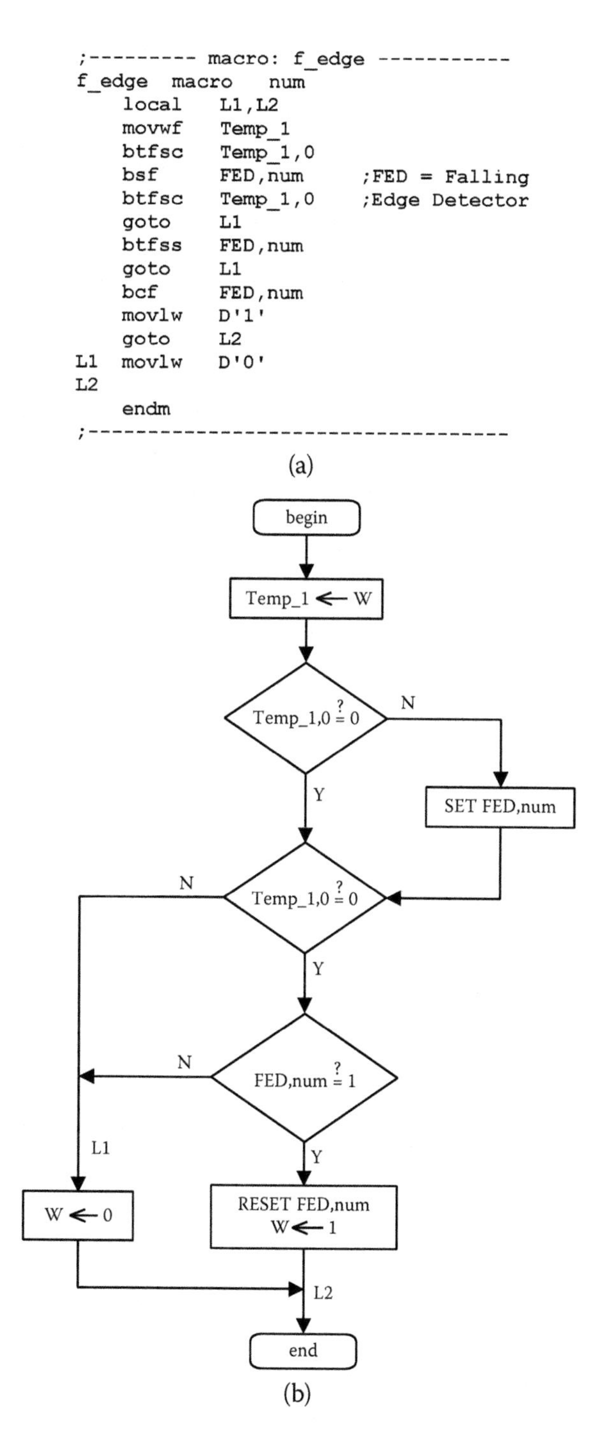

```
;--------- macro: f_edge -----------
f_edge  macro   num
    local   L1,L2
    movwf   Temp_1
    btfsc   Temp_1,0
    bsf     FED,num     ;FED = Falling
    btfsc   Temp_1,0    ;Edge Detector
    goto    L1
    btfss   FED,num
    goto    L1
    bcf     FED,num
    movlw   D'1'
    goto    L2
L1  movlw   D'0'
L2
    endm
;-----------------------------------
```

(a)

(b)

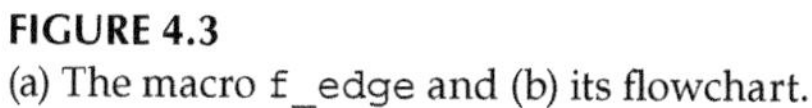

FIGURE 4.3
(a) The macro `f_edge` and (b) its flowchart.

TABLE 4.3

Symbol of the Macro `latch1` and Its Truth Table

Symbol

latch1

regi,biti — D Q — rego,bito

W — EN

EN : W,
D : regi,biti
Q : rego,bito

Truth Table

EN	D	Q_t	Q_{t+1}	Comment
0	×	Q_t	Q_t	No change
1	0	×	0	Reset
1	1	×	1	Set

× : don't care.

4.3 Macro latch1 (D Latch with Active High Enable)

The symbol of the macro `latch1` and its truth table are depicted in Table 4.3. Figure 4.4 shows the macro `latch1` and its flowchart. The macro `latch1` defines a D latch function with active high enable. Unlike the edge triggered flip-flops and the edge detector macros, in which eight functions are described, this function defines only one D latch function. However, we are free to use this macro as much as we need with different input/output variables. The macro `latch1` has two Boolean input variables, namely, EN, passed into the macro through W, and D (regi,biti), and a single Boolean output variable, Q (rego,bito). The input signal EN (active high enable input) should be loaded into W before this macro is run. When the active high enable input EN is OFF (0), no state change is issued for the output Q and it holds its current state. When the active high enable input EN is ON (1), the output Q is loaded with the state of the input D. Operands for the instruction `latch1` are shown in Table 4.4.

4.4 The Macro latch0 (D Latch with Active Low Enable)

The symbol of the macro `latch0` and its truth table are depicted in Table 4.5. Figure 4.5 shows the macro `latch0` and its flowchart. The macro `latch0` defines a D latch function with active low enable. Unlike the edge triggered flip-flops and the edge detector macros, in which eight functions are described,

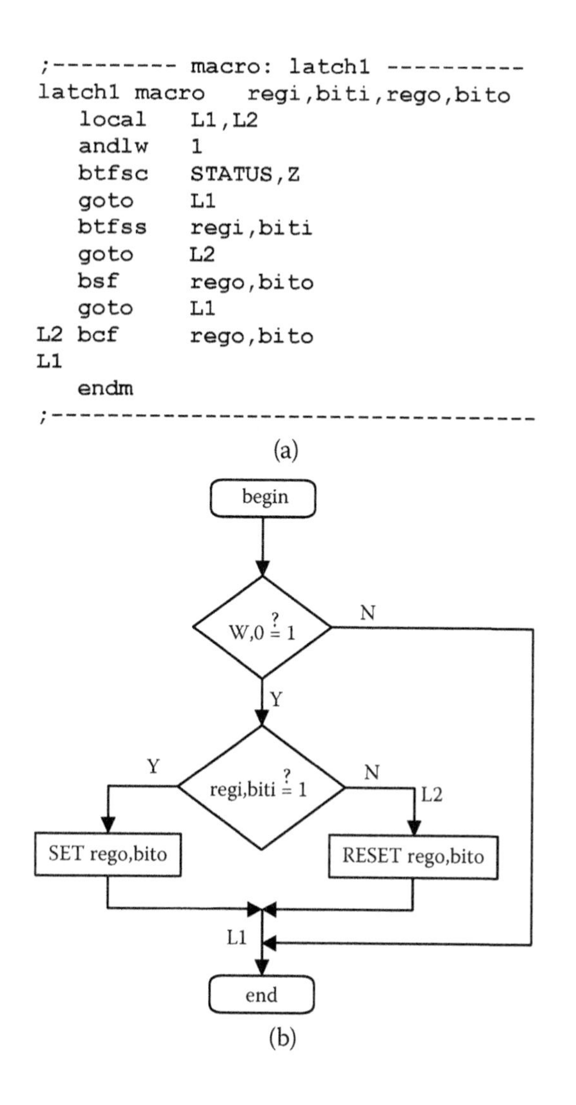

```
;--------- macro: latch1 ----------
latch1 macro   regi,biti,rego,bito
   local    L1,L2
   andlw    1
   btfsc    STATUS,Z
   goto     L1
   btfss    regi,biti
   goto     L2
   bsf      rego,bito
   goto     L1
L2 bcf      rego,bito
L1
   endm
;-----------------------------------
```


FIGURE 4.4
(a) The macro `latch1` and (b) its flowchart.

TABLE 4.4

Operands for the Instruction `latch1`

Input/Output	Data Type	Operands
D regi,biti (Bit)	BOOL	I, Q, M, TON8_Q, TOF8_Q, TP8_Q, TOS8_Q, CTU8_Q, CTD8_Q, CTUD8_Q, LOGIC1, LOGIC0, FRSTSCN, SCNOSC
Q rego,bito (Bit)	BOOL	Q, M, TON8_Q, TOF8_Q, TP8_Q, TOS8_Q, CTU8_Q, CTD8_Q, CTUD8_Q

TABLE 4.5

Symbol of Macro `latch0` and Its Truth Table

Symbol
latch0 regi,biti — D Q — rego,bito W —o EN **EN** : W **D** : regi,biti **Q** : rego,bito
Truth Table

EN	D	Q_t	Q_{t+1}	Comment
1	×	Q_t	Q_t	No change
0	0	×	0	Reset
0	1	×	1	Set

× : don't care.

this function defines only one D latch function. However, we are free to use this macro as much as we need with different input/output variables. The macro `latch0` has two Boolean input variables, namely, EN, passed into the macro through W and D (regi,biti), and a single Boolean output variable, Q (rego,bito). The input signal EN (active low enable input) should be loaded into W before this macro is run. When the active low enable input EN is ON (1), no state change is issued for the output Q and it holds its current state. When the active low enable input EN is OFF (0), the output Q is loaded with the state of the input D. Operands for the instruction `latch0` are shown in Table 4.6.

4.5 Macro `dff_r` (Rising Edge Triggered D Flip-Flop)

The symbol of the macro `dff_r` and its truth table are depicted in Table 4.7. Figure 4.6 shows the macro `dff_r` and its flowchart. The macro `dff_r` defines eight rising edge triggered D flip-flop functions selected with the

TABLE 4.6

Operands for the Instruction `latch0`

Input/Output	Data Type	Operands
D regi,biti (Bit)	BOOL	I, Q, M, TON8_Q, TOF8_Q, TP8_Q, TOS8_Q, CTU8_Q, CTD8_Q, CTUD8_Q, LOGIC1, LOGIC0, FRSTSCN, SCNOSC
Q rego,bito (Bit)	BOOL	Q, M, TON8_Q, TOF8_Q, TP8_Q, TOS8_Q, CTU8_Q, CTD8_Q, CTUD8_Q

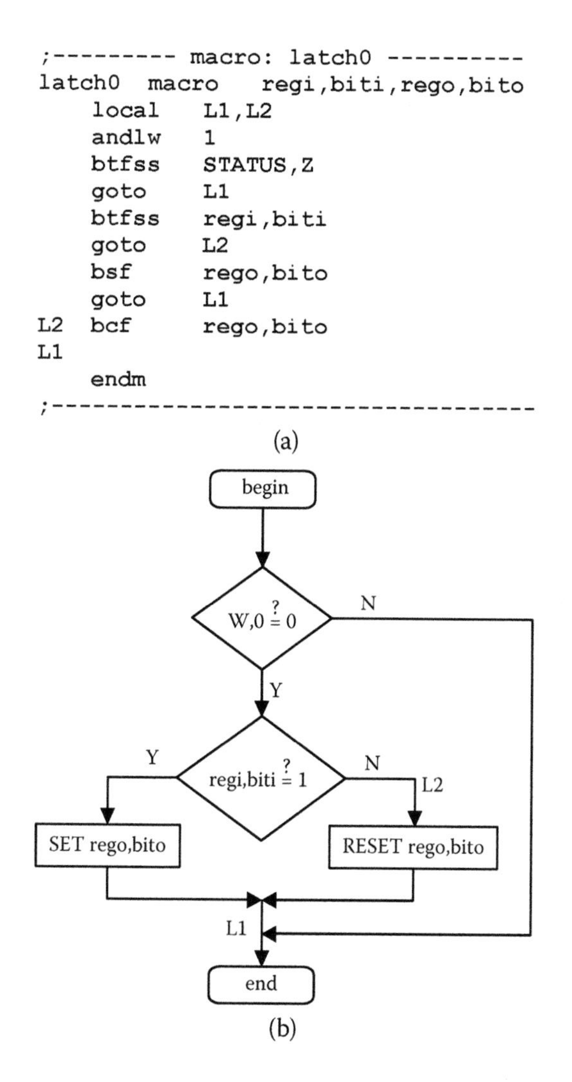

```
;--------- macro: latch0 ----------
latch0  macro   regi,biti,rego,bito
    local   L1,L2
    andlw   1
    btfss   STATUS,Z
    goto    L1
    btfss   regi,biti
    goto    L2
    bsf     rego,bito
    goto    L1
L2  bcf     rego,bito
L1
    endm
;-----------------------------------
```

(a)

(b)

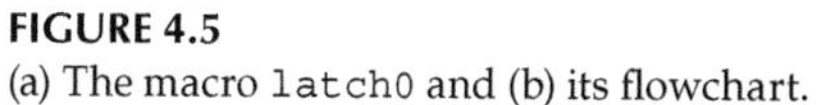

FIGURE 4.5
(a) The macro `latch0` and (b) its flowchart.

num = 0, 1, …, 7. It has two Boolean input variables, namely, clock input C, passed into the macro through W, and data input D (regi,biti), and a single Boolean output variable, flip-flop output Q (rego,bito). The clock input signal C should be loaded into W before this macro is run. When the clock input signal C is ON (1) or OFF (0), or changes its state from ON to OFF (↓), no state change is issued for the output Q and it holds its current state. When the state of clock input signal C is changed from OFF to ON (↑), the output Q is loaded with the state of the input D. Operands for the instruction `dff_r` are shown in Table 4.8.

TABLE 4.7

Symbol of the Macro `dff_r` and Its Truth Table

Symbol
dff_r; regi,biti — D, W — C, — num; Q — rego,bito. **C** : W, **D** : regi,biti, **Q** : rego,bito, **num** = 0, 1, ..., 7

Truth Table

D	C	Q_t	Q_{t+1}	Comment
×	0	Q_t	Q_t	No change
×	1	Q_t	Q_t	No change
×	↓	Q_t	Q_t	No change
0	↑	×	0	Reset
1	↑	×	1	Set

× : don't care.

TABLE 4.8

Operands for the Instruction `dff_r`

Input/Output	Data Type	Operands
D regi,biti (Bit)	BOOL	I, Q, M, TON8_Q, TOF8_Q, TP8_Q, TOS8_Q, CTU8_Q, CTD8_Q, CTUD8_Q, LOGIC1, LOGIC0, FRSTSCN, SCNOSC
Q rego,bito (Bit)	BOOL	Q, M, TON8_Q, TOF8_Q, TP8_Q, TOS8_Q, CTU8_Q, CTD8_Q, CTUD8_Q

```
;--------- macro: dff_r ------------
dff_r   macro   num, regi,biti,rego,bito
        local   L1,L2
        movwf   Temp_1
        btfss   Temp_1,0
        bsf     DFF_RED,num ;DFF_RED = Rising Edge
        btfss   Temp_1,0    ;Detector for rising edge
        goto    L1          ;triggered D flip-flop
        btfss   DFF_RED,num
        goto    L1
        bcf     DFF_RED,num
        btfss   regi,biti
        goto    L2
        bsf     rego,bito
        goto    L1
L2      bcf     rego,bito
L1
        endm
;-----------------------------------
```

(a)

FIGURE 4.6
(a) The macro `dff_r` and (b) its flowchart. (*Continued*)

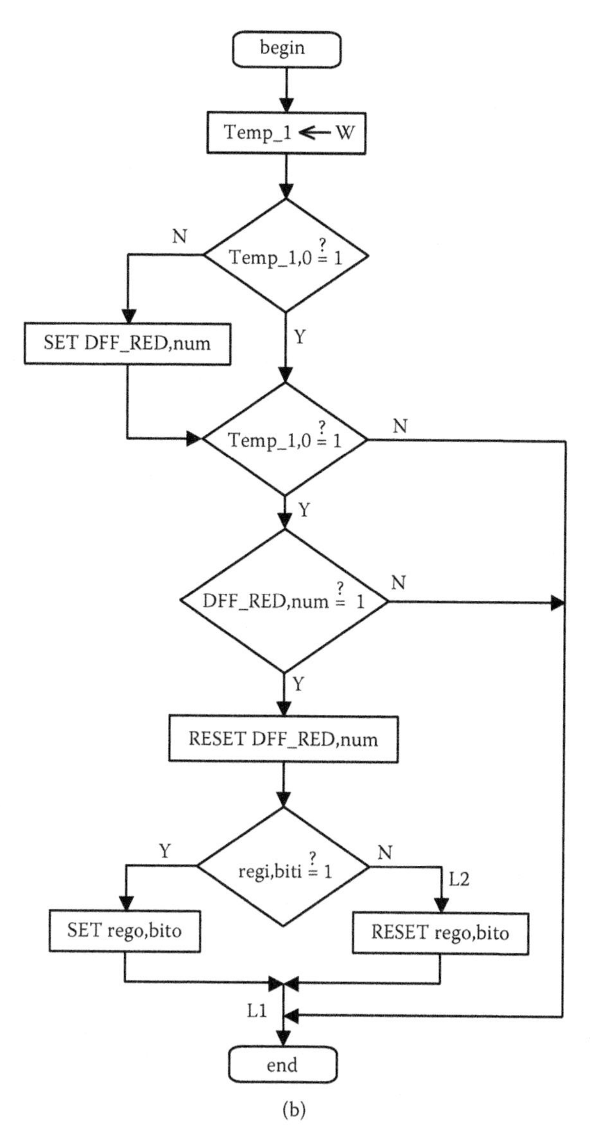

FIGURE 4.6 *(Continued)*
(a) The macro `dff_r` and (b) its flowchart.

4.6 Macro `dff_f` (Falling Edge Triggered D Flip-Flop)

The symbol of the macro `dff_f` and its truth table are depicted in Table 4.9. Figure 4.7 shows the macro `dff_f` and its flowchart. The macro `dff_f` defines eight falling edge triggered D flip-flop functions selected with the num = 0, 1, …, 7. It has two Boolean input variables, namely, clock input C, passed into the

TABLE 4.9

Symbol of the Macro `dff_f` and Its Truth Table

Symbol
dff_f; regi,biti — D; W —o> C; — num; Q — rego,bito **C** : W **D** : regi,biti **Q** : rego,bito **num** = 0, 1, …, 7

Truth Table

D	C	Q_t	Q_{t+1}	Comment
×	0	Q_t	Q_t	No change
×	1	Q_t	Q_t	No change
×	↑	Q_t	Q_t	No change
0	↓	×	0	Reset
1	↓	×	1	Set

× : don't care.

macro through W, and data input D (regi,biti), and a single Boolean output variable, flip-flop output Q (rego,bito). The clock input signal C should be loaded into W before this macro is run. When the clock input signal C is ON (1) or OFF (0), or changes its state from OFF to ON (↑), no state change is issued for the output Q and it holds its current state. When the state of clock input signal C is changed from ON to OFF (↓), the output Q is loaded with the state of the input D. Operands for the instruction `dff_f` are shown in Table 4.10.

```
;--------- macro: dff_f ------------
dff_f   macro   num,regi,biti,rego,bito
    local   L1,L2
    movwf   Temp_1
    btfsc   Temp_1,0
    bsf     DFF_FED,num ;DFF_FED = Falling Edge
    btfsc   Temp_1,0    ;Detector for falling edge
    goto    L1          ;triggered D flip-flop
    btfss   DFF_FED,num
    goto    L1
    bcf     DFF_FED,num
    btfss   regi,biti
    goto    L2
    bsf     rego,bito
    goto    L1
L2  bcf     rego,bito
L1
    endm
;-----------------------------------
```

(a)

FIGURE 4.7
(a) The macro `dff_f` and (b) its flowchart. (*Continued*)

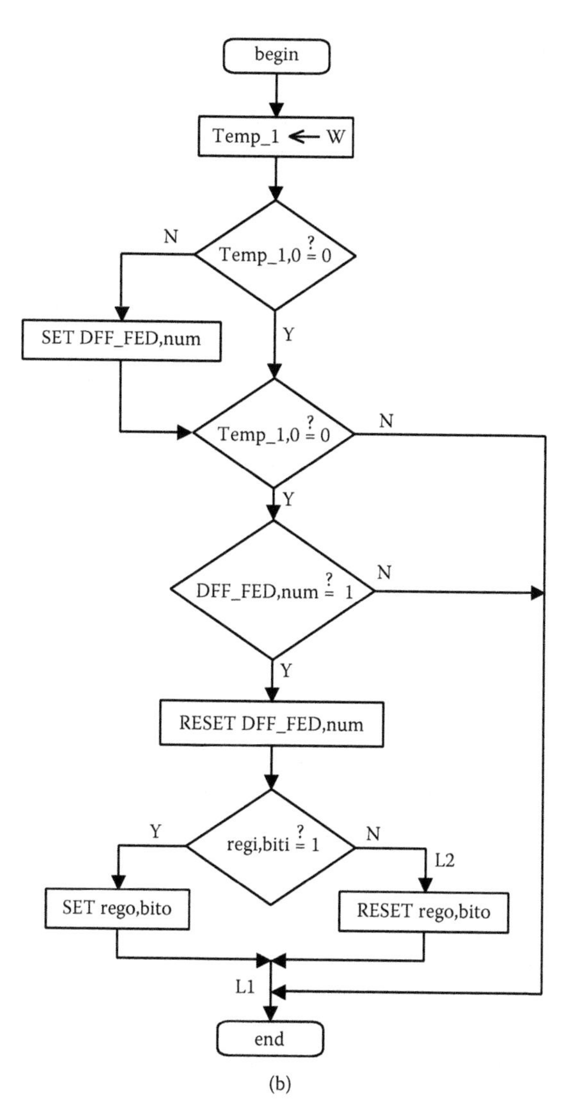

(b)

FIGURE 4.7 (*Continued*)
(a) The macro `dff_f` and (b) its flowchart.

TABLE 4.10

Operands for the Instruction `dff_f`

Input/Output	Data Type	Operands
D regi,biti (Bit)	BOOL	I, Q, M, TON8_Q, TOF8_Q, TP8_Q, TOS8_Q, CTU8_Q, CTD8_Q, CTUD8_Q, LOGIC1, LOGIC0, FRSTSCN, SCNOSC
Q rego,bito (Bit)	BOOL	Q, M, TON8_Q, TOF8_Q, TP8_Q, TOS8_Q, CTU8_Q, CTD8_Q, CTUD8_Q

TABLE 4.11

Symbol of the Macro `tff_r` and Its Truth Table

Symbol

Truth Table

T	C	Q_t	Q_{t+1}	Comment
×	0	Q_t	Q_t	No change
×	1	Q_t	Q_t	No change
×	↓	Q_t	Q_t	No change
0	↑	Q_t	Q_t	No change
1	↑	Q_t	$\overline{Q_t}$	Toggle

× : don't care.

4.7 Macro `tff_r` (Rising Edge Triggered T Flip-Flop)

The symbol of the macro `tff_r` and its truth table are depicted in Table 4.11. Figure 4.8 shows the macro `tff_r` and its flowchart. The macro `tff_r` defines eight rising edge triggered T flip-flop functions selected with the

```
;--------- macro: tff_r ------------
tff_r   macro   num,regi,biti,rego,bito
        local   L1,L2
        movwf   Temp_1
        btfss   Temp_1,0
        bsf     TFF_RED,num ;TFF_RED=Rising Edge
        btfss   Temp_1,0    ;Detector for rising edge
        goto    L1          ;triggered T flip-flop
        btfss   TFF_RED,num
        goto    L1
        bcf     TFF_RED,num
        btfss   regi,biti
        goto    L1
        btfsc   rego,bito
        goto    L2
        bsf     rego,bito
        goto    L1
L2      bcf     rego,bito
L1
        endm
;-----------------------------------
```

(a)

FIGURE 4.8
(a) The macro `tff_r` and (b) its flowchart. (*Continued*)

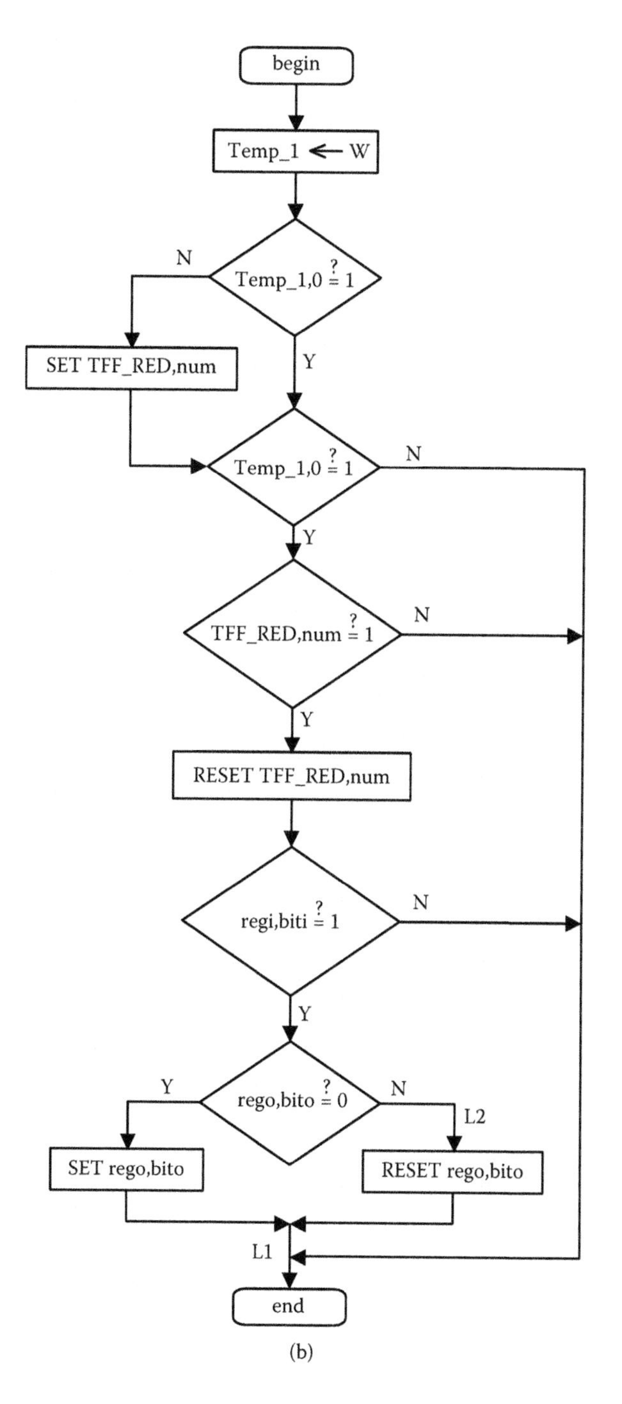

FIGURE 4.8 (*Continued*)
(a) The macro `tff_r` and (b) its flowchart.

TABLE 4.12

Operands for the Instruction `tff_r`

Input/Output	Data Type	Operands
T regi,biti (Bit)	BOOL	I, Q, M, TON8_Q, TOF8_Q, TP8_Q, TOS8_Q, CTU8_Q, CTD8_Q, CTUD8_Q, LOGIC1, LOGIC0, FRSTSCN, SCNOSC
Q rego,bito (Bit)	BOOL	Q, M, TON8_Q, TOF8_Q, TP8_Q, TOS8_Q, CTU8_Q, CTD8_Q, CTUD8_Q

num = 0, 1, …, 7. It has two Boolean input variables, namely, clock input C, passed into the macro through W, and toggle input T (regi,biti), and a single Boolean output variable, flip-flop output Q (rego,bito). The clock input signal C should be loaded into W before this macro is run. When the clock input signal C is ON (1) or OFF (0), or changes its state from ON to OFF (↓), no state change is issued for the output Q and it holds its current state. When the state of clock input signal C is changed from OFF to ON (↑), if T = 0, then no state change is issued for the output Q and it holds its current state. When the state of clock input signal C is changed from OFF to ON (↑), if T = 1, then the output Q is toggled. Operands for the instruction `tff_r` are shown in Table 4.12.

4.8 Macro `tff_f` (Falling Edge Triggered T Flip-Flop)

The symbol of the macro `tff_f` and its truth table are depicted in Table 4.13. Figure 4.9 shows the macro `tff_f` and its flowchart. The macro `tff_f` defines eight falling edge triggered T flip-flop functions selected with the num = 0, 1, …, 7. It has two Boolean input variables, namely, clock input C, passed into the macro through W, and toggle input T (regi,biti), and a single Boolean output variable, flip-flop output Q (rego,bito). The clock input signal C should be loaded into W before this macro is run. When the clock input signal C is ON (1) or OFF (0), or changes state from OFF to ON (↑), no state change is issued for the output Q and it holds its current state. When the state of clock input signal C is changed from ON to OFF (↓): if T = 0, then no state change is issued for the output Q; if T = 1, then the output Q is toggled. Operands for the instruction `tff_f` are shown in Table 4.14.

4.9 Macro `jkff_r` (Rising Edge Triggered JK Flip-Flop)

The symbol of the macro `jkff_r` and its truth table are depicted in Table 4.15. Figure 4.10 shows the macro `jkff_r` and its flowchart. The macro `jkff_r` defines eight rising edge triggered JK flip-flop functions selected with the num = 0, 1, …, 7. It has three Boolean input variables, namely, clock input C, passed into the macro through W, and data inputs J (regj,bitj) and K (regk,bitk),

TABLE 4.13

Symbol of the Macro `tff_f` and Its Truth Table

Symbol

Truth Table

T	C	Q_t	Q_{t+1}	Comment
×	0	Q_t	Q_t	No change
×	1	Q_t	Q_t	No change
×	↑	Q_t	Q_t	No change
0	↓	Q_t	Q_t	No change
1	↓	Q_t	$\overline{Q_t}$	Toggle

× : don't care.

```
;--------- macro: tff_f ------------
tff_f   macro   num,regi,biti,rego,bito
    local   L1,L2
    movwf   Temp_1
    btfsc   Temp_1,0
    bsf     TFF_FED,num ;TFF_FED = Falling Edge
    btfsc   Temp_1,0    ;Detector for falling edge
    goto    L1          ;triggered T flip-flop
    btfss   TFF_FED,num
    goto    L1
    bcf     TFF_FED,num
    btfss   regi,biti
    goto    L1
    btfsc   rego,bito
    goto    L2
    bsf     rego,bito
    goto    L1
L2  bcf     rego,bito
L1
     endm
;------------------------------------
```

(a)

FIGURE 4.9
(a) The macro `tff_f` and (b) its flowchart. (*Continued*)

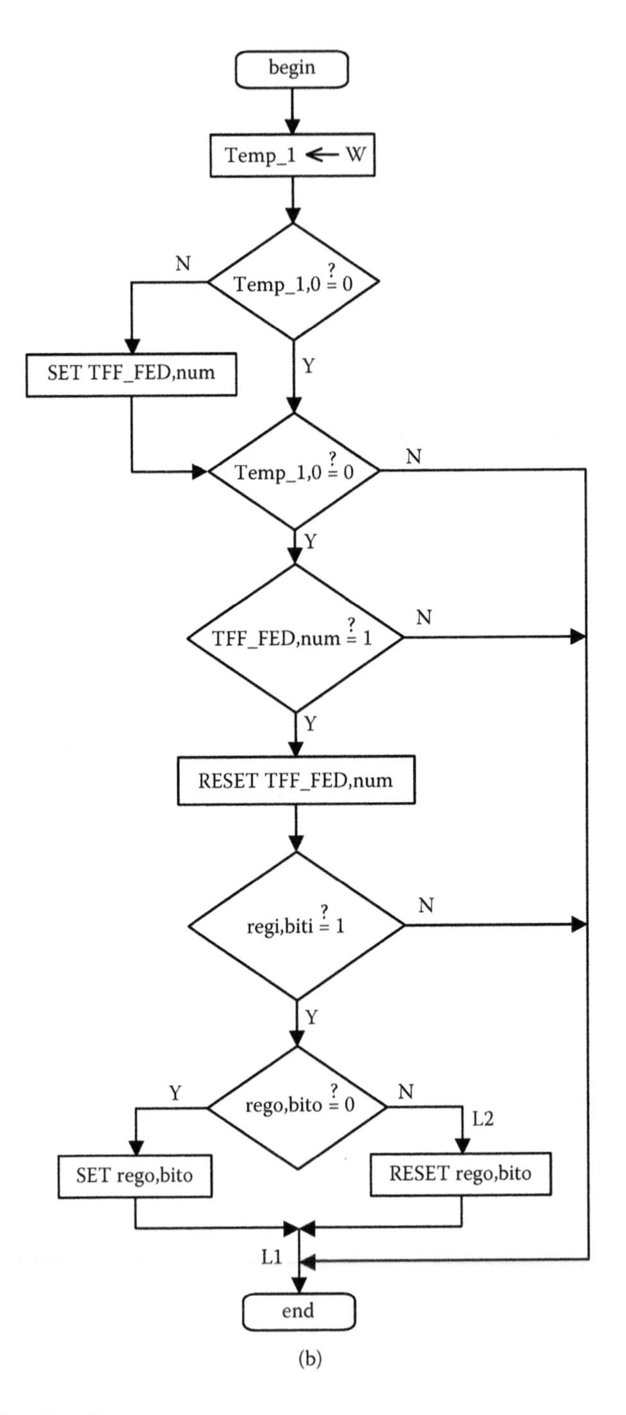

FIGURE 4.9 (*Continued*)
(a) The macro `tff_f` and (b) its flowchart.

TABLE 4.14

Operands for the Instruction `tff_f`

Input/Output	Data Type	Operands
T regi,biti (Bit)	BOOL	I, Q, M, TON8_Q, TOF8_Q, TP8_Q, TOS8_Q, CTU8_Q, CTD8_Q, CTUD8_Q, LOGIC1, LOGIC0, FRSTSCN, SCNOSC
Q rego,bito (Bit)	BOOL	Q, M, TON8_Q, TOF8_Q, TP8_Q, TOS8_Q, CTU8_Q, CTD8_Q, CTUD8_Q

and a single Boolean output variable, flip-flop output Q (rego,bito). The clock input signal C should be loaded into W before this macro is run. When the clock input signal C is ON (1) or OFF (0), or changes state from ON to OFF (↓), no state change is issued for the output Q and it holds its current state. When the state of clock input signal C is changed from OFF to ON (↑): if JK = 00, then no state change is issued; if JK = 01, then Q is reset; if JK = 10, then Q is set; and finally if JK = 11, then Q is toggled. Operands for the instruction `jkff_r` are shown in Table 4.16.

TABLE 4.15

Symbol of the Macro `jkff_r` and Its Truth Table

Symbol

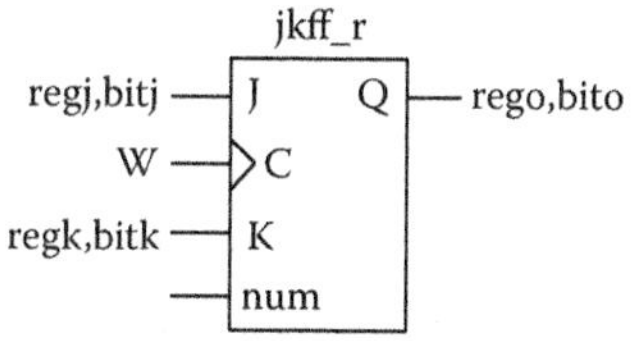

jkff_r

regj,bitj — J Q — rego,bito

W — >C

regk,bitk — K

— num

C: W
J: regj,bitj
K: regk,bitk
Q: rego,bito
num= 0, 1, ..., 7

Truth Table

J	K	C	Q_t	Q_{t+1}	Comment
×	×	0	Q_t	Q_t	No change
×	×	1	Q_t	Q_t	No change
×	×	↓	Q_t	Q_t	No change
0	0	↑	Q_t	Q_t	No change
0	1	↑	×	0	Reset
1	0	↑	×	1	Set
1	1	↑	Q_t	$\overline{Q_t}$	Toggle

× : don't care.

```
;--------- macro: jkff_r -----------
jkff_r macro    num,regj,bitj,regk,bitk,rego,bito
     local    L1,L2,L3,L4
     movwf    Temp_1
     btfss    Temp_1,0
     bsf      JKFF_RED,num        ;JKFF_RED = Rising Edge
     btfss    Temp_1,0            ;Detector for rising edge
     goto     L1                  ;triggered JK flip-flop
     btfss    JKFF_RED,num
     goto     L1
     bcf      JKFF_RED,num
     btfss    regj,bitj
     goto     L4                  ;if j=0 then goto L4
     btfss    regk,bitk
     goto     L3                  ;if j=1&k=0 then SET rego,bito (goto L3)
     btfsc    rego,bito           ;if j=1&k=1
     goto     L2                  ;then TOGGLE
     goto     L3                  ;rego,bito
L4   btfss    regk,bitk
     goto     L1                  ;if j=0&k=0 then NO CHANGE (goto L1)
     goto     L2                  ;if j=0&k=1 then RESET rego,bito
L3   bsf      rego,bito
     goto     L1
L2   bcf      rego,bito
L1
     endm
;-----------------------------------
```

(a)

FIGURE 4.10
(a) The macro `jkff_r` and (b) its flowchart. (*Continued*)

4.10 Macro `jkff_f` (Falling Edge Triggered JK Flip-Flop)

The symbol of the macro `jkff_f` and its truth table are depicted in Table 4.17. Figure 4.11 shows the macro `jkff_f` and its flowchart. The macro `jkff_f` defines eight falling edge triggered JK flip-flop functions selected with the num = 0, 1, …, 7. It has three Boolean input variables, namely, clock input C, passed into the macro through W, and data inputs J (regj,bitj) and K (regk,bitk), and a single Boolean output variable, flip-flop output Q (rego,bito). The clock input signal C should be loaded into W before this macro is run. When the clock input signal C is ON (1) or OFF (0), or changes state from OFF to ON (↑), no state change is issued for the output Q and it holds its current state. When the state of clock input signal C is changed from ON to OFF (↓): if JK = 00, then no state change is issued; if JK = 01, then Q is reset; if JK = 10, then Q is set; and finally if JK = 11, then Q is toggled. Operands for the instruction `jkff_f` are shown in Table 4.18.

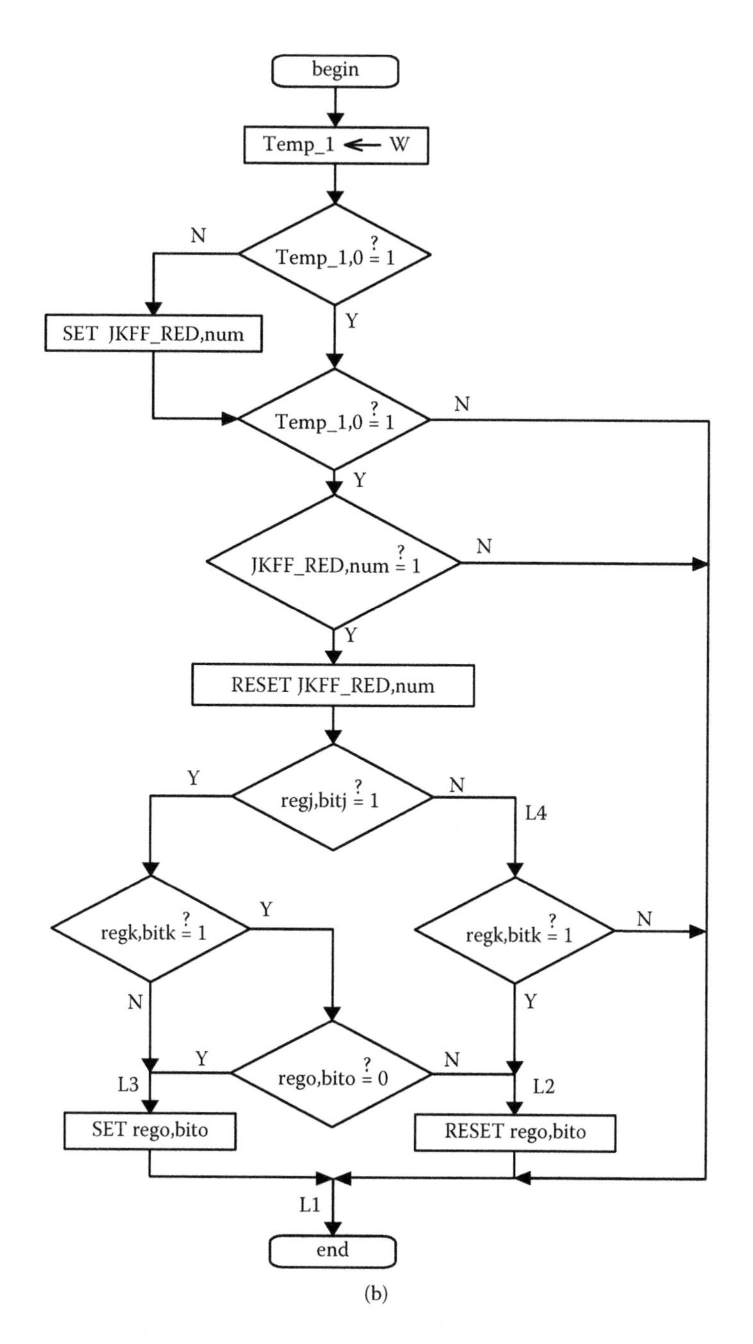

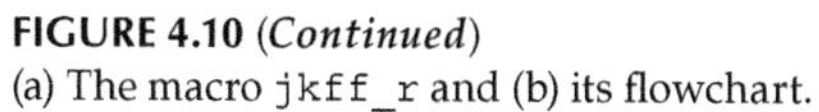

FIGURE 4.10 (*Continued*)
(a) The macro `jkff_r` and (b) its flowchart.

TABLE 4.16

Operands for the Instruction `jkff_r`

Input/Output	Data Type	Operands
J,K regj,bitj regk,bitk (Bit)	BOOL	I, Q, M, TON8_Q, TOF8_Q, TP8_Q, TOS8_Q, CTU8_Q, CTD8_Q, CTUD8_Q, LOGIC1, LOGIC0, FRSTSCN, SCNOSC
Q rego,bito (Bit)	BOOL	Q, M, TON8_Q, TOF8_Q, TP8_Q, TOS8_Q, CTU8_Q, CTD8_Q, CTUD8_Q

4.11 Examples for Flip-Flop Macros

In this section, we will consider two examples, UZAM_plc_16i16o_ex5.asm and UZAM_plc_16i16o_ex6.asm, to show the usage of flip-flop macros. In order to test the respective example please take the files from the CD-ROM attached to this book and then open the respective program by MPLAB IDE and compile it. After that, by using the PIC programmer software, take the compiled file UZAM_plc_16i16o_ex5.hex or UZAM_plc_16i16o_ex6.hex, and by your PIC programmer hardware send it to the program memory of

TABLE 4.17

Symbol of the Macro `jkff_f` and Its Truth Table

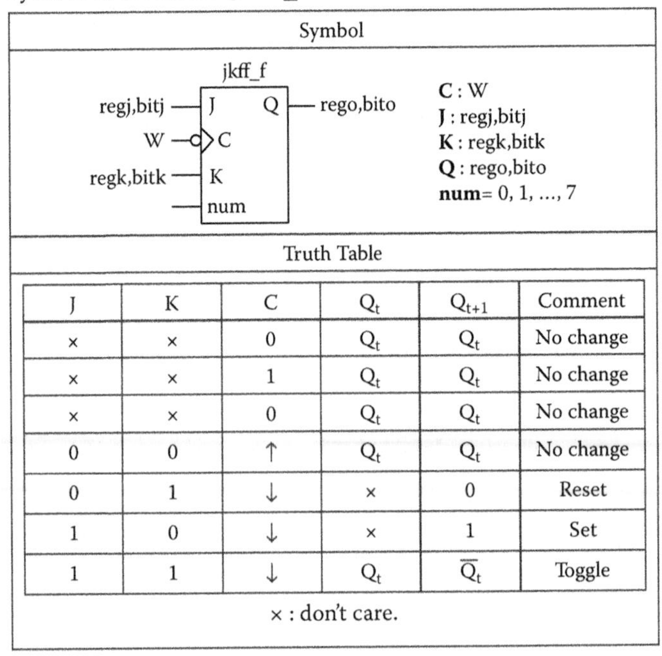
Symbol

Truth Table

J	K	C	Q_t	Q_{t+1}	Comment
×	×	0	Q_t	Q_t	No change
×	×	1	Q_t	Q_t	No change
×	×	0	Q_t	Q_t	No change
0	0	↑	Q_t	Q_t	No change
0	1	↓	×	0	Reset
1	0	↓	×	1	Set
1	1	↓	Q_t	$\overline{Q_t}$	Toggle

× : don't care.

```
;--------- macro: jkff_f -----------
jkff_f  macro   num,regj,bitj,regk,bitk,rego,bito
        local   L1,L2,L3,L4
        movwf   Temp_1
        btfsc   Temp_1,0
        bsf     JKFF_FED,num        ;JKFF_FED = Falling Edge
        btfsc   Temp_1,0            ;Detector for falling edge
        goto    L1                  ;triggered JK flip-flop
        btfss   JKFF_FED,num
        goto    L1
        bcf     JKFF_FED,num
        btfss   regj,bitj
        goto    L4                  ;if j=0 then goto L4
        btfss   regk,bitk
        goto    L3                  ;if j=1&k=0 then SET rego,bito (goto L3)
        btfsc   rego,bito           ;if j=1&k=1
        goto    L2                  ;then TOGGLE
        goto    L3                  ;rego,bito
L4      btfss   regk,bitk
        goto    L1                  ;if j=0&k=0 then NO CHANGE (goto L1)
        goto    L2                  ;if j=0&k=1 then RESET rego,bito
L3      bsf     rego,bito
        goto    L1
L2      bcf     rego,bito
L1
        endm
;-----------------------------------
```

(a)

FIGURE 4.11
(a) The macro `jkff_f` and (b) its flowchart. (*Continued*)

PIC16F648A microcontroller within the PIC16F648A-based PLC. To do this, switch the 4PDT in PROG position and the power switch in OFF position. After loading the UZAM_plc_16i16o_ex5.hex or UZAM_plc_16i16o_ex6.hex, switch the 4PDT in RUN and the power switch in ON position. Please check each program's accuracy by cross-referencing it with the related macros.

Let us now consider these two example programs: The first example program, UZAM_plc_16i16o_ex5.asm, is shown in Figure 4.12. It shows the usage of the following flip-flop macros: `r_edge`, `f_edge`, `latch1`, `latch0`, `dff_r`, `dff_f`. The ladder and schematic diagrams of the user program of UZAM_plc_16i16o_ex5.asm, shown in Figure 4.12, are depicted in Figure 4.13(a) and (b), respectively. It may not possible to observe the effects of `r_edge` and `f_edge` shown in rungs 1 and 2 due to the time delays caused by the macro `HC595`, explained in the Chapter 2. On the other hand, you can observe their effects from rungs 5 and 6, respectively, where `r_edge` and `f_edge` are both used together with the macro `latch1`. Observe that in rung 5 we obtain a rising edge triggered D flip-flop by using an `r_edge` and a `latch1`. Similarly, in rung 6 we obtain a falling edge triggered D flip-flop by using an `f_edge` and a `latch1`. Note that in this example, `_set` and `_reset` functions are both used as asynchronous SET and RESET inputs for the D type flip-flops.

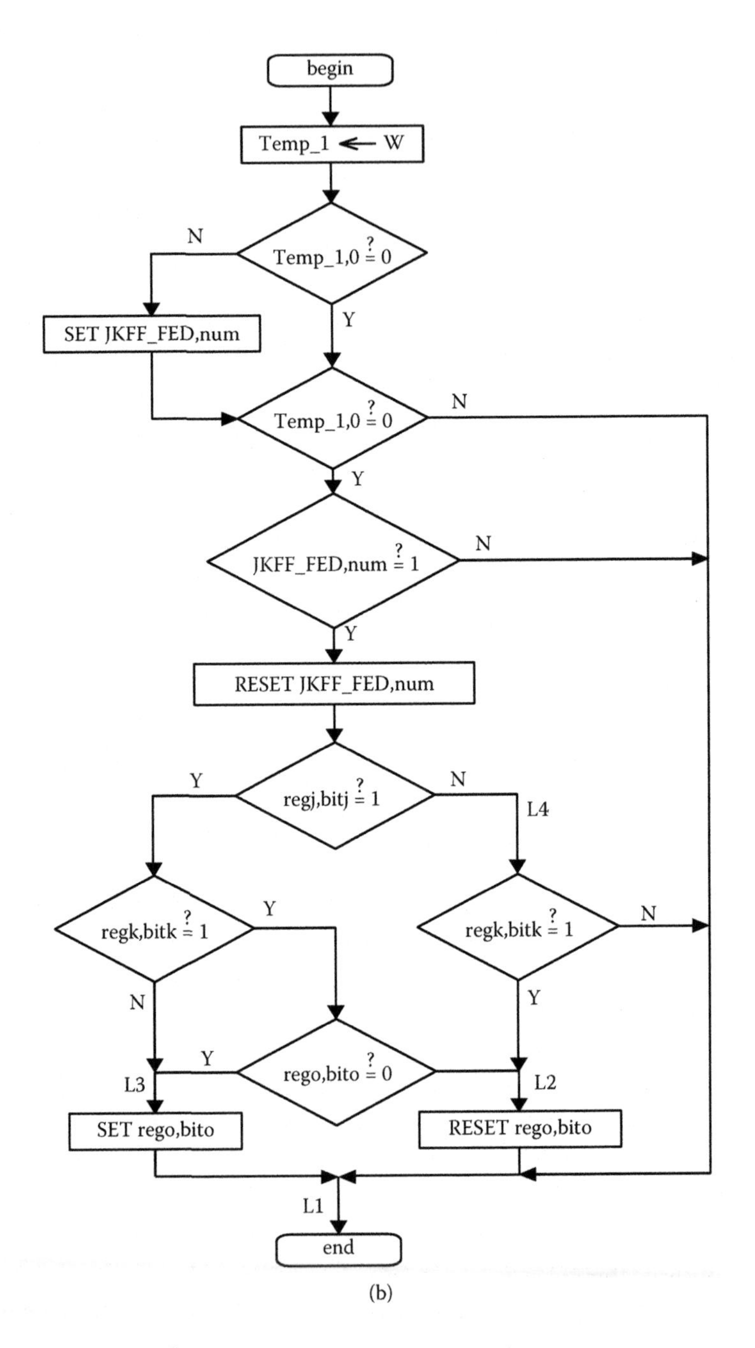

FIGURE 4.11 *(Continued)*
(a) The macro `jkff_f` and (b) its flowchart.

TABLE 4.18

Operands for the Instruction `jkff_f`

Input/Output	Data Type	Operands
J,K regj,bitj regk,bitk (Bit)	BOOL	I, Q, M, TON8_Q, TOF8_Q, TP8_Q, TOS8_Q, CTU8_Q, CTD8_Q, CTUD8_Q, LOGIC1, LOGIC0, FRSTSCN, SCNOSC
Q rego,bito (Bit)	BOOL	Q, M, TON8_Q, TOF8_Q, TP8_Q, TOS8_Q, CTU8_Q, CTD8_Q, CTUD8_Q

```
.
;--------------- user program starts here --
      ld          I0.0          ;rung 1
      r_edge      0
      out         Q0.0

      ld          I0.1          ;rung 2
      f_edge      0
      out         Q0.1

      ld          I0.2          ;rung 3
      latch1      I0.3,Q0.2

      ld          I0.4          ;rung 4
      latch0      I0.5,Q0.3

      ld          I0.0          ;rung 5
      r_edge      1
      latch1      I0.1,Q0.4

      ld          I0.6          ;rung 6
      f_edge      1
      latch1      I0.7,Q0.7

      ld          I1.0          ;rung 7
      dff_r       0,I1.1,Q1.0

      ld          I1.2          ;rung 8
      _set        Q1.0

      ld          I1.3          ;rung 9
      _reset      Q1.0

      ld          I1.4          ;rung 10
      dff_f       0,I1.5,Q1.7

      ld          I1.6          ;rung 11
      _set        Q1.7

      ld          I1.7          ;rung 12
      _reset      Q1.7
;--------------- user program ends here ----
.
```

FIGURE 4.12
The user program of UZAM_plc_16i16o_ex5.asm.

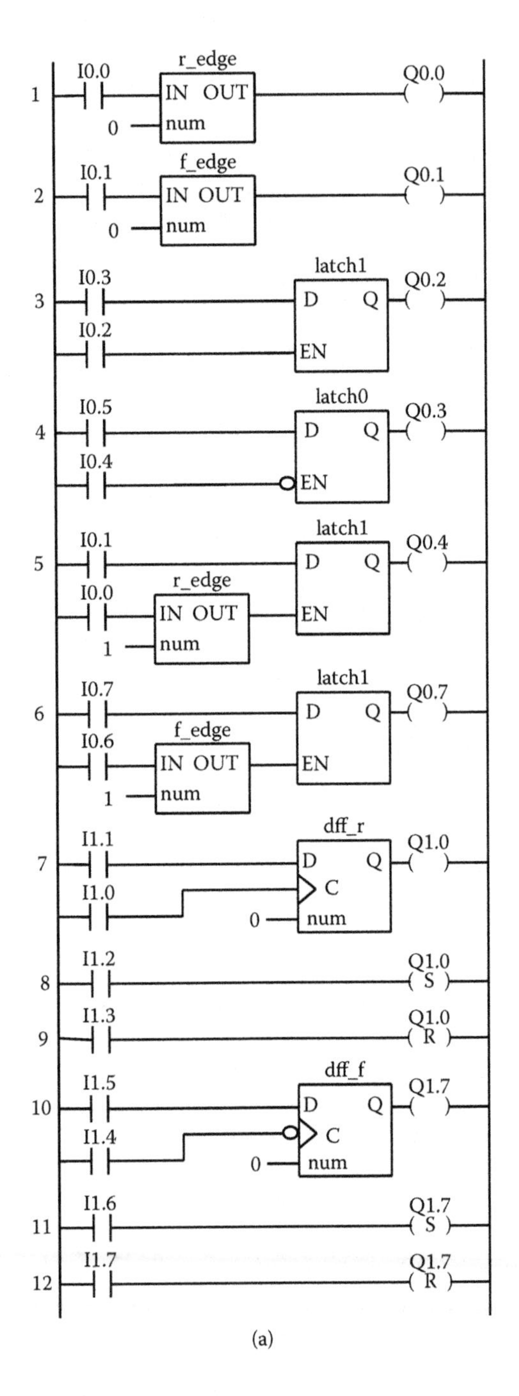

(a)

FIGURE 4.13
The user program of UZAM_plc_16i16o_ex5.asm: (a) ladder diagram. (*Continued*)

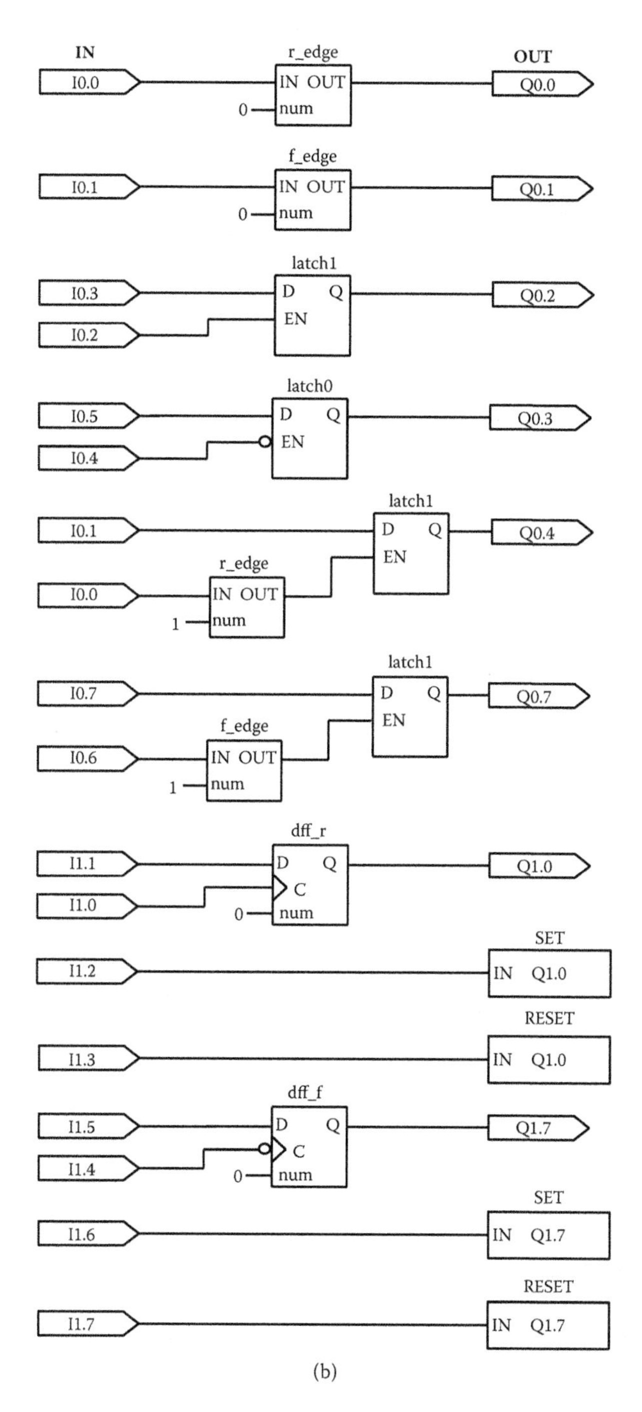

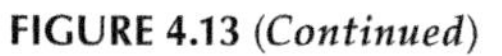
FIGURE 4.13 (*Continued*)
The user program of UZAM_plc_16i16o_ex5.asm: (b) schematic diagram.

```
.
;--------------- user program starts here --
     ld          I0.0          ;rung 1
     tff_r       0,I0.1,Q0.0

     ld          I0.2          ;rung 2
     _set        Q0.0

     ld          I0.3          ;rung 3
     _reset      Q0.0

     ld          I0.4          ;rung 4
     tff_f       0,I0.5,Q0.7

     ld          I0.6          ;rung 5
     _set        Q0.7

     ld          I0.7          ;rung 6
     _reset      Q0.7

     ld          I1.0          ;rung 7
     jkff_r      0,I1.1,I1.2,Q1.0

     ld          I1.3          ;rung 8
     _set        Q1.0

     ld          I1.4          ;rung 9
     _reset      Q1.0

     ld          I1.5          ;rung 10
     jkff_f      0,I1.6,I1.7,Q1.7
;--------------- user program ends here ----
.
```

FIGURE 4.14
The user program of UZAM_plc_16i16o_ex6.asm.

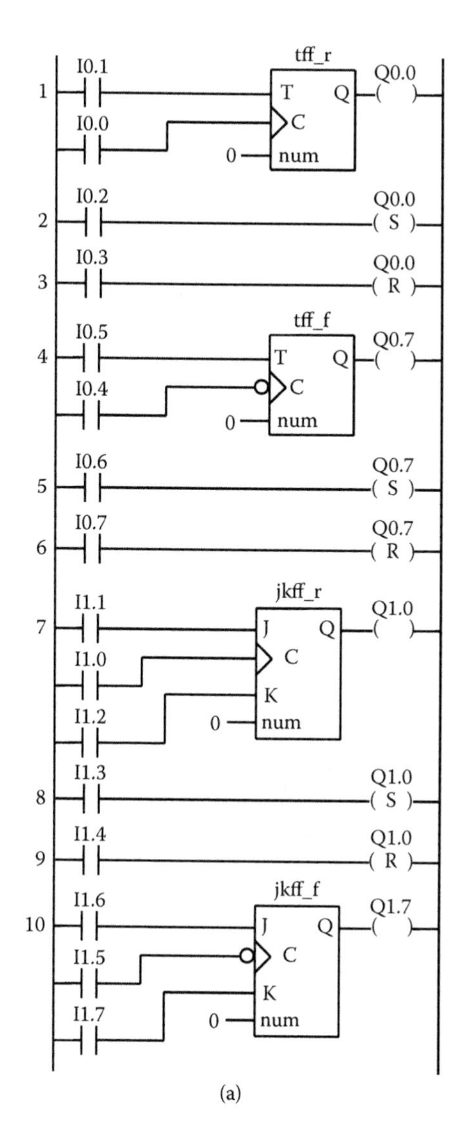

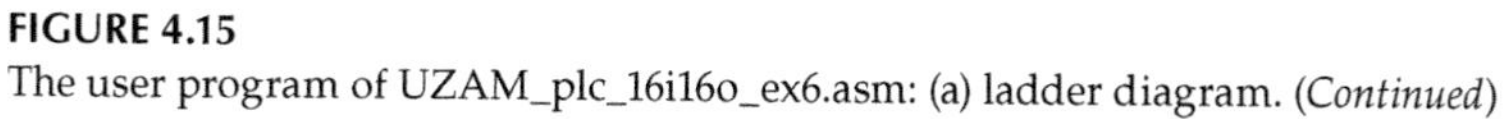

FIGURE 4.15
The user program of UZAM_plc_16i16o_ex6.asm: (a) ladder diagram. (*Continued*)

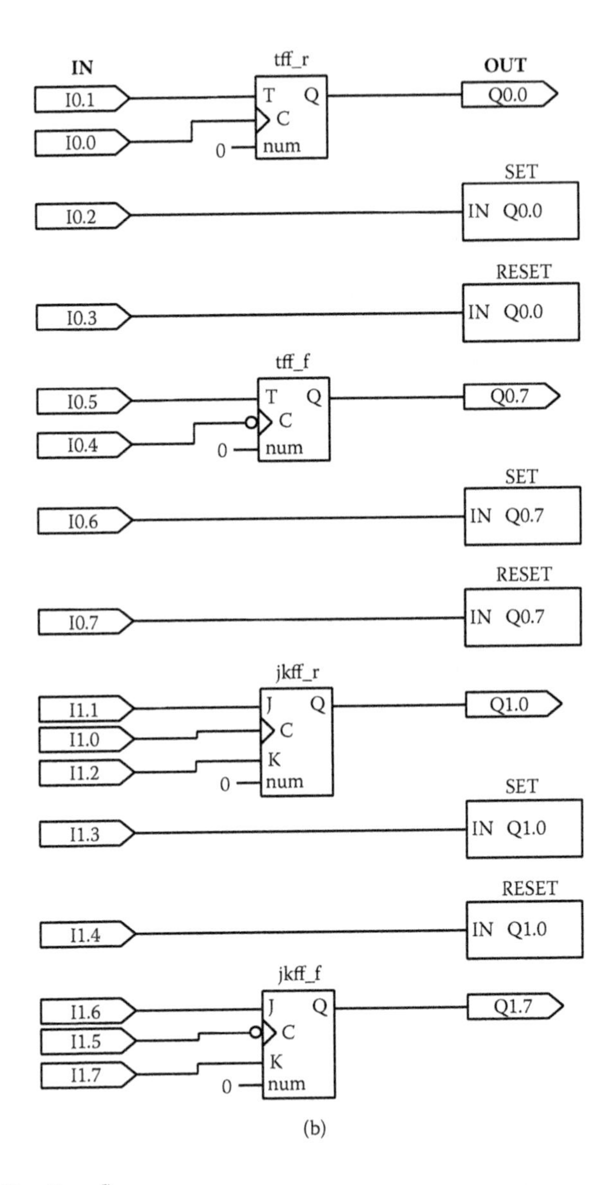

FIGURE 4.15 *(Continued)*
The user program of UZAM_plc_16i16o_ex6.asm: (b) schematic diagram.

The second example program, UZAM_plc_16i16o_ex6.asm, is shown in Figure 4.14. It shows the usage of the following flip-flop macros: `tff_r`, `tff_f`, `jkff_r`, and `jkff_f`. The ladder and schematic diagrams of the user program of UZAM_plc_16i16o_ex6.asm, shown in Figure 4.14, are depicted in Figure 4.15(a) and (b), respectively. Note that in this example, `_set` and `_reset` functions are both used as asynchronous SET and RESET inputs for the T and JK type flip-flops.

5

Timer Macros

In this chapter, the following timer macros are described:

`TON_8` (on-delay timer)
`TOF_8` (off-delay timer)
`TP_8` (pulse timer)
`TOS_8` (oscillator timer)

Timers can be used in a wide range of applications where a time delay function is required based on an input signal. The definition of 8-bit variables to be used for the timer macros, and their allocation in BANK 0 of SRAM data memory are shown in Figure 5.1(a) and (b), respectively. The status bits, which will be explained in the next sections, of all timers are defined as shown in Figure 5.2(a). All 8-bit variables defined for timers must be cleared at the beginning of the PLC operation for a proper operation. Therefore, all variables of timer macros are initialized within the macro `initialize`, as shown in Figure 5.2(b). The file definitions.inc, included within the CD-ROM attached to this book, contains all timer macros defined for the PIC16F648A-based PLC.

Let us now consider the timer macros. In the following, first, a general description is given for the considered timer function, and then its 8-bit implementation in the PIC16F648A-based PLC is provided.

5.1 On-Delay Timer (TON)

The on-delay timer can be used to delay setting an output true (ON—1) for a fixed period of time after an input signal becomes true (ON—1). The symbol and timing diagram of the on-delay timer (TON) are both shown in Figure 5.3. As the input signal IN goes true (ON—1), the timing function is started, and therefore the elapsed time ET starts to increase. When the elapsed time ET reaches the time specified by the preset time input PT, the output Q goes true (ON—1) and the elapsed time is held. The output Q remains true (ON—1) until the input signal IN goes false (OFF—0). If the input signal IN is not true (ON—1) longer than the delay time specified in

```
;--------------- VARIABLE DEFINITIONS ---
     CBLOCK 0x47
     TON8_Q,TOF8_Q,TP8_Q,TOS8_Q
     endc
     CBLOCK 0x4B
     TON8            ;TON8, TON8+1, ..., TON8+7
     endc
     CBLOCK 0x53
     TOF8            ;TOF8, TOF8+1, ..., TOF8+7
     endc
     CBLOCK 0x5B
     TP8             ;TP8, TP8+1, ..., TP8+7
     endc
     CBLOCK 0x63
     TOS8            ;TOS8, TOS8+1, ..., TOS8+7
     endc
     CBLOCK 0x6B
     TON8_RED,TOF8_RED,TP8_RED1,TP8_RED2,TOS8_RED
     endc
;----------------------------------------
```

(a)

FIGURE 5.1
(a) The definition of 8-bit variables to be used for the timer macros. (*Continued*)

PT, the output Q remains false (OFF—0). The following section explains the implementation of eight 8-bit on-delay timers for the PIC16F648A-based PLC.

5.2 Macro TON_8 (8-Bit On-Delay Timer)

The macro TON_8 defines eight on-delay timers selected with the num = 0, 1, …, 7. The macro TON_8 and its flowchart are shown in Figure 5.4. The symbol of the macro TON_8 is depicted in Table 5.1. IN (input signal), Q (output signal = timer status bit), and CLK (free-running timing signals—ticks: T0.0, T0.1, …, T0.7, T1.0, T1.1, …, T1.7) are all defined as Boolean variables. The time constant tcnst is an integer constant (here, for 8-bit resolution, it is chosen as any number in the range 1–255) and is used to define preset time PT, which is obtained by the formula PT = tcnst × CLK, where CLK should be used as the period of the free-running timing signals—ticks. The on-delay timer outputs are represented by the status bits: TON8_Q,num (num = 0, 1, …, 7), namely, TON8_Q0, TON8_Q1, …, TON8_Q7, as shown in Figure 5.2(a). A Boolean variable, TON8_RED,num (num = 0, 1, …, 7), is used as a rising edge detector for identifying the rising edges of the chosen CLK. An 8-bit integer variable TON8+num (num = 0, 1, …, 7) is used to count the rising edges of the CLK. The count value of TON8+num (num = 0, 1, …, 7) defines the elapsed time ET as follows: ET = CLK × count value of

Address	BANK 0
47h	TON8_Q
48h	TOF8_Q
49h	TP8_Q
4Ah	TOS8_Q
4Bh	TON8
4Ch	TON8+1
4Dh	TON8+2
4Eh	TON8+3
4Fh	TON8+4
50h	TON8+5
51h	TON8+6
52h	TON8+7
53h	TOF8
54h	TOF8+1
55h	TOF8+2
56h	TOF8+3
57h	TOF8+4
58h	TOF8+5
59h	TOF8+6
5Ah	TOF8+7
5Bh	TP8
5Ch	TP8+1
5Dh	TP8+2
5Eh	TP8+3
5Fh	TP8+4
60h	TP8+5
61h	TP8+6
62h	TP8+7
63h	TOS8
64h	TOS8+1
65h	TOS8+2
66h	TOS8+3
67h	TOS8+4
68h	TOS8+5
69h	TOS8+6
6Ah	TOS8+7
6Bh	TON8_RED
6Ch	TOF8_RED
6Dh	TP8_RED1
6Eh	TP8_RED2
6Fh	TOS8_RED

BANK 0

(b)

FIGURE 5.1 (*Continued*)
(b) Their allocation in BANK 0 of SRAM data memory.

```
;- defining on delay timer outputs -
#define TON8_Q0 TON8_Q,0
#define TON8_Q1 TON8_Q,1
#define TON8_Q2 TON8_Q,2
#define TON8_Q3 TON8_Q,3
#define TON8_Q4 TON8_Q,4
#define TON8_Q5 TON8_Q,5
#define TON8_Q6 TON8_Q,6
#define TON8_Q7 TON8_Q,7

;- defining off delay timer outputs -
#define TOF8_Q0 TOF8_Q,0
#define TOF8_Q1 TOF8_Q,1
#define TOF8_Q2 TOF8_Q,2
#define TOF8_Q3 TOF8_Q,3
#define TOF8_Q4 TOF8_Q,4
#define TOF8_Q5 TOF8_Q,5
#define TOF8_Q6 TOF8_Q,6
#define TOF8_Q7 TOF8_Q,7

;- defining puls timer outputs -----
#define TP8_Q0 TP8_Q,0
#define TP8_Q1 TP8_Q,1
#define TP8_Q2 TP8_Q,2
#define TP8_Q3 TP8_Q,3
#define TP8_Q4 TP8_Q,4
#define TP8_Q5 TP8_Q,5
#define TP8_Q6 TP8_Q,6
#define TP8_Q7 TP8_Q,7

;- defining osilator timer outputs -
#define TOS8_Q0 TOS8_Q,0
#define TOS8_Q1 TOS8_Q,1
#define TOS8_Q2 TOS8_Q,2
#define TOS8_Q3 TOS8_Q,3
#define TOS8_Q4 TOS8_Q,4
#define TOS8_Q5 TOS8_Q,5
#define TOS8_Q6 TOS8_Q,6
#define TOS8_Q7 TOS8_Q,7
```

(a)

FIGURE 5.2
(a) The definition of status bits of timer macros. (*Continued*)

TON8+num (num = 0, 1, …, 7). Let us now briefly consider how the macro `TON_8` works. First, preset time PT is defined by means of a reference timing signal CLK = `t_reg,t_bit` and a time constant `tcnst`. If the input signal IN, taken into the macro by means of W, is false (OFF—0), then the output signal TON8_Q,num (num = 0, 1, …, 7) is forced to be false (OFF—0), and the counter TON8+num (num = 0, 1, …, 7) is loaded with 00h. If the input signal IN is true (ON—1) and the output signal Q, i.e., the status bit TON8_Q,num (num = 0, 1, …, 7), is false (OFF—0), then with each rising

```
;------------------ macro: initialize ---------------
initialize    macro
    local   L1
    BANK1                   ;goto BANK1
    movlw b'00000111'       ;W<--b'00000111':Fosc/4-->TMR0,PS=256
    movwf OPTION_REG        ;pull-up on PORTB, OPTION_REG <-- W
    movlw b'00000001'       ;PORTB is both input and output port
    movwf TRISB             ;TRISB <-- b'00000001'
    BANK0                   ;goto BANK0
    clrf  PORTA             ;Clear PortA
    clrf  PORTB             ;Clear PortB
    clrf  TMR0              ;Clear TMR0
    movlw h'20'             ;initialize the pointer
    movwf FSR               ;to RAM
L1  clrf  INDF              ;clear INDF register
    incf  FSR,f             ;increment pointer
    btfss FSR,7             ;all done?
    goto  L1                ;if not goto L1
                            ;if yes carry on
    movlw 06h               ;W <--- 06h
    movwf Temp_2            ;Temp_2 <--- W(06h)
    endm
;-----------------------------------------------------
```

(b)

FIGURE 5.2 (*Continued*)
(b) The initialization of all variables of timer macros within the macro `initialize`.

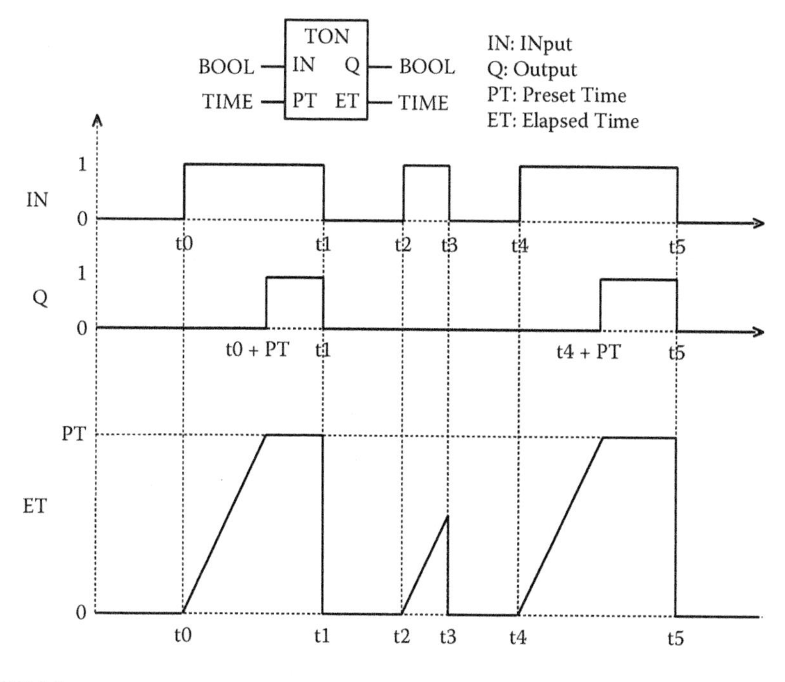

FIGURE 5.3
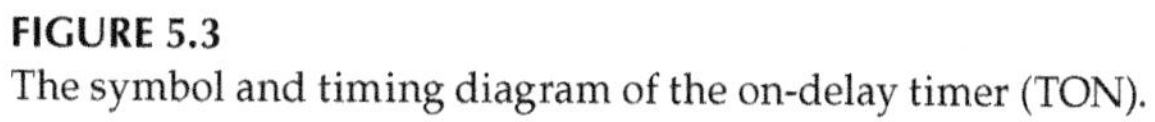
The symbol and timing diagram of the on-delay timer (TON).

```
;--------- macro: TON_8 -----------
TON_8 macro   num,t_reg,t_bit,tcnst
      local   L1,L2
      movwf   Temp_1
      btfsc   Temp_1,0
      goto    L2
      movlw   00h
      movwf   TON8+num
      bcf     TON8_Q,num
      goto    L1
L2    btfsc   TON8_Q,num
      goto    L1
      btfss   t_reg,t_bit
      bsf     TON8_RED,num
      btfss   t_reg,t_bit
      goto    L1
      btfss   TON8_RED,num
      goto    L1
      bcf     TON8_RED,num
      incf    TON8+num,f
      movfw   TON8+num
      xorlw   tcnst
      skpnz
      bsf     TON8_Q,num
L1
      endm
;-----------------------------------
```

(a)

FIGURE 5.4
(a) The macro `TON_8` and (b) its flowchart. (*Continued*)

edge of the reference timing signal CLK = `t_reg,t_bit` the related counter TON8+num is incremented by one. In this case, when the count value of TON8+num is equal to the number `tcnst`, then state change from 0 to 1 is issued for the output signal (timer status bit) TON8_Q,num (num = 0, 1, …, 7). If the input signal IN and the output signal Q, i.e., the status bit TON8_Q,num (num = 0, 1, …, 7) are both true (ON—1), then no action is taken and the elapsed time ET is held. In this macro a previously defined 8-bit variable Temp_1 is also utilized.

5.3 Off-Delay Timer (TOF)

The off-delay timer can be used to delay setting an output false (OFF—0) for a fixed period of time after an input signal goes false (OFF—0); i.e., the output is held ON for a given period longer than the input. The symbol and

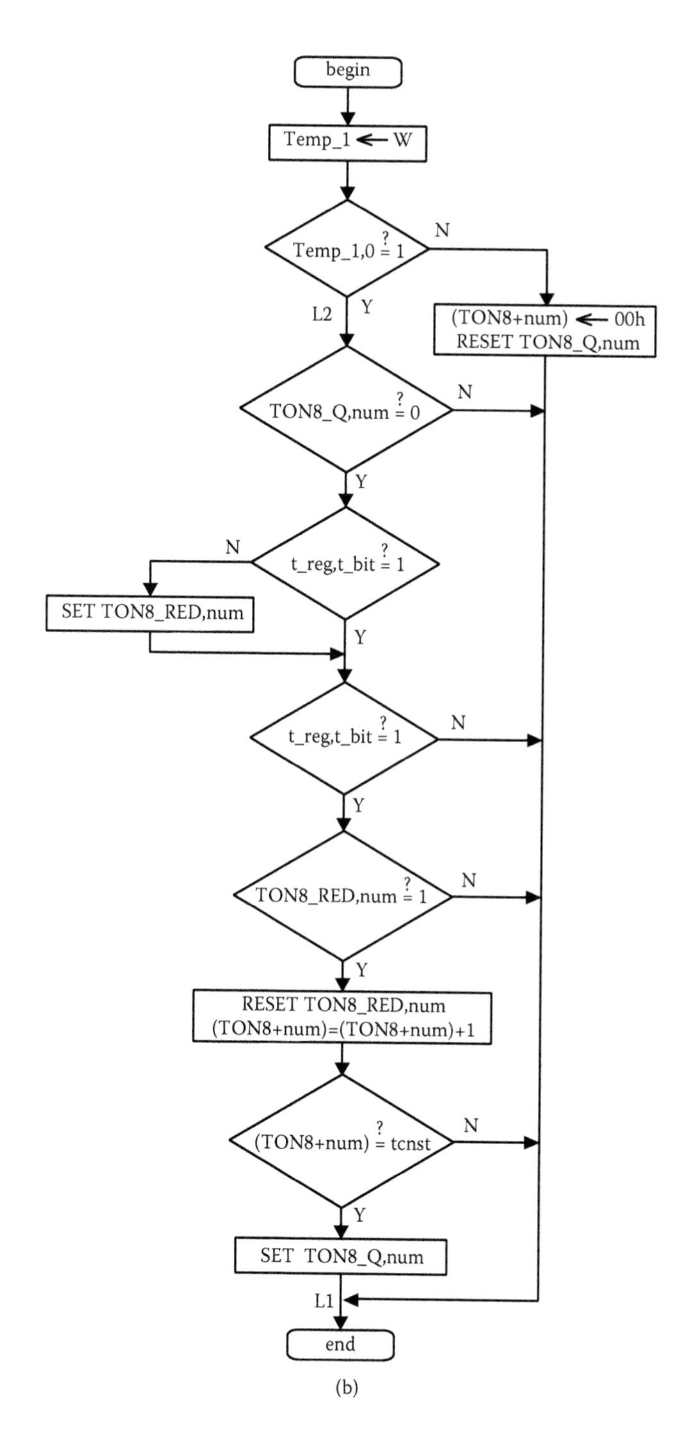

FIGURE 5.4 (*Continued*)
(a) The macro `TON_8` and (b) its flowchart.

TABLE 5.1

Symbol of the Macro TON_8

TON_8 symbol	Parameters
TON_8 IN Q CLK tcnst num PT = tcnst × CLK	**IN** (through W) = 0 or 1 **CLK** (t_reg,t_bit) = T0.0(1.024 ms), ..., T1.7(3355.4432 ms) **tcnst** (8bit) = 1, 2, ..., 255 **num** = 0, 1, ..., 7 **Q** = TON8_Q,num (num= 0, 1, ..., 7)

timing diagram of the off-delay timer (TOF) are both shown in Figure 5.5. As the input signal IN goes true (ON—1), the output Q follows and remains true (ON—1), until the input signal IN is false (OFF—0) for the period specified in preset time input PT. As the input signal IN goes false (OFF—0), the elapsed time ET starts to increase. It continues to increase until it reaches the preset time input PT, at which point the output Q is set false (OFF—0) and the elapsed time is held. If the input signal IN is only false (OFF—0) for a period shorter than the input PT, the output Q remains true (ON—1). The following section explains the implementation of eight 8-bit off-delay timers for the PIC16F648A-based PLC.

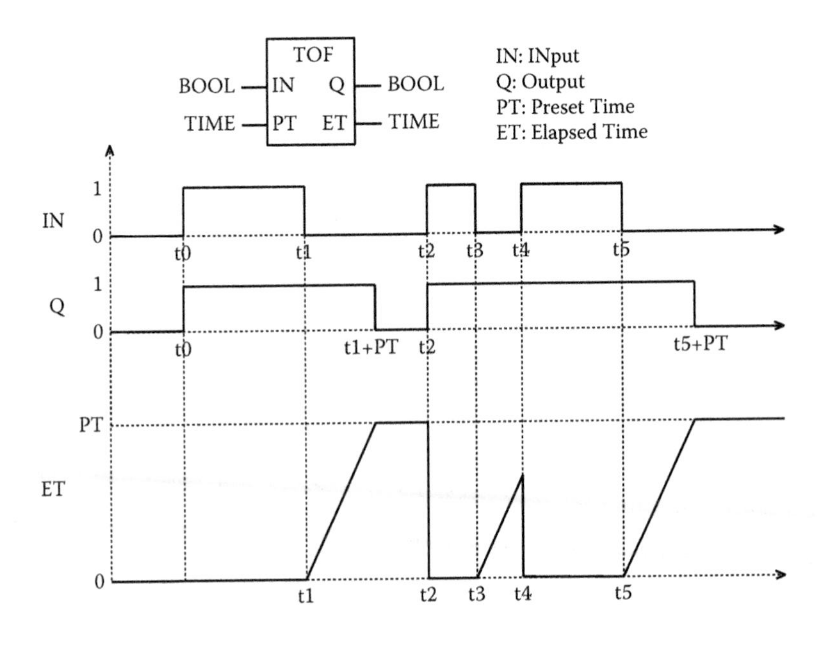

FIGURE 5.5
The symbol and timing diagram of the off-delay timer (TOF).

5.4 Macro `TOF_8` (8-Bit Off-Delay Timer)

The macro `TOF_8` defines eight off-delay timers selected with the num = 0, 1, …, 7. The macro `TOF_8` and its flowchart are shown in Figure 5.6. The symbol of the macro `TOF_8` is depicted in Table 5.2. IN (input signal), Q (output signal = timer status bit), and CLK (free-running timing signals—ticks: T0.0, T0.1, …, T0.7, T1.0, T1.1, …, T1.7) are all defined as Boolean variables. The time constant `tcnst` is an integer constant (here, for 8-bit resolution, it is chosen as any number in the range 1–255) and is used to define preset time PT, which is obtained by the formula PT = `tcnst` × CLK, where CLK should be used as the period of the free-running timing signals—ticks. The off-delay timer outputs are represented by the status bits: TOF8_Q,num (num = 0, 1, …, 7), namely, TOF8_Q0, TOF8_Q1, …, TOF8_Q7, as shown in Figure 5.2(a). We use a Boolean variable, TOF8_RED,num (num = 0, 1, …, 7), as a rising edge detector for identifying the rising edges of the chosen CLK. An 8-bit integer variable TOF8+num (num = 0, 1, …, 7) is used to count the rising edges of the

```
;--------- macro: TOF_8 -----------
TOF_8 macro   num,t_reg,t_bit,tcnst
      local   L1,L2
      movwf   Temp_1
      btfss   Temp_1,0
      goto    L2
      movlw   00h
      movwf   TOF8+num
      bsf     TOF8_Q,num
      goto    L1
L2    btfss   TOF8_Q,num
      goto    L1
      btfss   t_reg,t_bit
      bsf     TOF8_RED,num
      btfss   t_reg,t_bit
      goto    L1
      btfss   TOF8_RED,num
      goto    L1
      bcf     TOF8_RED,num
      incf    TOF8+num,f
      movfw   TOF8+num
      xorlw   tcnst
      skpnz
      bcf     TOF8_Q,num
L1
      endm
;-----------------------------------
```

(a)

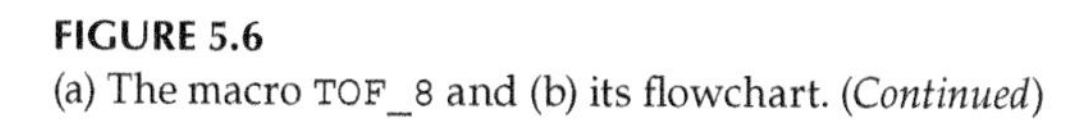

FIGURE 5.6
(a) The macro `TOF_8` and (b) its flowchart. (*Continued*)

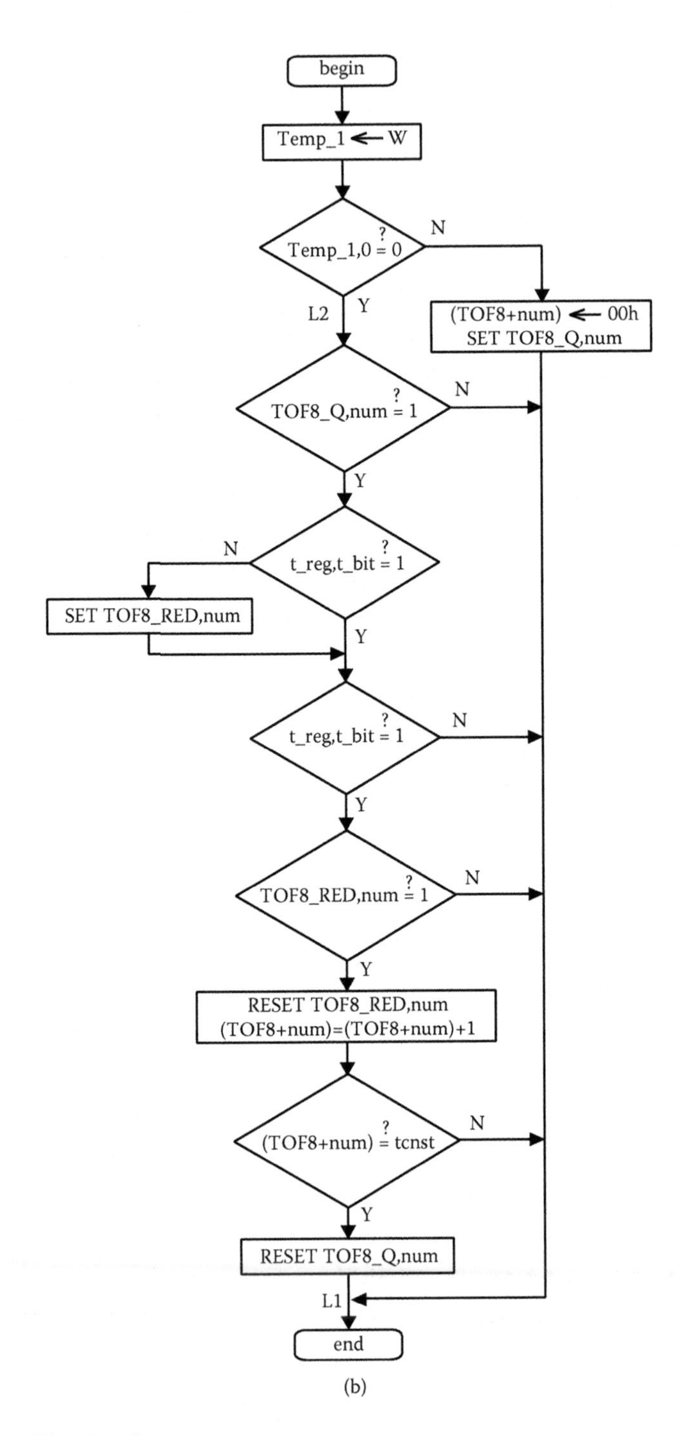

FIGURE 5.6 (*Continued*)
(a) The macro `TOF_8` and (b) its flowchart.

TABLE 5.2

Symbol of the Macro `TOF_8`

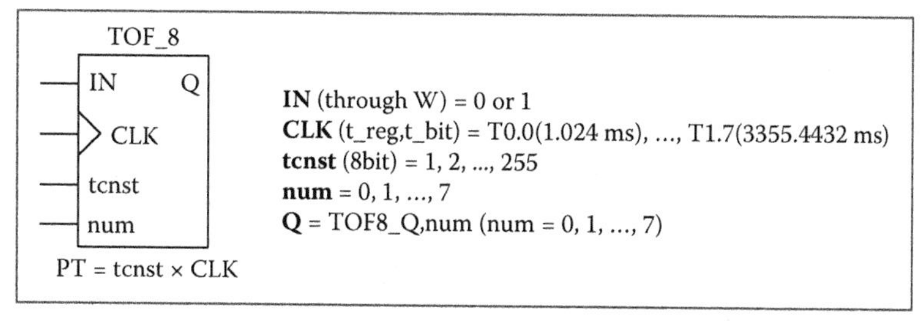

CLK. The count value of TOF8+num (num = 0, 1, …, 7) defines the elapsed time ET as follows: ET = CLK × count value of TOF8+num (num = 0, 1, …, 7). Let us now briefly consider how the macro `TOF_8` works. First, preset time PT is defined by means of a reference timing signal CLK = `t_reg,t_bit` and a time constant `tcnst`. If the input signal IN, taken into the macro by means of W, is true (ON—1), then the output signal TOF8_Q,num (num = 0, 1, …, 7) is forced to be true (ON—1), and the counter TOF8+num (num = 0, 1, …, 7) is loaded with 00h. When IN = 1 and TOF8_Q,num = 1, if IN goes false (OFF—0), then with each rising edge of the reference timing signal CLK = `t_reg,t_bit` the related counter TOF8+num is incremented by one. In this case, when the count value of TOF8+num is equal to the number `tcnst`, then state change from 1 to 0 is issued for the output signal (timer status bit) TOF8_Q,num (num = 0, 1, …, 7). In this macro a previously defined 8-bit variable Temp_1 is also utilized.

5.5 Pulse Timer (TP)

The pulse timer can be used to generate output pulses of a given time duration. The symbol and timing diagram of the pulse timer (TP) are both shown in Figure 5.7. As the input signal IN goes true (ON—1) (t0, t2, t4), the output Q follows and remains true (ON—1) for the pulse duration as specified by the preset time input PT. While the pulse output Q is true (ON—1), the elapsed time ET is increased (between t0 and t0 + PT, between t2 and t2 + PT, and between t4 and t4 + PT). On the termination of the pulse, the elapsed time ET is reset. The output Q will remain true (ON—1) until the pulse time has elapsed, irrespective of the state of the input signal IN. The following section explains the implementation of eight 8-bit pulse timers for the PIC16F648A-based PLC.

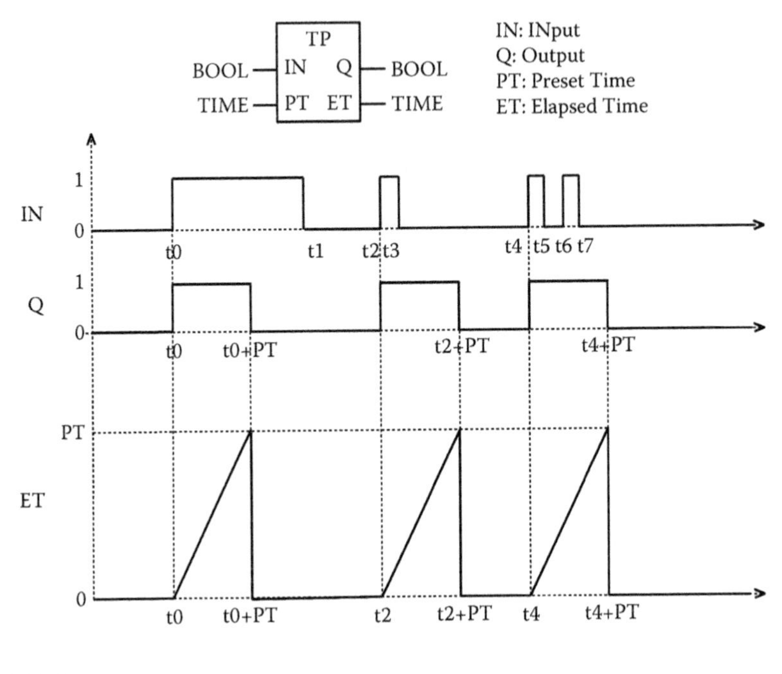

FIGURE 5.7
The symbol and timing diagram of the pulse timer (TP).

5.6 Macro `TP_8` (8-Bit Pulse Timer)

The macro `TP_8` defines eight pulse timers selected with the num = 0, 1, …, 7. The macro `TP_8` and its flowchart are shown in Figure 5.8. The symbol of the macro `TP_8` is depicted in Table 5.3. The macro `TP_8` defines eight pulse timers selected with the num = 0, 1, …, 7. IN (input signal), Q (output signal = timer status bit), and CLK (free-running timing signals—ticks: T0.0, T0.1, …, T0.7, T1.0, T1.1, …, T1.7) are all defined as Boolean variables. The time constant `tcnst` is an integer constant (here, for 8-bit resolution, it is chosen as any number in the range 1–255) and is used to define preset time PT, which is obtained by the formula PT = `tcnst` × CLK, where CLK should be used as the period of the free-running timing signals—ticks. The pulse timer outputs are represented by the status bits: TP8_Q,num (num = 0, 1, …, 7), namely, TP8_Q0, TP8_Q1, …, TP8_Q7, as shown in Figure 5.2(a). A Boolean variable, TP8_RED1,num (num = 0, 1, …, 7), is used as a rising edge detector for identifying the rising edges of the chosen CLK. Similarly, another Boolean variable, TP8_RED2,num (num = 0, 1, …, 7), is used as a rising edge detector for identifying the rising edges of the input signal IN, taken into the macro by means of W. An 8-bit integer variable TP8+num (num = 0, 1, …, 7) is used to count the rising edges of the CLK. The count

```
;--------- macro: TP_8 -----------
TP_8 macro     num,t_reg,t_bit,tcnst
     local     L1,L2,L3,L4
     movwf     Temp_1
     btfss     Temp_1,0
     bsf       TP8_RED2,num
     btfss     Temp_1,0
     goto      L3
     btfss     TP8_RED2,num
     goto      L3
     bsf       TP8_Q,num
L3   btfsc     TP8_Q,num
     goto      L2
     btfss     Temp_1,0
     goto      L1
L2   btfss     t_reg,t_bit
     bsf       TP8_RED1,num
     btfss     t_reg,t_bit
     goto      L1
     btfss     TP8_RED1,num
     goto      L1
     bcf       TP8_RED1,num
     incf      TP8+num,f
     movfw     TP8+num
     xorlw     tcnst
     skpz
     goto      L1
     bcf       TP8_Q,num
     bcf       TP8_RED2,num
L1   btfss     TP8_Q,num
     clrf      TP8+num
     endm
;---------------------------------
```

(a)

FIGURE 5.8
(a) The macro `TP_8` and (b) its flowchart. (*Continued*)

value of TP8+num (num = 0, 1, …, 7) defines the elapsed time ET as follows: ET = CLK × count value of TP8+num (num = 0, 1, …, 7). Let us now briefly consider how the macro `TP_8` works. First, preset time PT is defined by means of a reference timing signal CLK = `t_reg,t_bit` and a time constant `tcnst`. If the rising edge of the input signal IN is detected, by means of TP8_RED2,num, then the output signal TP8_Q,num (num = 0, 1, …, 7) is forced to be true (ON—1). After the output becomes true, i.e., TP8_Q,num = 1, the related counter TP8+num is incremented by one with each rising edge of the reference timing signal CLK = `t_reg,t_bit` detected by means of TP8_RED1,num. When the count value of TP8+num is equal to the number `tcnst`, then state change from 1 to 0 is issued for the output signal (timer

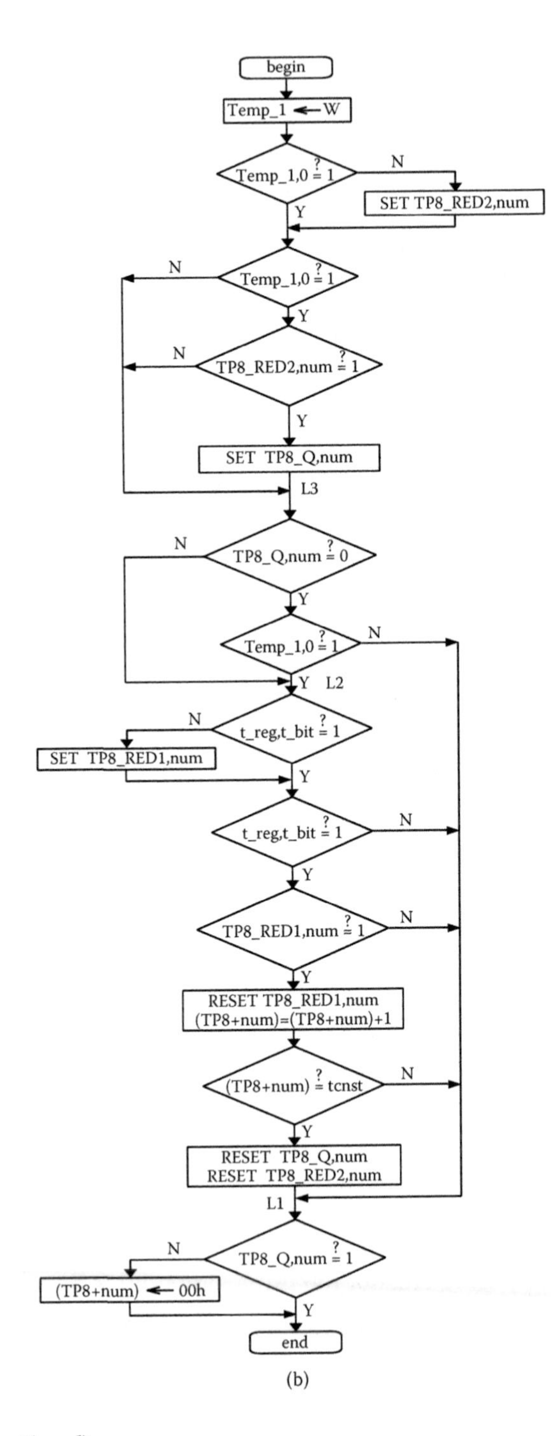

(b)

FIGURE 5.8 (*Continued*)
(a) The macro `TP_8` and (b) its flowchart.

TABLE 5.3

Symbol of the Macro `TP_8`

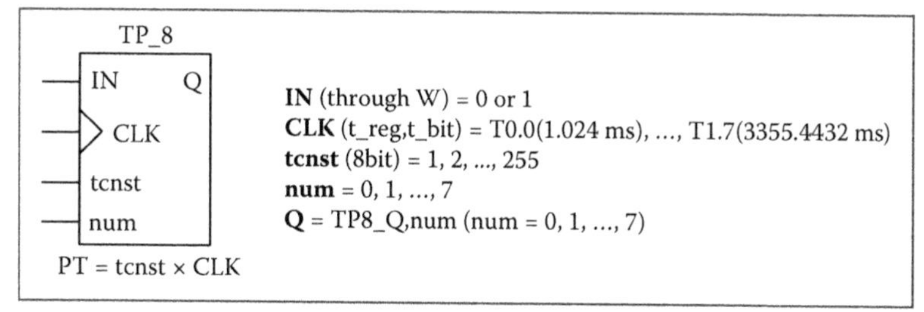

status bit) TP8_Q,num (num = 0, 1, …, 7), and at the same time the counter TP8+num (num = 0, 1, …, 7) is cleared. In this macro a previously defined 8-bit variable Temp_1 is also utilized.

5.7 Oscillator Timer (TOS)

The oscillator timer can be used to generate pulse trains with given durations for true (ON) and false (OFF) times. Therefore, the oscillator timer can be used in pulse width modulation (PWM) applications. The symbol and timing diagram of the oscillator timer (TOS) are both shown in Figure 5.9. PT0 (respectively PT1) defines the false (OFF) time (respectively true (ON) time) of the pulse. As the input signal IN goes and remains true (ON—1), the OFF timing function is started, and therefore the elapsed time ET0 is increased. When the elapsed time ET0 reaches the time specified by the preset time input PT0, the output Q goes true (ON—1) and ET0 is cleared. At the same time, as long as the input signal IN remains true (ON—1), the ON timing function is started, and therefore the elapsed time ET1 is increased. When the elapsed time ET1 reaches the time specified by the preset time input PT1, the output Q goes false (OFF—1) and ET1 is cleared. Then it is time for the next operation for OFF and ON times. This operation will carry on as long as the input signal IN remains true (ON—1), generating the pulse trains based on PT0 and PT1. If the input signal IN goes and remains false (OFF—0), then the output Q is forced to be false (OFF—0). The following section explains the implementation of eight 8-bit oscillator timers (TOS) for the PIC16F648A-based PLC.

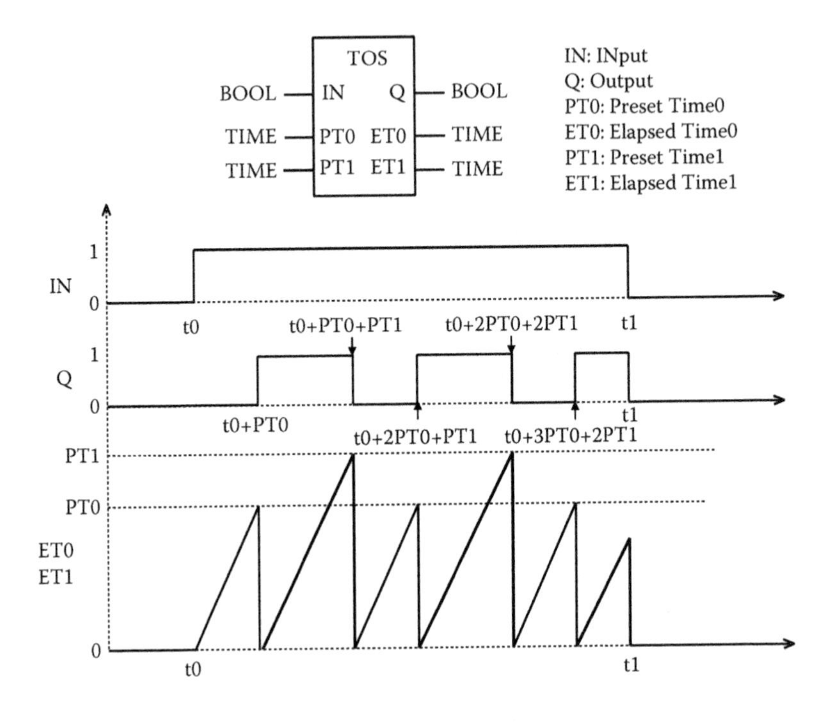

FIGURE 5.9
Symbol and timing diagram of the oscillator timer (TOS).

5.8 Macro `TOS_8` (8-Bit Oscillator Timer)

The macro `TOS_8` defines eight oscillator timers selected with the num = 0, 1, …, 7. The macro `TOS_8` and its flowchart are shown in Figure 5.10. The symbol of the macro `TOS_8` is depicted in Table 5.4. IN (input signal), Q (output signal = timer status bit), and CLK (free-running timing signals—ticks: T0.0, T0.1, …, T0.7, T1.0, T1.1, …, T1.7) are all defined as Boolean variables. The time constant `tcnst0` is an integer constant (here, for 8-bit resolution, it is chosen as any number in the range 1–255) and is used to define preset time PT0, which is obtained by the formula PT0 = `tcnst0` × CLK, where CLK should be used as the period of the free-running timing signals—ticks. The time constant `tcnst1` is an integer constant (here, for 8-bit resolution, it is chosen as any number in the range 1–255) and is used to define preset time PT1, which is obtained by the formula PT1 = `tcnst1` × CLK, where CLK should be used as the period of the free-running timing signals—ticks. The oscillator timer outputs are represented by the status bits: TOS8_Q,num (num = 0, 1, …, 7), namely, TOS8_Q0, TOS8_Q1, …, TOS8_Q7, as shown in Figure 5.2(a). We use a Boolean variable, TOS8_RED,num (num = 0, 1, …, 7), as a rising edge detector for identifying the rising edges of the chosen CLK. An 8-bit

```
;--------- macro: TOS_8 --------------------
TOS_8 macro   num,t_reg,t_bit,tcnst0,tcnst1
      local   L1,L2,L3
      movwf   Temp_1
      btfsc   Temp_1,0
      goto    L3
      movlw   00h
      movwf   TOS8+num
      bcf     TOS8_Q,num
      goto    L1
L3    btfss   t_reg,t_bit
      bsf     TOS8_RED,num
      btfss   t_reg,t_bit
      goto    L1
      btfss   TOS8_RED,num
      goto    L1
      bcf     TOS8_RED,num
      incf    TOS8+num,f
      btfsc   TOS8_Q,num
      goto    L2
      movfw   TOS8+num
      xorlw   tcnst0
      skpz
      goto    L1
      bsf     TOS8_Q,num
      movlw   00h
      movwf   TOS8+num
      goto    L1
L2    movfw   TOS8+num
      xorlw   tcnst1
      skpz
      goto    L1
      bcf     TOS8_Q,num
      movlw   00h
      movwf   TOS8+num
L1
      endm
;-------------------------------------------
```

(a)

FIGURE 5.10
(a) The macro `TOS_8` and (b) its flowchart. (*Continued*)

integer variable TOS8+num (num = 0, 1, …, 7) is used to count the rising edges of the CLK. Note that we use the same counter TOS8+num (num = 0, 1, …, 7) to obtain the time delays for both OFF and ON times, as these durations are mutually exclusive. The count value of TOS8+num (num = 0, 1, …, 7) defines the elapsed time ET0 or ET1 as follows: ET(0 or 1) = CLK × count value of TOS8+num (num = 0, 1, …, 7). Let us now briefly consider how the macro TOS_8 works. First, preset time PT0 (respectively PT1) is defined by means of a reference timing signal CLK = `t_reg,t_bit` and a time

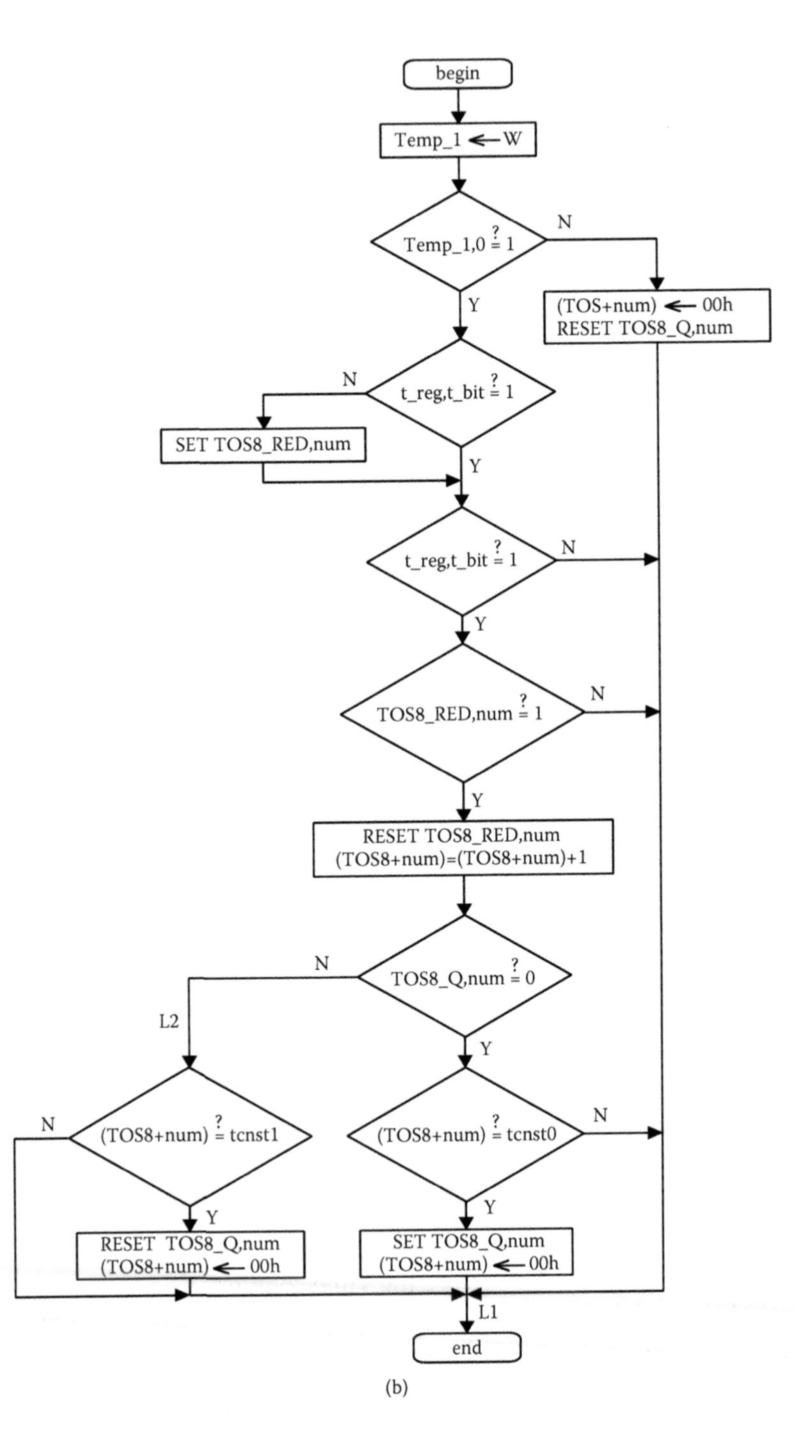

(b)

FIGURE 5.10 (*Continued*)
(a) The macro `TOS_8` and (b) its flowchart.

TABLE 5.4

Symbol of the Macro `TOS_8`

TOS_8	
IN Q	**IN** (through W) = 0 or 1
CLK	**CLK** (t_reg,t_bit) = T0.0(1.024 ms), ..., T1.7(3355.4432 ms)
tcnst0	**tcnst0** (8bit) = 1, 2, ..., 255
tcnst1	**tcnst1** (8bit) = 1, 2, ..., 255
num	**num** = 0, 1, ..., 7
	Q = TOS8_Q,num (num = 0, 1, ..., 7)
PT0 = tcnst0 × CLK	
PT1 = tcnst1 × CLK	

constant `tcnst0` (respectively `tcnst1`). If the input signal IN, taken into the macro by means of W, is false (OFF—0), then the output signal TOS8_Q,num (num = 0, 1, ..., 7) is forced to be false (OFF—0), and the counter TOS8+num (num = 0, 1, ..., 7) is loaded with 00h. If the input signal IN is true (ON—1) and the output signal Q, i.e., the status bit TON8_Q,num (num = 0, 1, ..., 7), is false (OFF—0), then with each rising edge of the reference timing signal CLK = `t_reg,t_bit` the related counter TON8+num is incremented by one. In this case, when the count value of TON8+num is equal to the number `tcnst0`, then TON8+num is cleared and a state change from 0 to 1 is issued for the output signal (timer status bit) TON8_Q,num (num = 0, 1, ..., 7). If both the input signal IN and the output signal Q, i.e., the status bit TON8_Q,num (num = 0, 1, ..., 7), are true (ON—1), then with each rising edge of the reference timing signal CLK = `t_reg,t_bit` the related counter TON8+num is incremented by one. In this case, when the count value of TON8+num is equal to the number `tcnst1`, then TON8+num is cleared and a state change from 1 to 0 is issued for the output signal (timer status bit) TON8_Q,num (num = 0, 1, ..., 7). This process will continue as long as the input signal IN remains true (ON—1). In this macro a previously defined 8-bit variable Temp_1 is also utilized.

5.9 Example for Timer Macros

In this section, we will consider an example, namely, UZAM_plc_16i16o_ex7.asm, to show the usage of timer macros. In order to test this example, please take the file from the CD-ROM attached to this book and then open the program by MPLAB IDE and compile it. After that, by using the PIC programmer software, take the compiled file UZAM_plc_16i16o_ex7.hex, and by your PIC programmer hardware, send it to the program memory of PIC16F648A microcontroller within the PIC16F648A-based PLC. To do this, switch the 4PDT in PROG position and the power switch in OFF position. After loading the UZAM_plc_16i16o_ex7.hex, switch the 4PDT in RUN and the power switch in the ON position. Please check the program's accuracy by cross-referencing it with the related macros.

Let us now consider this example program: The example program UZAM_plc_16i16o_ex7.asm is shown in Figure 5.11. It shows the usage of all timer macros described above. The ladder diagram of the user program of UZAM_plc_16i16o_ex7.asm, shown in Figure 5.11, is depicted in Figure 5.12.

In the first two rungs, an on-delay timer TON_8 is implemented as follows: the input signal IN is taken from I0.0 num = 0, and therefore we choose the first on-delay timer, whose timer status bit (or output Q) is TON8_Q0. The preset time PT = `tcnst` × CLK = 50 × 104.8576 ms (T1.2) = 5242.88 ms = 5.24288 s. As can be seen from the second rung, the timer status bit TON8_Q0 is sent to output Q0.0.

In rungs 3 and 4, an off-delay timer TOF_8 is implemented as follows: the input signal IN is taken from I0.2 num = 1, and therefore we choose the second off-delay timer, whose timer status bit (or output Q) is TOF8_Q1. The preset time PT = `tcnst` × CLK = 50 × 104.8576 ms (T1.2) = 5242.88 ms = 5.24288 s. As can be seen from rung 4, the timer status bit TOF8_Q1 is sent to output Q0.2.

In rungs 5 and 6, a pulse timer TP_8 is implemented as follows: the input signal IN is taken from I0.4 num = 2, and therefore we choose the third pulse timer, whose timer status bit (or output Q) is TP8_Q2. The preset time PT = `tcnst` × CLK = 50 × 104.8576 ms (T1.2) = 5242.88 ms = 5.24288 s. As can be seen from rung 6, the timer status bit TP8_Q2 is sent to output Q0.4.

In rungs 7 and 8, an oscillator timer TOS_8 is implemented as follows: the input signal IN is taken from I0.6 num = 3, and therefore we choose the fourth oscillator timer, whose timer status bit (or output Q) is TOS8_Q3. The preset time PT0 = `tcnst0` × CLK = 50 × 104.8576 ms (T1.2) = 5242.88 ms = 5.24288 s. The preset time PT1 = `tcnst1` × CLK = 50 × 104.8576 ms (T1.2) = 5242.88 ms = 5.24288 s. In this setup, the pulse trains we will obtain have a 50% duty cycle with the time period of T = 100 × 104.8576 ms = 10,485.76 ms = 10.48576 s. As can be seen from rung 8, the timer status bit TOS8_Q3 is sent to output Q0.6.

In rungs 9 and 10, another on-delay timer TON_8 is implemented as follows: the input signal IN is taken from I1.1 num = 4, and therefore we choose the fifth on-delay timer, whose timer status bit (or output Q) is TON8_Q4. The preset time PT = `tcnst` × CLK = 10 × 419.4304 ms (T1.4) = 4194.304 ms = 4.194304 s. As can be seen from rung 10, the timer status bit TON8_Q4 is sent to output Q1.1.

In rungs 11 and 12, another off-delay timer TOF_8 is implemented as follows: the input signal IN is taken from I1.3 num = 5, and therefore we choose the sixth off-delay timer, whose timer status bit (or output Q) is TOF8_Q5. The preset time PT = `tcnst` × CLK = 10 × 419.4304 ms

```
.
;--------------- user program starts here --
      ld     I0.0               ;rung 1
      TON_8 0,T1.2,.50

      ld     TON8_Q0            ;rung 2
      out    Q0.0

      ld     I0.2               ;rung 3
      TOF_8 1,T1.2,.50

      ld     TOF8_Q1            ;rung 4
      out    Q0.2

      ld     I0.4               ;rung 5
      TP_8  2,T1.2,.50

      ld     TP8_Q2             ;rung 6
      out    Q0.4

      ld     I0.6               ;rung 7
      TOS_8 3,T1.2,.50,.50

      ld     TOS8_Q3            ;rung 8
      out    Q0.6

      ld     I1.1               ;rung 9
      TON_8 4,T1.4,.10

      ld     TON8_Q4            ;rung 10
      out    Q1.1

      ld     I1.3               ;rung 11
      TOF_8 5,T1.4,.10

      ld     TOF8_Q5            ;rung 12
      out    Q1.3

      ld     I1.5               ;rung 13
      TP_8  6,T1.4,.10

      ld     TP8_Q6             ;rung 14
      out    Q1.5

      ld     I1.7               ;rung 15
      TOS_8 7,T1.4,.10,.10

      ld     TOS8_Q7            ;rung 16
      out    Q1.7
;--------------- user program ends here ----
.
```

FIGURE 5.11
The user program of UZAM_plc_16i16o_ex7.asm.

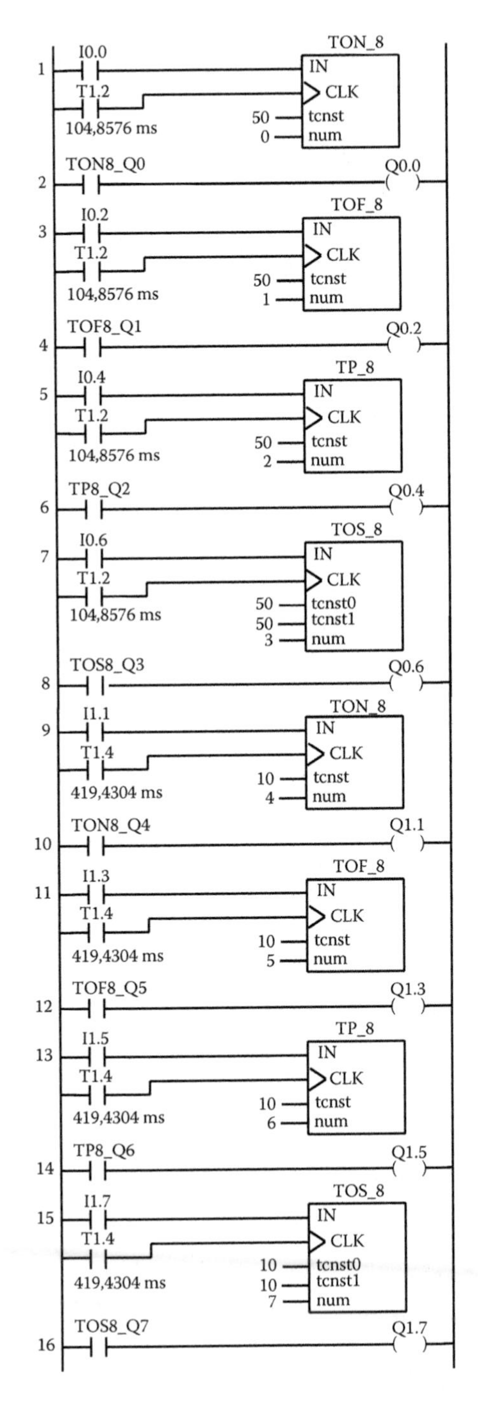

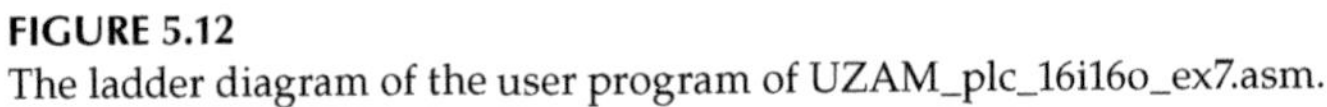

FIGURE 5.12
The ladder diagram of the user program of UZAM_plc_16i16o_ex7.asm.

(T1.4) = 4194.304 ms = 4.194304 s. As can be seen from rung 12, the timer status bit TOF8_Q5 is sent to output Q1.3.

In rungs 13 and 14, another pulse timer TP_8 is implemented as follows: the input signal IN is taken from I1.5 num = 6, and therefore we choose the seventh pulse timer, whose timer status bit (or output Q) is TP8_Q6. The preset time PT = `tcnst` × CLK = 10 × 419.4304 ms (T1.4) = 4194.304 ms = 4.194304 s. As can be seen from rung 14, the timer status bit TP8_Q6 is sent to output Q1.5.

In rungs 15 and 16, another oscillator timer TOS_8 is implemented as follows: the input signal IN is taken from I1.7 num = 7, and therefore we choose the eighth oscillator timer, whose timer status bit (or output Q) is TOS8_Q7. The preset time PT0 = `tcnst0` × CLK = 10 × 419.4304 ms (T1.4) = 4194.304 ms = 4.194304 s. The preset time PT1 = `tcnst1` × CLK = 10 × 419.4304 ms (T1.4) = 4194.304 ms = 4.194304 s. In this setup, the pulse trains we will obtain have a 50% duty cycle with the time period of T = 20 × 419,4304 ms = 8,388608 s. As can be seen from rung 16, the timer status bit TOS8_Q7 is sent to output Q1.7.

6

Counter Macros

In this chapter, the following counter macros are described:

CTU_8 (up counter)
CTD_8 (down counter)
CTUD_8 (up/down counter)

In addition two macros, move_R and load_R, are also described for data transfer.

6.1 Move and Load Macros

In a PLC, numbers are often required to be moved from one location to another; a timer preset value may be required to be changed according to plant conditions, or the result of some calculations may be used in another part of a program. To satisfy this need for 8-bit variables, in the PIC16F648A-based PLC we define the macro move_R. Similarly, the macro load_R is also described to load an 8-bit number into an 8-bit variable.

The algorithm and the symbol of the macro move_R are depicted in Table 6.1. Figure 6.1 shows the macro move_R and its flowchart. In this macro, EN is a Boolean input variable taken into the macro through W, and ENO is a Boolean output variable sent out from the macro through W. Output ENO follows the input EN. This means that when EN = 0, ENO is forced to be 0, and when EN = 1, ENO is forced to be 1. This is especially useful if we want to carry out more than one operation based on a single input condition. When EN = 1, the macro move_R transfers the data from the 8-bit input variable IN to the 8-bit output variable OUT.

The algorithm and the symbol of the macro load_R are depicted in Table 6.2. Figure 6.2 shows the macro load_R and its flowchart. In this macro, EN is a Boolean input variable taken into the macro through W, and ENO is a Boolean output variable sent out from the macro through W. Output ENO follows the input EN. This means that when EN = 0, ENO is forced to be 0, and when EN = 1, ENO is forced to be 1. When EN = 1, the macro load_R transfers the 8-bit constant data IN, within the 8-bit output variable OUT.

TABLE 6.1

Algorithm and Symbol of the Macro `move_R`

Algorithm	Symbol

if EN = 1 then
OUT = IN;
ENO = 1;
else ENO = 0;
end if;
	move_R
	W — EN ENO — W
	— IN OUT —
	IN, OUT (8 bit register)
	EN (through W) = 0 or 1
	ENO (through W) = 0 or 1

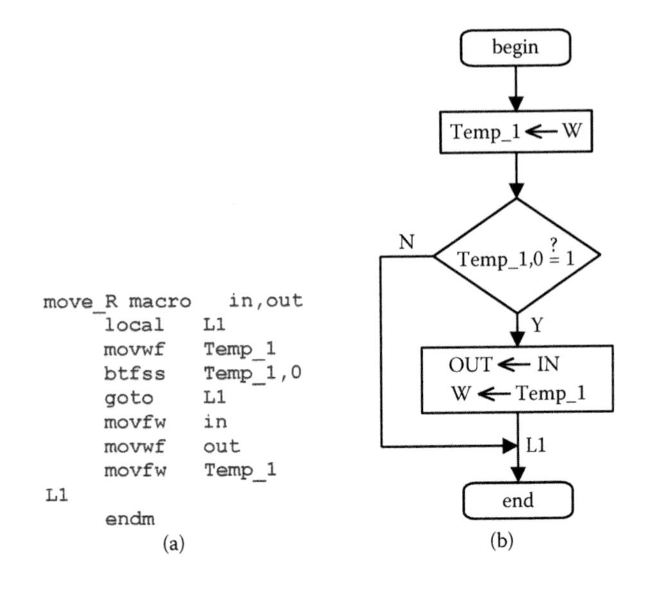

FIGURE 6.1
(a) The macro `move_R` and (b) its flowchart.

TABLE 6.2

Algorithm and Symbol of the Macro `load_R`

Algorithm	Symbol

if EN = 1 then
OUT = IN;
ENO = 1;
else ENO = 0;
end if;
	load_R
	W — EN ENO — W
	— IN OUT —
	IN (8 bit constant)
	OUT (8 bit register)
	EN (through W) = 0 or 1
	ENO (through W) = 0 or 1

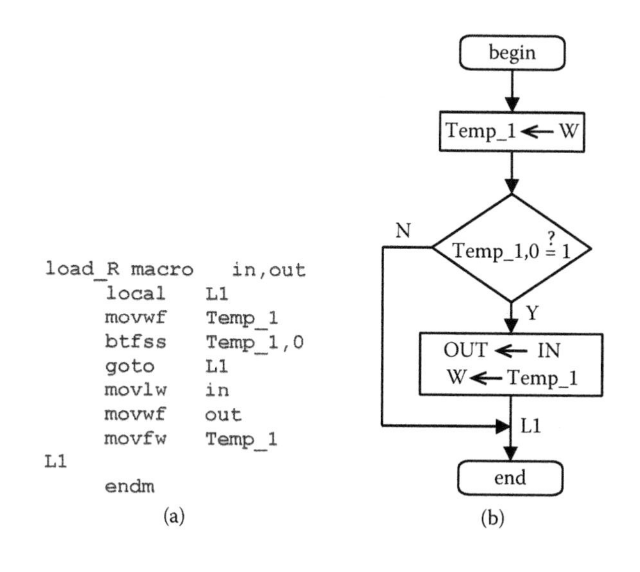

FIGURE 6.2
(a) The macro `load_R` and (b) its flowchart.

The file definitions.inc, included within the CD-ROM attached to this book, contains these two macros.

6.2 Counter Macros

Counters can be used in a wide range of applications. In this chapter, three counter functions, up counter, down counter, and up/down counter, are described. The definition of 8-bit variables to be used for the counter macros, and their allocation in BANK 0 of SRAM data memory are both shown in Figure 6.3(a) and (b), respectively. Here, it is important to note that as we restrict ourselves to use the BANK 0, where there are not enough registers left, we cannot define different sets of 8-bit variables to be used in the counting process for each counter type. Rather, we define eight 8-bit variables and share them for each counter type. As a result, in total we can define eight different counters at most, irrespective of the counter type. The status bits, which will be explained in the next sections, of all counters are defined as shown in Figure 6.4(a). All the 8-bit variables defined for counters must be cleared at the beginning of the PLC operation for a proper operation. Therefore, all variables of counter macros are initialized within the macro `initialize`, as shown in Figure 6.4(b). The file definitions.inc, included within the CD-ROM attached to this book, contains all counter macros defined for the PIC16F648A-based PLC.

```
;--------------- VARIABLE DEFINITIONS ---
     CBLOCK 0x70
     CTU8_Q,CTD8_Q,CTUD8_Q
     endc
     CBLOCK 0x73
     CV_8              ;CV8, CV8+1, ..., CV8+7
     endc
     CBLOCK 0x7B
     CTU8_RED,CTD8_RED,CTUD8_RED
     endc
;----------------------------------------
```

(a)

70h	CTU8_Q
71h	CTD8_Q
72h	CTUD8_Q
73h	CV_8
74h	CV_8+1
75h	CV_8+2
76h	CV_8+3
77h	CV_8+4
78h	CV_8+5
79h	CV_8+6
7Ah	CV_8+7
7Bh	CTU8_RED
7Ch	CTD8_RED
7Dh	CTUD8_RED
7Eh	
7Fh	

BANK 0

(b)

FIGURE 6.3
(a) Definition of 8-bit variables to be used for the counter macros. (b) Their allocation in BANK 0 of SRAM data memory.

Let us now consider the counter macros. In the following, first, a general description will be given for the considered counter function, and then its implementation in the PIC16F648A-based PLC will be provided.

6.3 Up Counter (CTU)

The up counter (CTU) can be used to signal when a count has reached a maximum value. The symbol of the up counter (CTU) is shown in Figure 6.5, while its truth table is given in Table 6.3. The up counter counts

```
;- defining Up Counter
;-(CTU) outputs -
#define CTU8_Q0 CTU8_Q,0
#define CTU8_Q1 CTU8_Q,1
#define CTU8_Q2 CTU8_Q,2
#define CTU8_Q3 CTU8_Q,3
#define CTU8_Q4 CTU8_Q,4
#define CTU8_Q5 CTU8_Q,5
#define CTU8_Q6 CTU8_Q,6
#define CTU8_Q7 CTU8_Q,7

;- defining Down Counter
;-(CTD) outputs -
#define CTD8_Q0 CTD8_Q,0
#define CTD8_Q1 CTD8_Q,1
#define CTD8_Q2 CTD8_Q,2
#define CTD8_Q3 CTD8_Q,3
#define CTD8_Q4 CTD8_Q,4
#define CTD8_Q5 CTD8_Q,5
#define CTD8_Q6 CTD8_Q,6
#define CTD8_Q7 CTD8_Q,7

;- defining Up/Down Counter
;-(CTUD) outputs -
#define CTUD8_Q0 CTUD8_Q,0
#define CTUD8_Q1 CTUD8_Q,1
#define CTUD8_Q2 CTUD8_Q,2
#define CTUD8_Q3 CTUD8_Q,3
#define CTUD8_Q4 CTUD8_Q,4
#define CTUD8_Q5 CTUD8_Q,5
#define CTUD8_Q6 CTUD8_Q,6
#define CTUD8_Q7 CTUD8_Q,7
```

(a)

```
;------------------- macro: initialize ----------------
initialize      macro
     local    L1
     BANK1                   ;goto BANK1
     movlw b'00000111'       ;W<--b'00000111':Fosc/4-->TMR0,PS=256
     movwf OPTION_REG        ;pull-up on PORTB, OPTION_REG <-- W
     movlw b'00000001'       ;PORTB is both input and output port
     movwf TRISB             ;TRISB <-- b'00000001'
     BANK0                   ;goto BANK0
     clrf  PORTA             ;Clear PortA
     clrf  PORTB             ;Clear PortB
     clrf  TMR0              ;Clear TMR0
     movlw h'20'             ;initialize the pointer
     movwf FSR               ;to RAM
L1   clrf  INDF              ;clear INDF register
     incf  FSR,f             ;increment pointer
     btfss FSR,7             ;all done?
     goto  L1                ;if not goto L1
                             ;if yes carry on
     movlw 06h               ;W <--- 06h
     movwf Temp_2            ;Temp_2 <--- W(06h)
     endm
;-----------------------------------------------------
```

(b)

FIGURE 6.4
(a) Definition of status bits of counter macros. (b) The initialization of all variables of counter macros within the macro `initialize`.

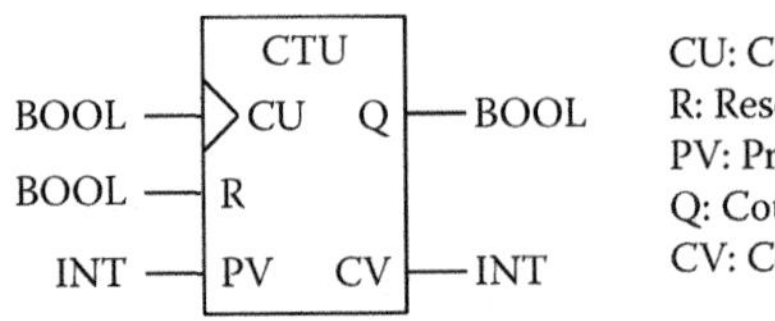

CU: Count Up Input
R: Reset Input
PV: Preset Value
Q: Counter Output
CV: Count Value

FIGURE 6.5
The up counter (CTU).

TABLE 6.3
Truth Table of the Up Counter (CTU)

CU	R	Operation
×	1	1. Set the output Q false (OFF – LOW) 2. Clear the count value CV to zero
0	0	NOP (No Operation is done)
1	0	NOP
↓	0	NOP
↑	0	If CV < PV, then increment CV (i.e. CV = CV + 1). If CV = PV, then hold CV and set the output Q true (ON – HIGH).

the number of rising edges (↑) detected at the input CU. PV defines the maximum value for the counter. Each time the counter is called with a new rising edge (↑) on CU, the count value CV is incremented by one. When the counter reaches the PV value, the counter output Q is set true (ON—1) and the counting stops. The reset input R can be used to set the output Q false (OFF—0) and clear the count value CV to zero. The following section explains the implementation of eight 8-bit up counters for the PIC16F648A-based PLC.

6.4 Macro `CTU_8` (8-Bit Up Counter)

The macro `CTU_8` defines eight up counters selected with the num = 0, 1, …, 7. Table 6.4 shows the symbol of the macro `CTU_8`. The macro `CTU_8` and its flowchart are depicted in Figure 6.6. CU (count up input), Q (output signal = counter status bit), and R (reset input) are all defined as Boolean variables. The PV (preset value) is an integer constant (here, for 8-bit resolution, it is chosen as any number in the range 1–255) and is used to define a maximum count value for the counter. The counter outputs are represented by the counter status bits: CTU8_Q,num (num = 0, 1, …, 7), namely, CTU8_Q0, CTU8_Q1, …, CTU8_Q7, as shown in Figure 6.4(a). We

TABLE 6.4

Symbol of the Macro CTU_8

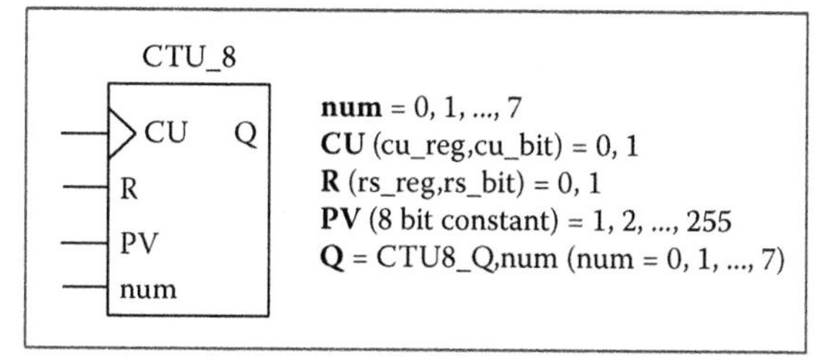

use a Boolean variable, CTU8_RED,num (num = 0, 1, …, 7), as a rising edge detector for identifying the rising edges of the CU. An 8-bit integer variable CV_8+num (num = 0, 1, …, 7) is used to count the rising edges of the CU. Let us now briefly consider how the macro CTU_8 works. If the input signal R is true (ON—1), then the output signal CTU8_Q,num (num = 0, 1, …, 7) is forced to be false (OFF—0), and the counter CV_8+num (num = 0, 1, …, 7) is loaded with 00h. If the input signal R is false (OFF—0), then with each rising edge of the CU, the related counter CV_8+num is incremented by one. In this case, when the count value of CV_8+num is equal to the PV, then state change from 0 to 1 is issued for the output

```
;--------- macro: CTU_8 -------------------------
CTU_8  macro    num,cu_reg,cu_bit,rs_reg,rs_bit,PV
       local    L1,L2
       btfss    rs_reg,rs_bit
       goto     L2
       movlw    00h
       movwf    CV_8+num
       bcf      CTU8_Q,num
       goto     L1
L2     btfsc    CTU8_Q,num
       goto     L1
       btfss    cu_reg,cu_bit
       bsf      CTU8_RED,num
       btfss    cu_reg,cu_bit
       goto     L1
       btfss    CTU8_RED,num
       goto     L1
       bcf      CTU8_RED,num
       incf     CV_8+num,f
       movfw    CV_8+num
       xorlw    PV
       skpnz
       bsf      CTU8_Q,num
L1
       endm
;-----------------------------------------------
```

(a)

FIGURE 6.6
(a) The macro CTU_8 and (b) its flowchart. (*Continued*)

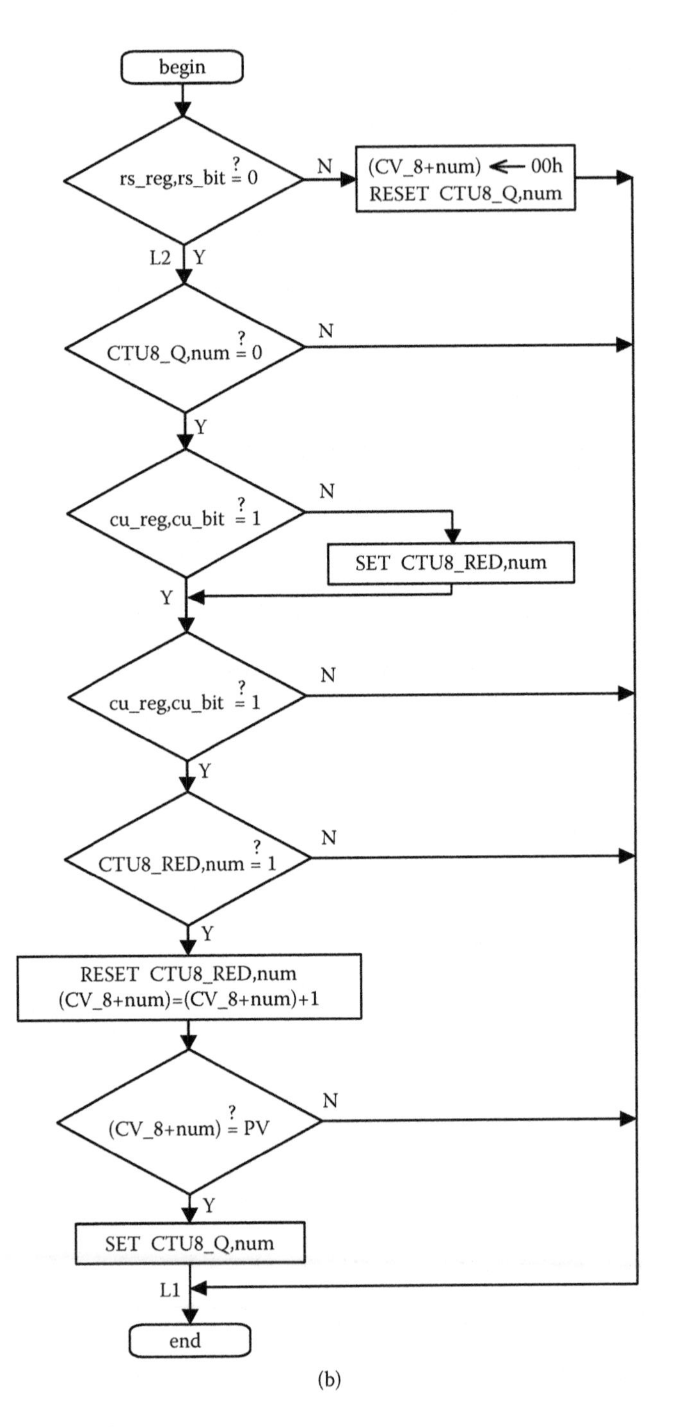

FIGURE 6.6 (*Continued*)
(a) The macro CTU_8 and (b) its flowchart.

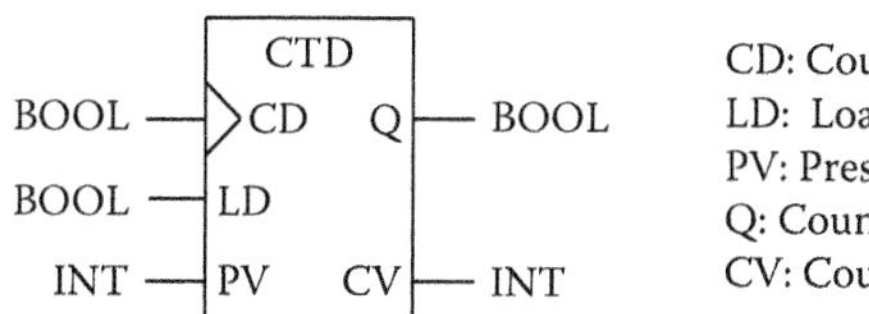

FIGURE 6.7
The down counter (CTD).

signal (counter status bit) CTU8_Q,num (num = 0, 1, …, 7) and the counting stops.

6.5 Down Counter (CTD)

The down counter (CTD) can be used to signal when a count has reached zero, on counting down from a preset value. The symbol of the down counter (CTD) is shown in Figure 6.7, while its truth table is given in Table 6.5. The down counter counts down the number of rising edges (↑) detected at the input CD. PV defines the starting value for the counter. Each time the counter is called with a new rising edge (↑) on CD, the count value CV is decremented by one. When the counter reaches zero, the counter output Q is set true (ON—1) and the counting stops. The load input LD can be used to clear the output Q to false (OFF—0) and load the count value CV with the preset value PV. The following section explains the implementation of eight 8-bit down counters for the PIC16F648A-based PLC.

TABLE 6.5
Truth Table of the Down Counter (CTD)

CD	LD	Operation
×	1	1. Clear the output **Q** to false (OFF – LOW) 2. Load the count value **CV** with the preset value **PV**
0	0	NOP (No Operation is done)
1	0	NOP
↓	0	NOP
↑	0	If CV > 0, then decrement CV (i.e., CV = CV – 1). If CV = 0, then hold CV and set the output **Q** true (ON – HIGH).

TABLE 6.6

Symbol of the Macro CTD_8

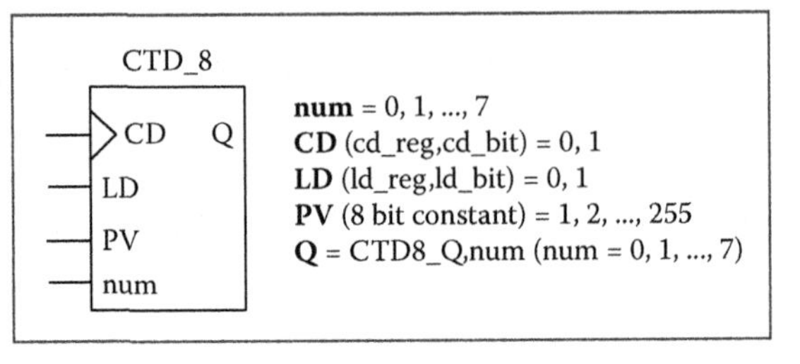

6.6 Macro CTD_8 (8-Bit Down Counter)

The macro CTD_8 defines eight down counters selected with the num = 0, 1, ..., 7. Table 6.6 shows the symbol of the macro CTD_8. The macro CTD_8 and its flowchart are depicted in Figure 6.8. CD (count down input), Q (output signal = counter status bit), and LD (load input) are all defined as Boolean

```
;--------- macro: CTD_8 ------------------------
CTD_8  macro    num,cd_reg,cd_bit,ld_reg,ld_bit,PV
       local    L1,L2
       btfss    ld_reg,ld_bit
       goto     L2
       movlw    PV
       movwf    CV_8+num
       bcf      CTD8_Q,num
       goto     L1
L2     btfsc    CTD8_Q,num
       goto     L1
       btfss    cd_reg,cd_bit
       bsf      CTD8_RED,num
       btfss    cd_reg,cd_bit
       goto     L1
       movfw    CV_8+num
       xorlw    0
       skpnz
       goto     L1
       btfss    CTD8_RED,num
       goto     L1
       bcf      CTD8_RED,num
       decf     CV_8+num,f
       movfw    CV_8+num
       xorlw    0
       skpnz
       bsf      CTD8_Q,num
L1
       endm
;-----------------------------------------------
```

(a)

FIGURE 6.8
(a) The macro CTD_8 and (b) its flowchart. (*Continued*)

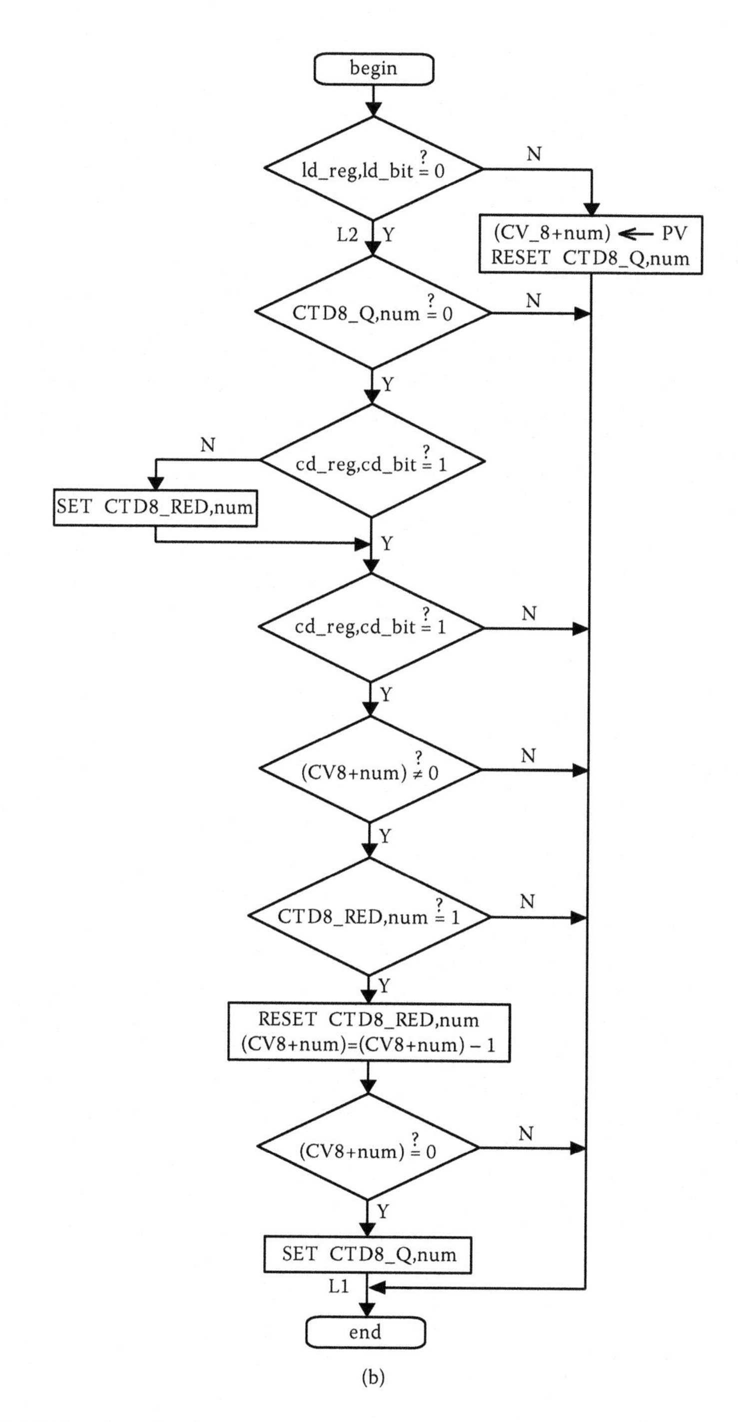

(b)

FIGURE 6.8 (*Continued*)
(a) The macro `CTD_8` and (b) its flowchart.

variables. The PV (preset value) is an integer constant (here, for 8-bit resolution, it is chosen as any number in the range 1–255) and is used to define a starting value for the counter. The counter outputs are represented by the counter status bits: CTD8_Q,num (num = 0, 1, …, 7), namely, CTD8_Q0, CTD8_Q1, …, CTD8_Q7, as shown in Figure 6.4(a). We use a Boolean variable, CTD8_RED,num (num = 0, 1, …, 7), as a rising edge detector for identifying the rising edges of the CD. An 8-bit integer variable CV_8+num (num = 0, 1, …, 7) is used to count the rising edges of the CD. Let us now briefly consider how the macro `CTD_8` works. If the input signal LD is true (ON—1), then the output signal CTU8_Q,num (num = 0, 1, …, 7) is forced to be false (OFF—0), and the counter CV_8+num (num = 0, 1, …, 7) is loaded with PV. If the input signal LD is false (OFF—0), then with each rising edge of the CD, the related counter CV_8+num is decremented by one. In this case, when the count value of CV_8+num is equal to zero, then state change from 0 to 1 is issued for the output signal (counter status bit) CTU8_Q,num (num = 0, 1, …, 7) and the counting stops.

6.7 Up/Down Counter (CTUD)

The up/down counter (CTUD) has two inputs CU and CD. It can be used to both count up on one input and count down on the other. The symbol of the up/down counter (CTUD) is shown in Figure 6.9, while its truth table is given in Table 6.7. The up/down counter counts up the number of rising edges (↑) detected at the input CU. The up/down counter counts down the number of rising edges (↑) detected at the input CD. PV defines the maximum value for the counter. When the counter reaches the PV value, the counter output Q is set true (ON—1) and the counting up stops. The reset input R can be used to set the output Q false (OFF—0) and clear the count value CV to zero. The load input LD can be used to load the count value CV with the preset value PV. When the counter reaches zero, the counting down stops. The following

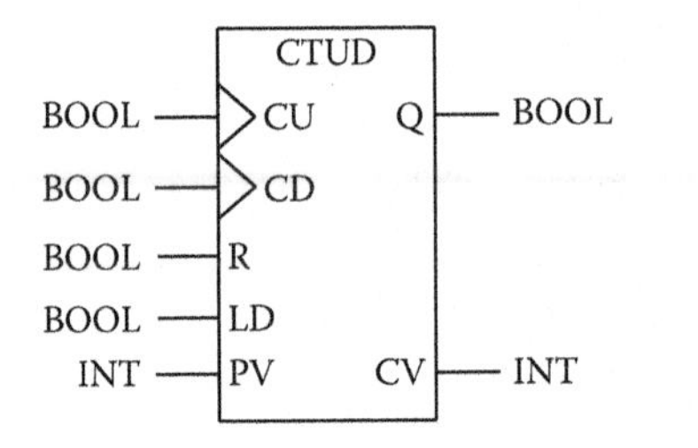

CU: Count Up Input
CD: Count Down Input
R: Reset Input
LD: Load Input
PV: Preset Value
Q: Counter Output
CV: Count Value

FIGURE 6.9
The up/down counter (CTUD).

TABLE 6.7
Truth Table of the Up/Down Counter (CTUD)

CU	CD	R	LD	Operation
×	×	1	×	1. Set the output **Q** false (OFF – LOW) 2. Clear the count value **CV** to zero
×	×	0	1	Load the count value **CV** with the preset value **PV**
0	0	0	0	NOP (No Operation is done)
0	1	0	0	NOP
1	0	0	0	NOP
1	1	0	0	NOP
1	↑	0	0	NOP
↑	1	0	0	NOP
×	↓	0	0	NOP
↓	×	0	0	NOP
↑	0	0	0	If CV < PV, then increment CV. If CV = PV, then hold CV and set the output **Q** true (ON – HIGH).
0	↑	0	0	If CV > 0, then decrement CV.

section explains the implementation of eight 8-bit up/down counters for the PIC16F648A-based PLC.

6.8 Macro CTUD_8 (8-Bit Up/Down Counter)

The macro CTUD_8 defines eight up/down counters selected with the num = 0, 1, …, 7. Table 6.8 shows the symbol of the macro CTUD_8. The macro CTUD8 and its flowchart are depicted in Figure 6.10. CU (count up input), CD (count down input), Q (output signal = counter status bit), R (reset input), and LD

TABLE 6.8
Symbol of the Macro CTUD8

CTUD_8

CD Q
CD
R
LD
PV
num

num = 0, 1, ..., 7
CU (cu_reg,cu_bit) = 0, 1
CD (cd_reg,cd_bit) = 0, 1
R (rs_reg,rs_bit) = 0, 1
LD (ld_reg,ld_bit) = 0, 1
PV (8 bit constant) = 1, 2, ..., 255
Q = CTUD8_Q,num (num = 0, 1, ..., 7)

```
;--------- macro: CTUD_8 ---------------------------
CTUD_8 macro   num,cu_reg,cu_bit,cd_reg,cd_bit,
rs_reg,rs_bit,ld_reg,ld_bit,PV
     local   L1,L2,L3,L4
     btfss   rs_reg,rs_bit
     goto    L4
     movlw   00h
     movwf   CV_8+num
     goto    L1
L4   btfss   ld_reg,ld_bit
     goto    L3
     movlw   PV
     movwf   CV_8+num
     goto    L1
L3   movlw   0
     btfsc   cu_reg,cu_bit
     movlw   1
     movwf   Temp_1
     movlw   0
     btfsc   cd_reg,cd_bit
     movlw   1
     iorwf   Temp_1,W
     movwf   Temp_1
     btfss   Temp_1,0
     bsf     CTUD8_RED,num
     btfss   Temp_1,0
     goto    L1
     btfss   CTUD8_RED,num
     goto    L1
     bcf     CTUD8_RED,num
     btfss   cu_reg,cu_bit
     goto    L2
     btfsc   CTUD8_Q,num       ;--- count up---
     goto    L1
     incf    CV_8+num,f
     goto    L1
L2   movfw   CV_8+num          ;---count down---
     xorlw   00h
     btfsc   STATUS,Z          ;skip if no Zero
     goto    L1
     decf    CV_8+num,f
L1   bcf     CTUD8_Q,num
     movfw   CV_8+num
     xorlw   PV
     btfsc   STATUS,Z          ;skip if no Zero
     bsf     CTUD8_Q,num
     endm
;-------------------------------------------------
```

(a)

FIGURE 6.10
(a) The macro CTUD8 and (b) its flowchart. (*Continued*)

(load input) are all defined as Boolean variables. The PV (preset value) is an integer constant (here, for 8-bit resolution, it is chosen as any number in the range 1–255) and is used to define a maximum count value for the counter. The counter outputs are represented by the counter status bits: CTUD8_Q,num (num = 0, 1, …, 7), namely, CTUD8_Q0, CTUD8_Q1, …, CTUD8_Q7, as shown

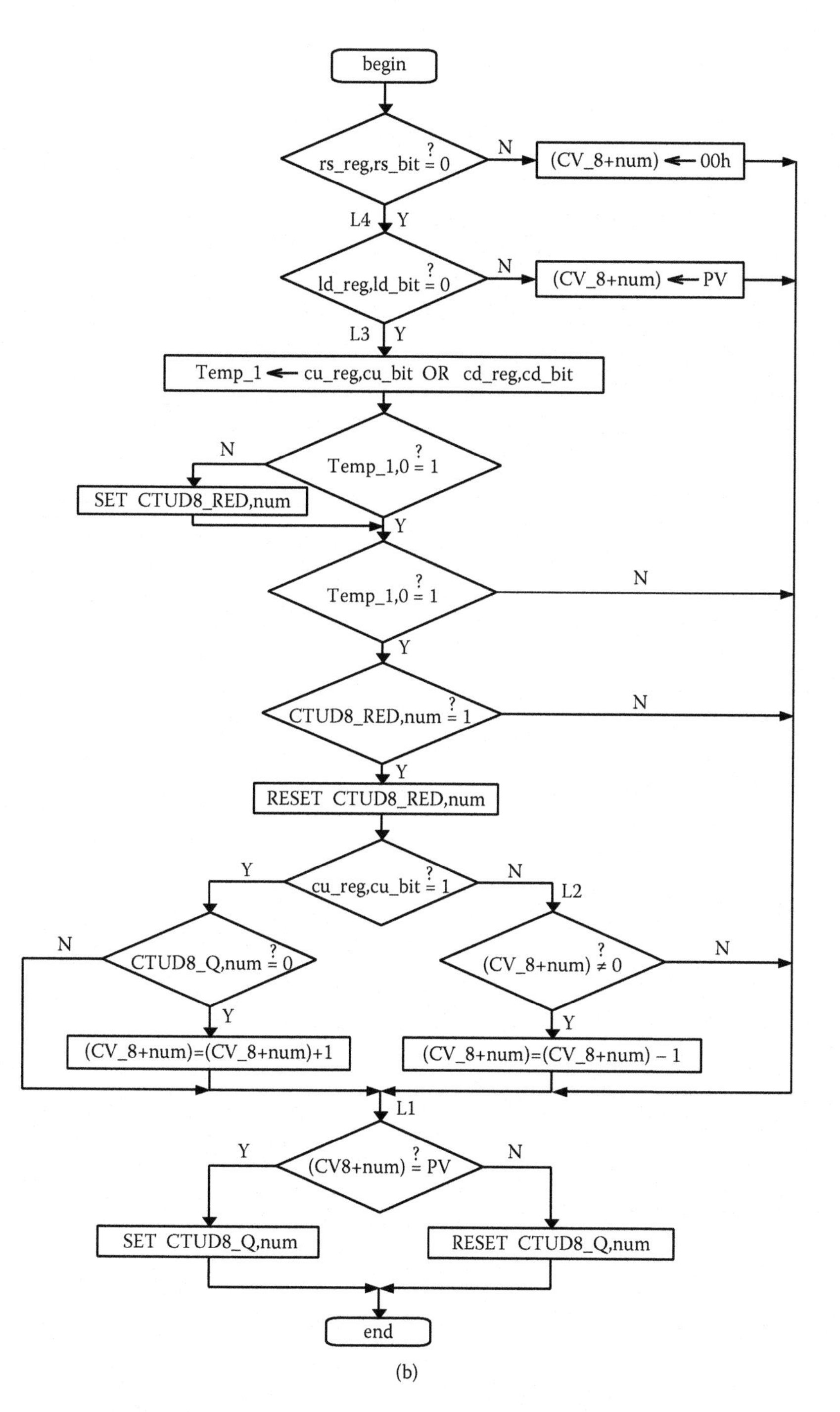

FIGURE 6.10 (*Continued*)
(a) The macro CTUD8 and (b) its flowchart.

in Figure 6.4(a). We use a Boolean variable, CTUD8_RED,num (num = 0, 1, …, 7), as a rising edge detector for identifying the rising edges of the inputs CU or CD. An 8-bit integer variable CV_8+num (num = 0, 1, …, 7) is used to count up the rising edges of the CU and count down the rising edges of the CD. Let us now briefly consider how the macro `CTUD_8` works. If the input signal R is true (ON—1), then the output signal CTU8_Q,num (num = 0, 1, …, 7) is forced to be false (OFF—0), and the counter CV_8+num (num = 0, 1, …, 7) is loaded with 00h. If the input signal R is false (OFF—0) and the input signal LD is true (ON—1), then the counter CV_8+num (num = 0, 1, …, 7) is loaded with PV. If the input signal R is false (OFF—0), the input signal LD is false (OFF—0), and the CD is false (OFF—0), then with each rising edge of the CU, the related counter CV_8+num is incremented by one. In this case, when the count value of CV_8+num is equal to the PV, then state change from 0 to 1 is issued for the output signal (counter status bit) CTU8_Q,num (num = 0, 1, …, 7) and the counting up stops. If the input signal R is false (OFF—0), the input signal LD is false (OFF—0), and the CU is false (OFF—0), then with each rising edge of the CD, the related counter CV_8+num is decremented by one. The counting down stops when the CV reaches zero.

6.9 Examples for Counter Macros

In this section, we will consider four examples, namely, UZAM_plc_16i16o_exX.asm (X = 8, 9, 10, 11), to show the usage of counter macros. In order to test one of these examples, please take the related file UZAM_plc_16i16o_exX.asm (X = 8, 9, 10, 11) from the CD-ROM attached to this book, and then open the program by MPLAB IDE and compile it. After that, by using the PIC programmer software, take the compiled file UZAM_plc_16i16o_exX.hex (X = 8, 9, 10, 11), and by your PIC programmer hardware, send it to the program memory of PIC16F648A microcontroller within the PIC16F648A-based PLC. To do this, switch the 4PDT in PROG position and the power switch in OFF position. After loading the file UZAM_plc_16i16o_exX.hex (X = 8, 9, 10, 11), switch the 4PDT in RUN and the power switch in ON position. Please check the program's accuracy by cross-referencing it with the related macros.

Let us now consider these example programs: The first example program, UZAM_plc_16i16o_ex8.asm, is shown in Figure 6.11. It shows the usage of the macro `CTU_8`. The ladder diagram of the user program of UZAM_plc_16i16o_ex8.asm, shown in Figure 6.11, is depicted in Figure 6.12. In the first two rungs, an up counter `CTU_8` is implemented as follows: the count up input CU is taken from I0.0, while the reset input R is taken from I0.1 num = 0, and therefore we choose the first up counter, whose counter status bit (or output Q) is CTU8_Q0. The preset value PV = 15. As can be seen from the second rung, the state of the counter status bit CTU8_Q0 is sent to output Q0.0. In the third rung, by using the `move_R` function, the contents of the register

```
.
;--------------- user program starts here -
        CTU_8       0,I0.0,I0.1,.15    ;rung 1

        ld          CTU8_Q0            ;rung 2
        out         Q0.0

        ld          LOGIC1             ;rung 3
        move_R      CV_8,Q1
;--------------- user program ends here ---
.
```

FIGURE 6.11
The user program of UZAM_plc_16i16o_ex8.asm.

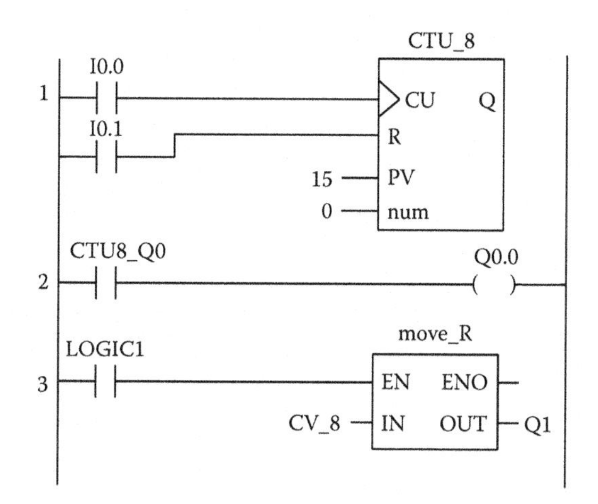

FIGURE 6.12
The ladder diagram of the user program of UZAM_plc_16i16o_ex8.asm.

`CV_8`, which keeps the current count value (CV) of the first up counter, are sent to the output register Q1.

The second example program, UZAM_plc_16i16o_ex9.asm, is shown in Figure 6.13. It shows the usage of the macro `CTD_8`. The ladder diagram of the user program of UZAM_plc_16i16o_ex9.asm, shown in Figure 6.13,

```
.
;--------------- user program starts here -
        CTD_8       4,I0.2,I0.3,.10    ;rung 1

        ld          CTD8_Q4            ;rung 2
        out         Q0.4

        ld          LOGIC1             ;rung 3
        move_R      CV_8+4,Q1
;--------------- user program ends here ---
.
```

FIGURE 6.13
The user program of UZAM_plc_16i16o_ex9.asm.

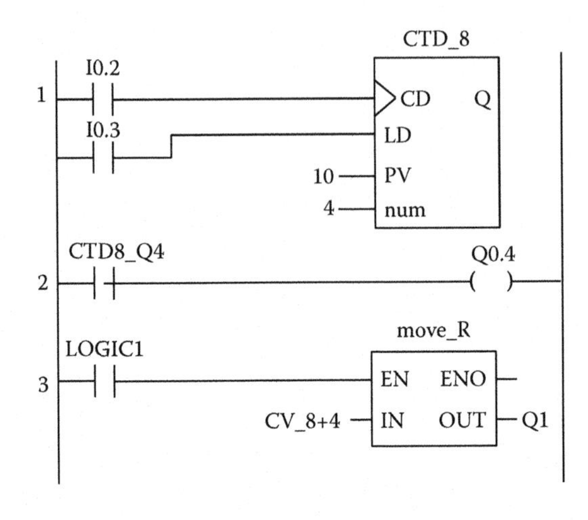

FIGURE 6.14
The ladder diagram of the user program of UZAM_plc_16i16o_ex9.asm.

is depicted in Figure 6.14. In the first two rungs, a down counter `CTD_8` is implemented as follows: the count down input CD is taken from I0.2, while the load input LD is taken from I0.3 num = 4, and therefore we choose the fifth down counter, whose counter status bit (or output Q) is CTD8_Q4. The preset value PV = 10. As can be seen from the second rung, the state of the counter status bit CTD8_Q4 is sent to output Q0.4. In the third rung, by using the `move_R` function the contents of the register `CV_8+4`, which keeps the current count value (CV) of the fifth down counter, are sent to the output register Q1.

The third example program, UZAM_plc_16i16o_ex10.asm, is shown in Figure 6.15. It shows the usage of the macro `CTUD_8`. The ladder diagram of the user program of UZAM_plc_16i16o_ex10.asm, shown in Figure 6.15, is depicted in Figure 6.16. In the first two rungs, an up/down counter `CTUD_8` is implemented as follows: the count up input CU is taken from I0.4, the count down input CD is taken from I0.5, while the reset input R is taken

```
;
;--------------- user program starts here -------------
        CTUD_8      7,I0.4,I0.5,I0.6,I0.7,.20        ;rung 1

        ld          CTUD8_Q7                         ;rung 2
        out         Q0.7

        ld          LOGIC1                           ;rung 3
        move_R      CV_8+7,Q1
;--------------- user program ends here ---------------
;
```

FIGURE 6.15
The user program of UZAM_plc_16i16o_ex10.asm.

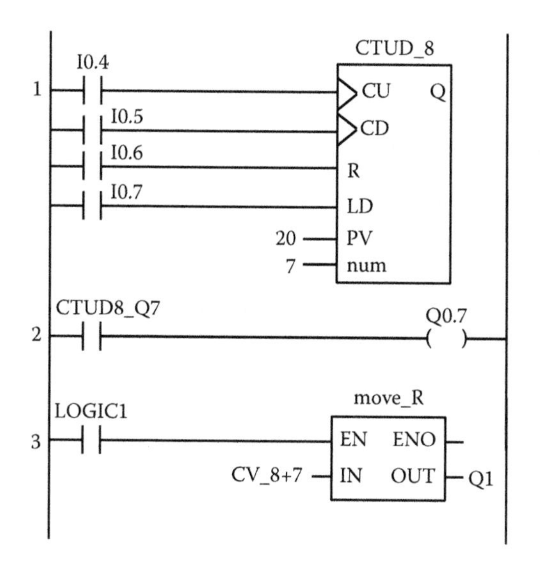

FIGURE 6.16
The ladder diagram of the user program of UZAM_plc_16i16o_ex10.asm.

from I0.6 and the load input LD is taken from I0.7 num = 7, and therefore we choose the eighth up/down counter, whose counter status bit (or output Q) is CTUD8_Q7. The preset value PV = 20. As can be seen from the second rung, the state of the counter status bit CTUD8_Q7 is sent to output Q0.7. In the third rung, by using the `move_R` function the contents of the register `CV_8+7`, which keeps the current count value (CV) of the eighth up/down counter, are sent to the output register Q1.

The fourth and last example program, UZAM_plc_16i16o_ex11.asm, is shown in Figure 6.17. It shows the usage of all counter macros. The ladder diagram of the user program of UZAM_plc_16i16o_ex11.asm, shown in Figure 6.17, is depicted in Figure 6.18. This example contains the previous three examples in one program.

In the first two rungs, an up counter `CTU_8` is implemented as follows: the count up input CU is taken from I0.0, while the reset input R is taken from I0.1. As num = 0, the first up counter is chosen, whose counter status bit (or output Q) is CTU8_Q0. The preset value PV = 15. As can be seen from the second rung, the state of the counter status bit CTU8_Q0 is sent to output Q0.0.

In rungs 3 and 4, a down counter `CTD_8` is implemented as follows: the count down input CD is taken from I0.2, while the load input LD is taken from I0.3. As num = 4, the fifth down counter is chosen, whose counter status bit (or output Q) is CTD8_Q4. The preset value PV = 10. As can be seen from the fourth rung, the state of the counter status bit CTD8_Q4 is sent to output Q0.4.

In rungs 5 and 6, an up/down counter `CTUD_8` is implemented as follows: the count up input CU is taken from I0.4, the count down input CD is taken

```
.
;--------------- user program starts here --------------
        CTU_8       0,I0.0,I0.1,.15                ;rung 1

        ld          CTU8_Q0                        ;rung 2
        out         Q0.0

        CTD_8       4,I0.2,I0.3,.10                ;rung 3

        ld          CTD8_Q4                        ;rung 4
        out         Q0.4

        CTUD_8      7,I0.4,I0.5,I0.6,I0.7,.20      ;rung 5

        ld          CTUD8_Q7                       ;rung 6
        out         Q0.7

        ld_not      I1.1                           ;rung 7
        and         I1.0
        out         M0.1

        ld          I1.1                           ;rung 8
        and_not     I1.0
        out         M0.2

        ld          I1.1                           ;rung 9
        and         I1.0
        out         M0.3

        ld          M0.1                           ;rung 10
        move_R      CV_8,Q1

        ld          M0.2                           ;rung 11
        move_R      CV_8+4,Q1

        ld          M0.3                           ;rung 12
        move_R      CV_8+7,Q1
;--------------- user program ends here ----------------
.
```

FIGURE 6.17
The user program of UZAM_plc_16i16o_ex11.asm.

from I0.5, while the reset input R is taken from I0.6 and the load input LD is taken from I0.7. As num = 7, the eighth up/down counter is chosen, whose counter status bit (or output Q) is CTUD8_Q7. The preset value PV = 20. As can be seen from the sixth rung, the state of the counter status bit CTUD8_Q7 is sent to output Q0.7.

In rungs 7 to 9, based on the input bits I1.1 and I1.0, one of three situations is chosen: If I1.1,I1.0 = 01, then M0.1 is activated. If I1.1,I1.0 = 10, then M0.2 is activated. Finally, if I1.1,I1.0 = 11, then M0.3 is activated.

In rung 10, if M0.1 = 1, then by using the `move_R` function, the contents of the register `CV_8`, which keeps the current count value (CV) of the first up counter, are sent to the output register Q1.

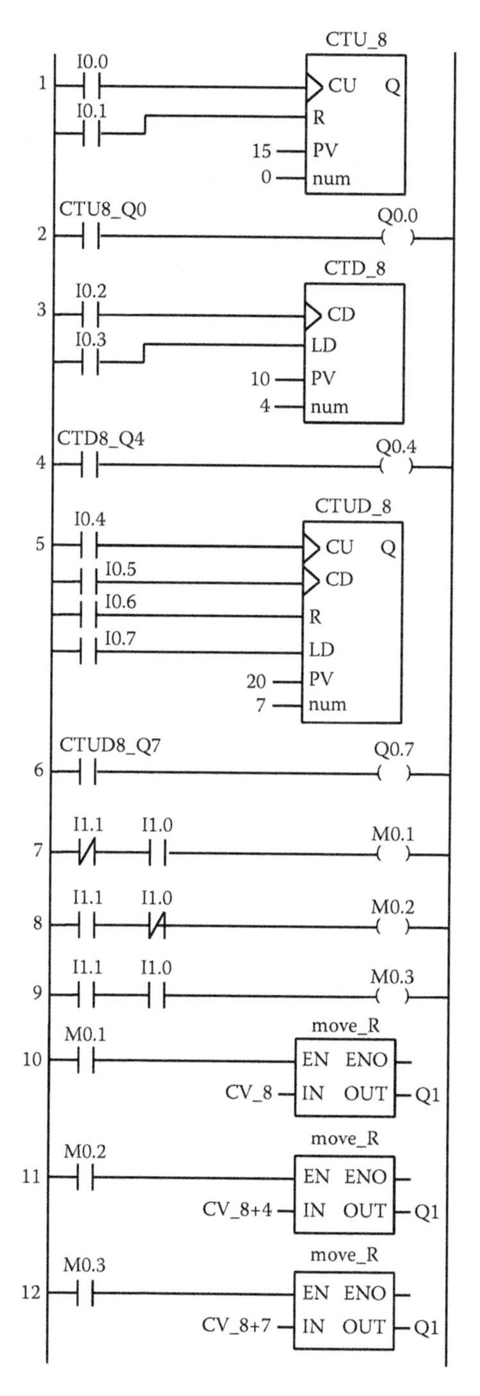

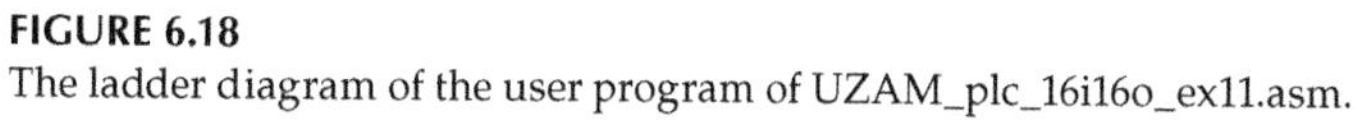
FIGURE 6.18
The ladder diagram of the user program of UZAM_plc_16i16o_ex11.asm.

In rung 11, if M0.2 = 1, then by using the `move_R` function, the contents of the register `CV_8+4`, which keeps the current count value (CV) of the fifth down counter, are sent to the output register Q1.

In rung 12, if M0.3 = 1, then by using the `move_R` function, the contents of the register `CV_8+7`, which keeps the current count value (CV) of the eighth up/down counter, are sent to the output register Q1.

7

Comparison Macros

Numerical values often need to be compared in PLC programs; typical examples are a batch counter saying the required number of items has been delivered, or alarm circuits indicating, for example, a temperature has gone above some safety level. These comparisons are performed by elements that have the generalized form of Figure 7.1, with two numerical inputs A and B corresponding to the values to be compared, and a Boolean (on/off) output that is true if the specified condition is met. The comparisons provided in this chapter are as follows:

A greater than B	(A > B)
A greater than or equal to B	(A > = B)
A equal to B	(A = B)
A less than B	(A < B)
A less than or equal to B	(A < = B)
A not equal to B	(A <> B)

where A and B are 8-bit numerical data.

In this chapter, two groups of comparison macros are described for the PIC16F648A-based PLC. In the former, the contents of two registers (R1 and R2) are compared according to the following:

GT (greater than, >)

GE (greater than or equal to, > =)

EQ (equal to, =)

LT (less than, <)

LE (less than or equal to, < =)

NE (not equal to, < >)

In the latter, similar comparison macros are also described for comparing the content of an 8-bit register (R) with an 8-bit constant (K). The file definitions.inc, included within the CD-ROM attached to this book, contains all comparison macros defined for the PIC16F648A-based PLC. Let us now consider these comparison macros in detail.

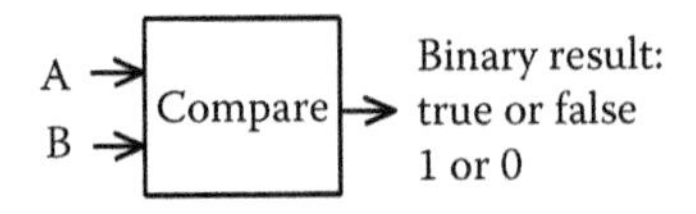

FIGURE 7.1
The generalized form of data comparison.

7.1 Macro R1_GT_R2

The definition, symbols, and algorithm of the macro R1_GT_R2 are depicted in Table 7.1. Figure 7.2 shows the macro R1_GT_R2 and its flowchart. The macro R1_GT_R2 has a Boolean input variable (active high enabling input), EN, passed into the macro through W, and a Boolean output variable, Q, passed out of the macro through W. This means that the input signal EN should be loaded into W before this macro is run, and the output signal Q will be provided within the W at the end of the macro. R1 and R2 are both 8-bit input variables. When EN = 0, no action is taken and the output Q (W) is forced to be 0. When EN = 1, if the content of R1 is greater than the content of R2 (R1 > R2), then the output Q (W) is forced to be 1. Otherwise, the output Q (W) is forced to be 0.

7.2 Macro R1_GE_R2

The definition, symbols, and algorithm of the macro R1_GE_R2 are depicted in Table 7.2. Figure 7.3 shows the macro R1_GE_R2 and its flowchart. The macro R1_GE_R2 has a Boolean input variable (active high enabling input), EN, passed into the macro through W, and a Boolean output variable, Q, passed out of the macro through W. This means that the input signal EN

TABLE 7.1

Definition, Symbols, and Algorithm of the Macro R1_GT_R2

Definition	Ladder Diagram Symbol	Schematic Symbol	Algorithm

is the content of register **R1** **G**reater **T**han the content of register **R2**?	R1
	W —| > |— W
	R2
		W — EN Q — W
		— R1 >
		— R2
		R1, R2 (8 bit register)
		EN (through W) = 0 or 1
		Q (through W) = 0 or 1
			if EN = 1 then
			if R1 > R2 then
			Q = 1;
			else Q = 0;
			end if;

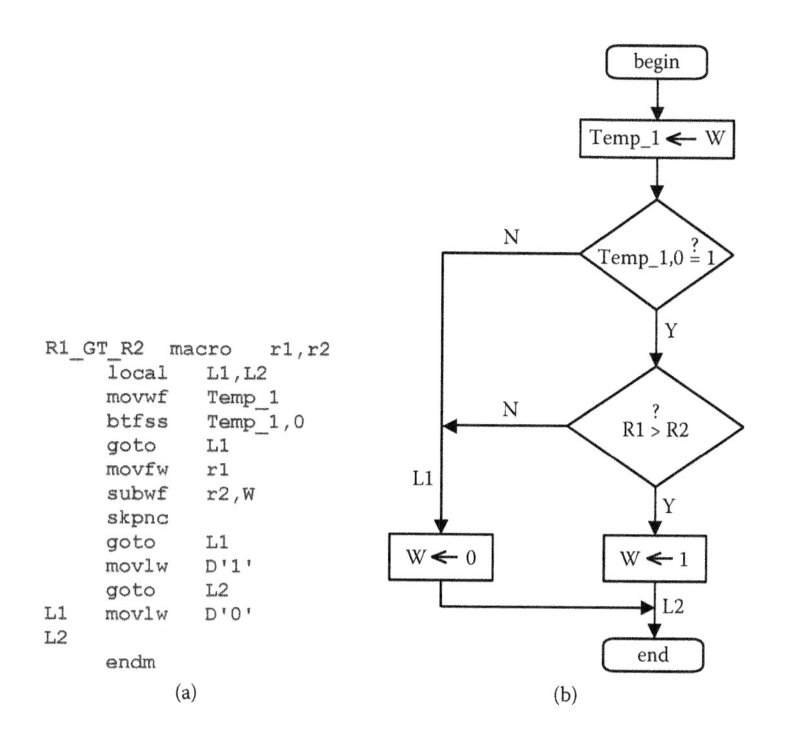

FIGURE 7.2
(a) The macro `R1_GT_R2` and (b) its flowchart.

should be loaded into W before this macro is run, and the output signal Q will be provided within the W at the end of the macro. R1 and R2 are both 8-bit input variables. When EN = 0, no action is taken and the output Q (W) is forced to be 0. When EN = 1, if the content of R1 is greater than or equal to the content of R2 (R1 ≥ R2), then the output Q (W) is forced to be 1. Otherwise, the output Q (W) is forced to be 0.

TABLE 7.2

Definition, Symbols, and Algorithm of the Macro `R1_GE_R2`

Definition	Ladder diagram symbol	Schematic symbol	Algorithm

is the content of register **R1** **G**reater than or **E**qual to the content of register **R2**?	W —| R1 >= R2 |— W	W — EN Q — W; R1; R2; >=
		R1, R2 (8 bit register)
		EN (through W) = 0 or 1
		Q (through W) = 0 or 1
			if EN = 1 then
			if R1 ≥ R2 then
			Q = 1;
			else Q = 0;
			end if;

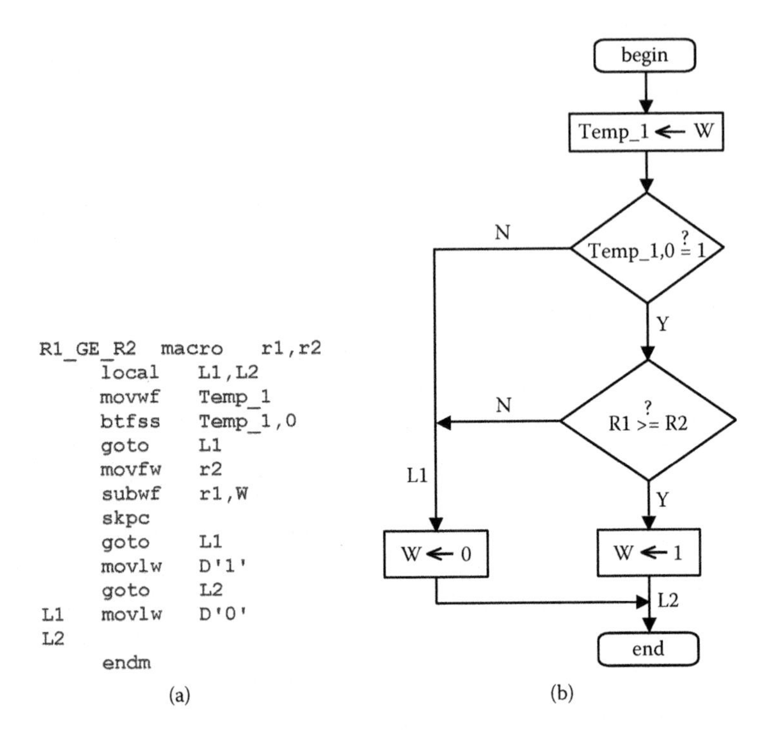

FIGURE 7.3
(a) The macro `R1_GE_R2` and (b) its flowchart.

7.3 Macro `R1_EQ_R2`

The definition, symbols, and algorithm of the macro `R1_EQ_R2` are depicted in Table 7.3. Figure 7.4 shows the macro `R1_EQ_R2` and its flowchart. The macro `R1_EQ_R2` has a Boolean input variable (active high enabling input), EN, passed into the macro through W, and a Boolean output variable, Q, passed out of the macro through W. This means that the input signal EN

TABLE 7.3

Definition, Symbols, and Algorithm of the Macro `R1_EQ_R2`

Definition	Ladder diagram symbol	Schematic symbol	Algorithm

is the content of register **R1 EQ**ual to the content of register **R2**?	W —| R1 = R2 |— W	W — EN Q — W; R1; R2; =
		 R1, R2 (8 bit register)
		 EN (through W) = 0 or 1
		 Q (through W) = 0 or 1
			if EN = 1 then
			 if R1 = R2 then
			 Q = 1;
			 else Q = 0;
			 end if;

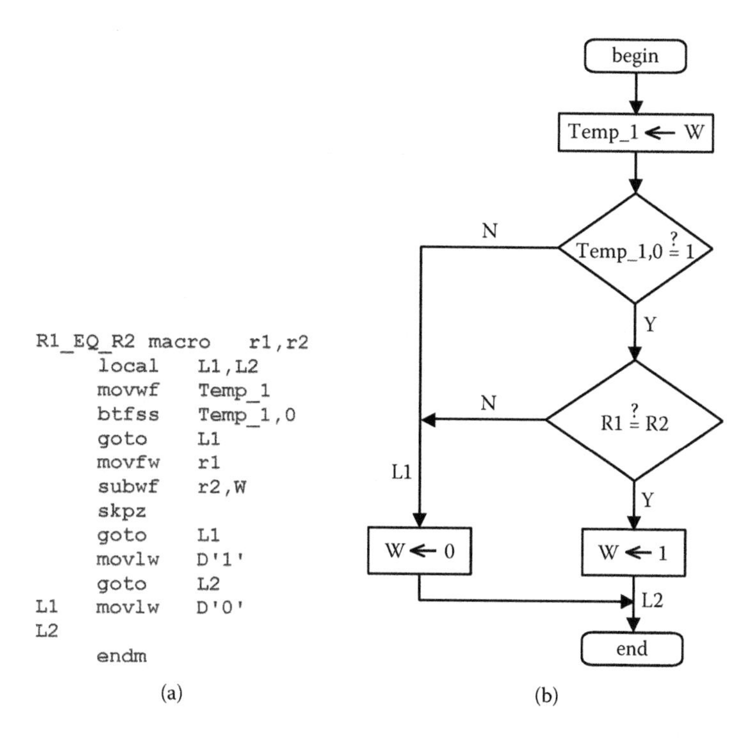

FIGURE 7.4
(a) The macro `R1_EQ_R2` and (b) its flowchart.

should be loaded into W before this macro is run, and the output signal Q will be provided within the W at the end of the macro. R1 and R2 are both 8-bit input variables. When EN = 0, no action is taken and the output Q (W) is forced to be 0. When EN = 1, if the content of R1 is equal to the content of R2 (R1 = R2), then the output Q (W) is forced to be 1. Otherwise, the output Q (W) is forced to be 0.

7.4 Macro `R1_LT_R2`

The definition, symbols, and algorithm of the macro `R1_LT_R2` are depicted in Table 7.4. Figure 7.5 shows the macro `R1_LT_R2` and its flowchart. The macro `R1_LT_R2` has a Boolean input variable (active high enabling input), EN, passed into the macro through W, and a Boolean output variable, Q, passed out of the macro through W. This means that the input signal EN should be loaded into W before this macro is run, and the output signal Q will be provided within the W at the end of the macro. R1 and R2 are both 8-bit input variables. When EN = 0, no action is taken and the output Q (W) is forced to be 0. When EN = 1, if the content of R1 is less than the content of

TABLE 7.4

Definition, Symbols, and Algorithm of the Macro `R1_LT_R2`

Definition	Ladder diagram symbol	Schematic symbol	Algorithm

is the content of register **R1** Less Than the content of register **R2**?	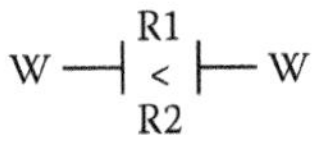 W ─┤ R1 < R2 ├─ W	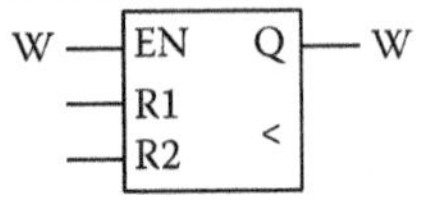 W ─ EN Q ─ W; ─ R1 <; ─ R2
		R1, R2 (8 bit register)
		EN (through W) = 0 or 1
		Q (through W) = 0 or 1
			if EN = 1 then
			if R1 < R2 then
			Q = 1;
			else Q = 0;
			end if;

```
R1_LT_R2  macro    r1,r2
          local    L1,L2
          movwf    Temp_1
          btfss    Temp_1,0
          goto     L1
          movfw    r2
          subwf    r1,W
          skpnc
          goto     L1
          movlw    D'1'
          goto     L2
L1        movlw    D'0'
L2
          endm
```

(a)

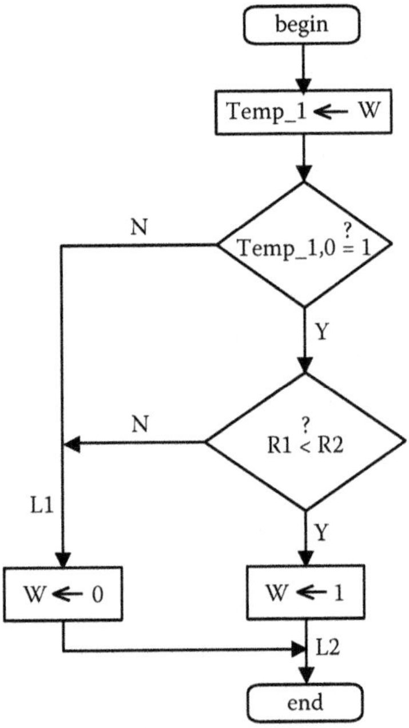

begin
Temp_1 ← W
Temp_1,0 ?= 1 (N → L1; Y)
R1 ?< R2 (N → L1; Y)
L1: W ← 0
W ← 1
L2
end

(b)

FIGURE 7.5
(a) The macro `R1_LT_R2` and (b) its flowchart.

R2 (R1 < R2), then the output Q (W) is forced to be 1. Otherwise, the output Q (W) is forced to be 0.

7.5 Macro `R1_LE_R2`

The definition, symbols, and algorithm of the macro `R1_LE_R2` are depicted in Table 7.5. Figure 7.6 shows the macro `R1_LE_R2` and its flowchart. The macro `R1_LE_R2` has a Boolean input variable (active high enabling input),

TABLE 7.5

Definition, Symbols, and Algorithm of the Macro `R1_LE_R2`

Definition	Ladder diagram symbol	Schematic symbol	Algorithm

is the content of register **R1** Less than or Equal to the content of register **R2**?	W —| R1 <= R2 |— W	W — EN Q — W; R1; R2; <=
		R1, R2 (8 bit register)
		EN (through W) = 0 or 1
		Q (through W) = 0 or 1
			if EN = 1 then
			if R1 ≤ R2 then
			Q = 1;
			else Q = 0;
			end if;

EN, passed into the macro through W, and a Boolean output variable, Q, passed out of the macro through W. This means that the input signal EN should be loaded into W before this macro is run, and the output signal Q will be provided within the W at the end of the macro. R1 and R2 are both 8-bit input variables. When EN = 0, no action is taken and the output Q (W) is forced to be 0. When EN = 1, if the content of R1 is less than or equal to the content of R2 (R1 ≤ R2), then the output Q (W) is forced to be 1. Otherwise, the output Q (W) is forced to be 0.

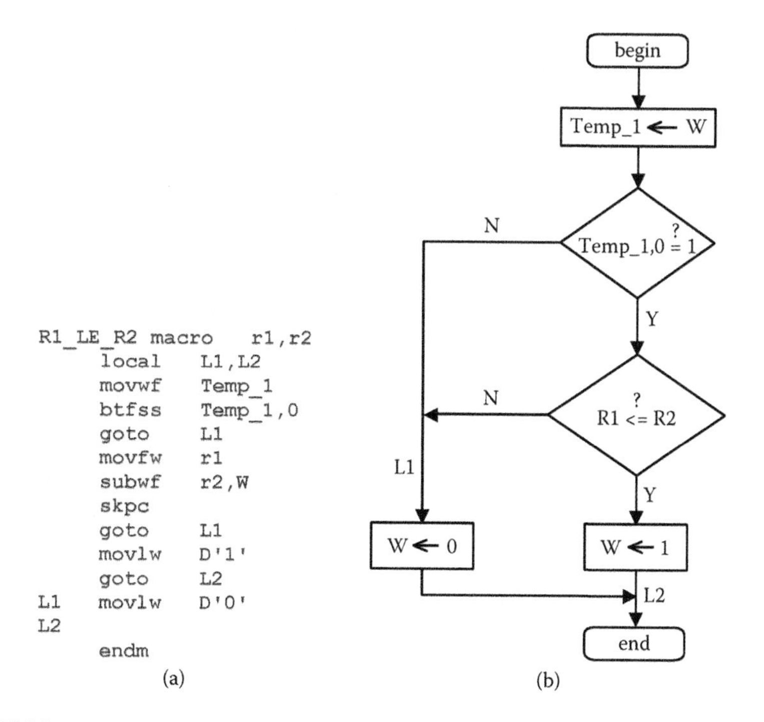

FIGURE 7.6
(a) The macro `R1_LE_R2` and (b) its flowchart.

TABLE 7.6

Definition, Symbols, and Algorithm of the Macro `R1_NE_R2`

Definition	Ladder diagram symbol	Schematic symbol	Algorithm

is the content of register **R1** Not Equal to the content of register **R2**?	W —┤ R1 <> R2 ├— W	W — EN Q — W; R1; R2; <>
		R1, R2 (8 bit register)
		EN (through W) = 0 or 1
		Q (through W) = 0 or 1
			if EN = 1 then
			if R1 ≠ R2 then
			Q = 1;
			else Q = 0;
			end if;

7.6 Macro `R1_NE_R2`

The definition, symbols, and algorithm of the macro `R1_NE_R2` are depicted in Table 7.6. Figure 7.7 shows the macro `R1_NE_R2` and its flowchart. The macro `R1_NE_R2` has a Boolean input variable (active high enabling input), EN, passed into the macro through W, and a Boolean output variable, Q, passed out of the macro through W. This means that the input signal EN

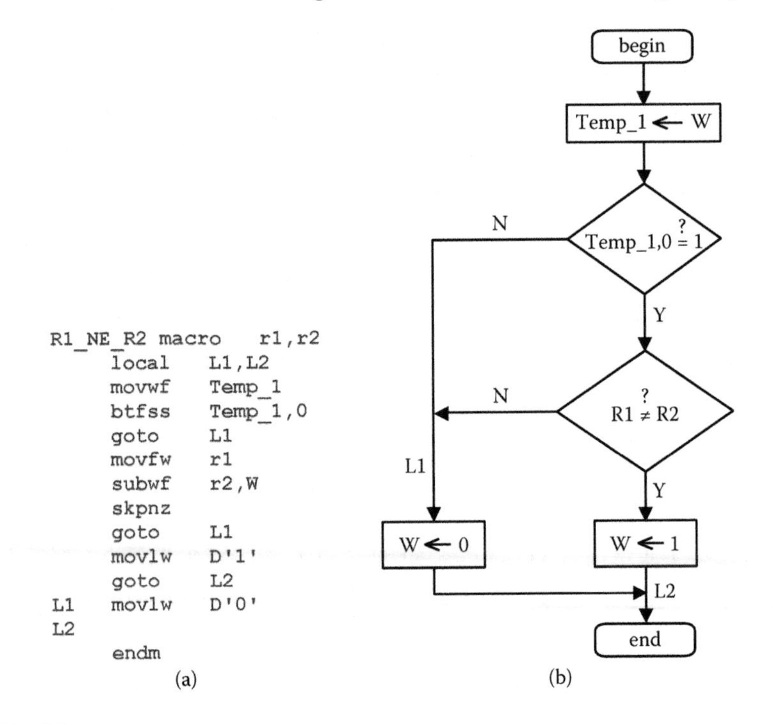

FIGURE 7.7
(a) The macro `R1_NE_R2` and (b) its flowchart.

TABLE 7.7

Definition, Symbols, and Algorithm of the Macro R_GT_K

Definition	Ladder Diagram Symbol	Schematic Symbol	Algorithm

is the content of register **R** **G**reater **T**han the constant **K**?	W —┤ R > K ├— W	W — EN Q — W; — R >; — K
		R (8 bit register)
		K (8 bit constant)
		EN (through W) = 0 or 1
		Q (through W) = 0 or 1
			if EN = 1 then
			if R > K then
			Q = 1;
			else Q = 0;
			end if;

should be loaded into W before this macro is run, and the output signal Q will be provided within the W at the end of the macro. R1 and R2 are both 8-bit input variables. When EN = 0, no action is taken and the output Q (W) is forced to be 0. When EN = 1, if the content of R1 is not equal to the content of R2 (R1 ≠ R2), then the output Q (W) is forced to be 1. Otherwise, the output Q (W) is forced to be 0.

7.7 Macro R_GT_K

The definition, symbols, and algorithm of the macro R_GT_K are depicted in Table 7.7. Figure 7.8 shows the macro R_GT_K and its flowchart. The macro R_GT_K has a Boolean input variable (active high enabling input), EN, passed into the macro through W, and a Boolean output variable, Q, passed out of the macro through W. This means that the input signal EN should be loaded into W before this macro is run, and the output signal Q will be provided within the W at the end of the macro. R is an 8-bit input variable, while K is an 8-bit constant value. When EN = 0, no action is taken and the output Q (W) is forced to be 0. When EN = 1, if the content of R is greater than the constant value K (R > K), then the output Q (W) is forced to be 1. Otherwise, the output Q (W) is forced to be 0.

7.8 Macro R_GE_K

The definition, symbols, and algorithm of the macro R_GE_K are depicted in Table 7.8. Figure 7.9 shows the macro R_GE_K and its flowchart. The macro R_GE_K has a Boolean input variable (active high enabling input), EN, passed

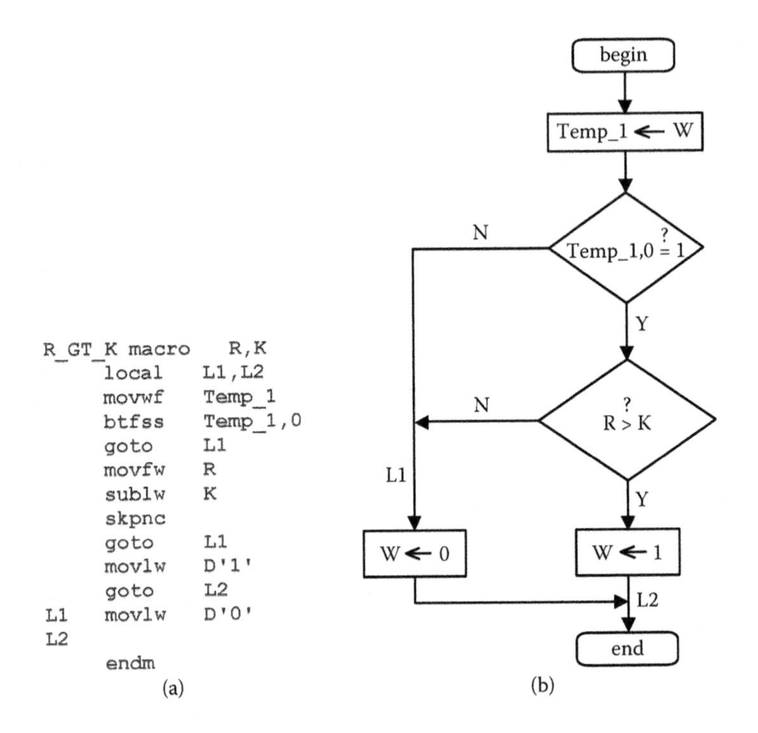

FIGURE 7.8
(a) The macro R_GT_K and (b) its flowchart.

into the macro through W, and a Boolean output variable, Q, passed out of the macro through W. This means that the input signal EN should be loaded into W before this macro is run, and the output signal Q will be provided within the W at the end of the macro. R is an 8-bit input variable, while K is an 8-bit constant value. When EN = 0, no action is taken and the output Q (W) is forced to be 0. When EN = 1, if the content of R is greater than or equal to the constant value K (R ≥ K), then the output Q (W) is forced to be 1. Otherwise, the output Q (W) is forced to be 0.

TABLE 7.8

Definition, Symbols, and Algorithm of the Macro R_GE_K

Definition	Ladder diagram symbol	Schematic symbol	Algorithm

is the content of register **R** **G**reater than or **E**qual to the constant **K**?	W —| R >= K |— W	W — EN Q — W; — R >=; — K
		R (8 bit register)
		K (8 bit constant)
		EN (through W) = 0 or 1
		Q (through W) = 0 or 1
			if EN = 1 then
			if R ≥ K then
			Q = 1;
			else Q = 0;
			end if;

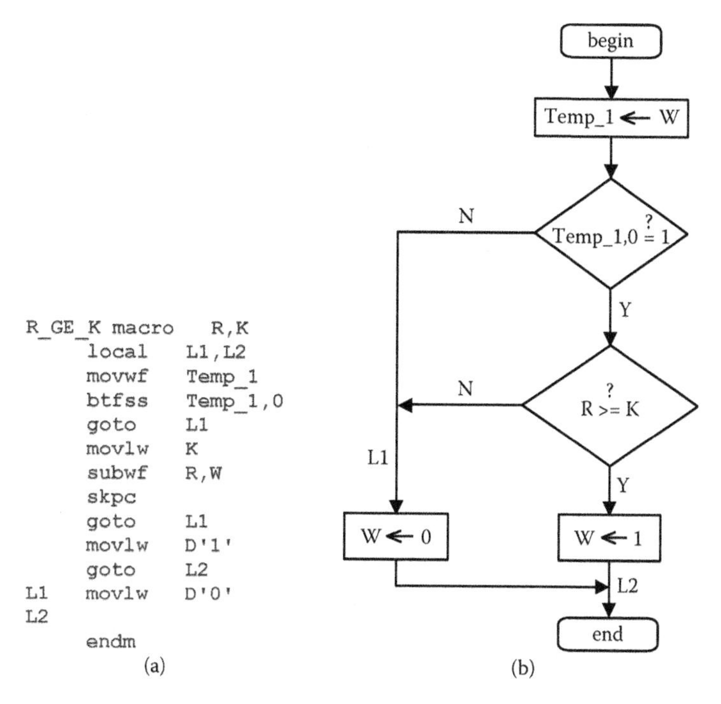

FIGURE 7.9
(a) The macro R_GE_K and (b) its flowchart.

7.9 Macro R_EQ_K

The definition, symbols, and algorithm of the macro R_EQ_K are depicted in Table 7.9. Figure 7.10 shows the macro R_EQ_K and its flowchart. The macro R_EQ_K has a Boolean input variable (active high enabling input), EN, passed into the macro through W, and a Boolean output variable, Q, passed out of the macro through W. This means that the input signal EN should

TABLE 7.9

Definition, Symbols, and Algorithm of the Macro R_EQ_K

Definition	Ladder diagram symbol	Schematic symbol	Algorithm

is the content of register **R** **EQ**ual to the constant **K**?	W ─┤ R = K ├─ W	W ─ EN Q ─ W; ─ R =; ─ K
		R (8 bit register)
		K (8 bit constant)
		EN (through W) = 0 or 1
		Q (through W) = 0 or 1
			if EN = 1 then
			if R = K then
			Q = 1;
			else Q = 0;
			end if;

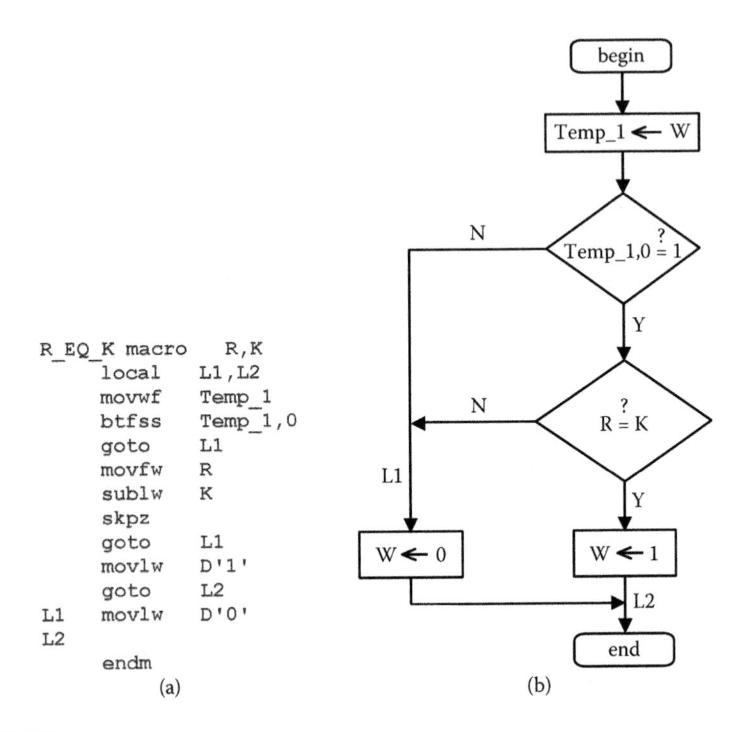

FIGURE 7.10
(a) The macro `R_EQ_K` and (b) its flowchart.

be loaded into W before this macro is run, and the output signal Q will be provided within the W at the end of the macro. R is an 8-bit input variable, while K is an 8-bit constant value. When EN = 0, no action is taken and the output Q (W) is forced to be 0. When EN = 1, if the content of R is equal to the constant value K (R = K), then the output Q (W) is forced to be 1. Otherwise, the output Q (W) is forced to be 0.

7.10 Macro `R_LT_K`

The definition, symbols, and algorithm of the macro `R_LT_K` are depicted in Table 7.10. Figure 7.11 shows the macro `R_LT_K` and its flowchart. The macro `R_LT_K` has a Boolean input variable (active high enabling input), EN, passed into the macro through W, and a Boolean output variable, Q, passed out of the macro through W. This means that the input signal EN should be loaded into W before this macro is run, and the output signal Q will be provided within the W at the end of the macro. R is an 8-bit input variable, while K is an 8-bit constant value. When EN = 0, no action is taken and the output Q (W) is forced to be 0. When EN = 1, if the content of R is less than the

TABLE 7.10

Definition, Symbols, and Algorithm of the Macro `R_LT_K`

Definition	Ladder diagram symbol	Schematic symbol	Algorithm

is the content of register **R** Less Than the constant **K**?	W —| R < K |— W	W — EN Q — W; — R; — K; <
		 R (8 bit register)
		 K (8 bit constant)
		 EN (through W) = 0 or 1
		 Q (through W) = 0 or 1
			if EN = 1 then
			 if R < K then
			 Q = 1;
			 else Q = 0;
			 end if;

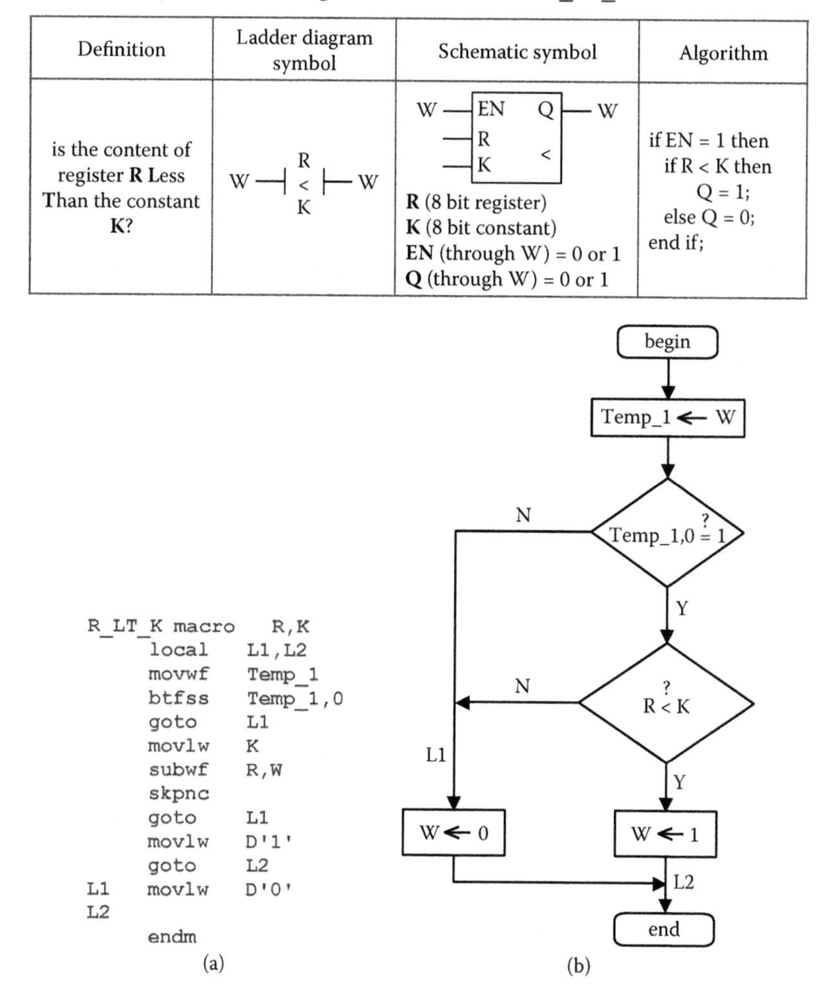

FIGURE 7.11
(a) The macro `R_LT_K` and (b) its flowchart.

constant value K (R < K), then the output Q (W) is forced to be 1. Otherwise, the output Q (W) is forced to be 0.

7.11 Macro `R_LE_K`

The definition, symbols, and algorithm of the macro `R_LE_K` are depicted in Table 7.11. Figure 7.12 shows the macro `R_LE_K` and its flowchart. The macro `R_LE_K` has a Boolean input variable (active high enabling input), EN,

TABLE 7.11

Definition, Symbols, and Algorithm of the Macro `R_LE_K`

Definition	Ladder diagram symbol	Schematic symbol	Algorithm

is the content of register **R** Less than or Equal to the constant **K**?	W —┤R <= K├— W	W — EN Q — W; — R <=; — K
		R (8 bit register)
		K (8 bit constant)
		EN (through W) = 0 or 1
		Q (through W) = 0 or 1
			if EN = 1 then
			if R ≤ K then
			Q = 1;
			else Q = 0;
			end if;

passed into the macro through W, and a Boolean output variable, Q, passed out of the macro through W. This means that the input signal EN should be loaded into W before this macro is run, and the output signal Q will be provided within the W at the end of the macro. R is an 8-bit input variable, while K is an 8-bit constant value. When EN = 0, no action is taken and the output Q (W) is forced to be 0. When EN = 1, if the content of R is less than or equal to the constant value K (R ≤ K), then the output Q (W) is forced to be 1. Otherwise, the output Q (W) is forced to be 0.

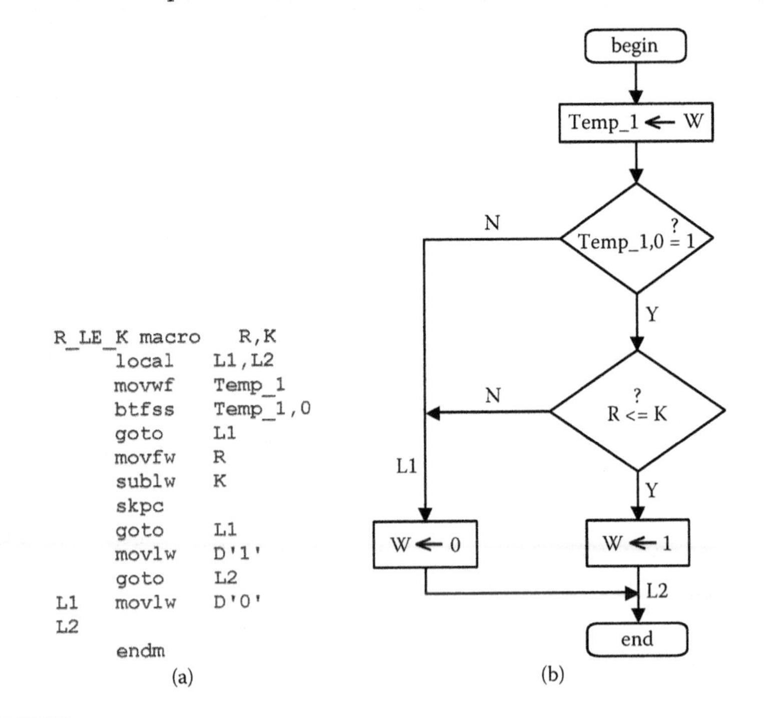

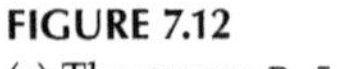

FIGURE 7.12
(a) The macro `R_LE_K` and (b) its flowchart.

TABLE 7.12

Definition, Symbols, and Algorithm of the Macro `R_NE_K`

Definition	Ladder diagram symbol	Schematic symbol	Algorithm

is the content of register **R** **N**ot **E**qual to the constant **K**?	W —┤ R <> K ├— W	W — EN Q — W; R; K; <>
		 R (8 bit register)
		 K (8 bit constant)
		 EN (through W) = 0 or 1
		 Q (through W) = 0 or 1
			if EN = 1 then
			 if R ≠ K then
			 Q = 1;
			 else Q = 0;
			 end if;

7.12 Macro `R_NE_K`

The definition, symbols, and algorithm of the macro `R_NE_K` are depicted in Table 7.12. Figure 7.13 shows the macro `R_NE_K` and its flowchart. The macro `R_NE_K` has a Boolean input variable (active high enabling input), EN, passed into the macro through W, and a Boolean output variable, Q, passed out of the macro through W. This means that the input signal EN should be loaded into W before this macro is run, and the output signal Q will be provided within the W at the end of the macro. R is an 8-bit input variable, while K is an 8-bit constant value. When EN = 0, no action is taken and the output

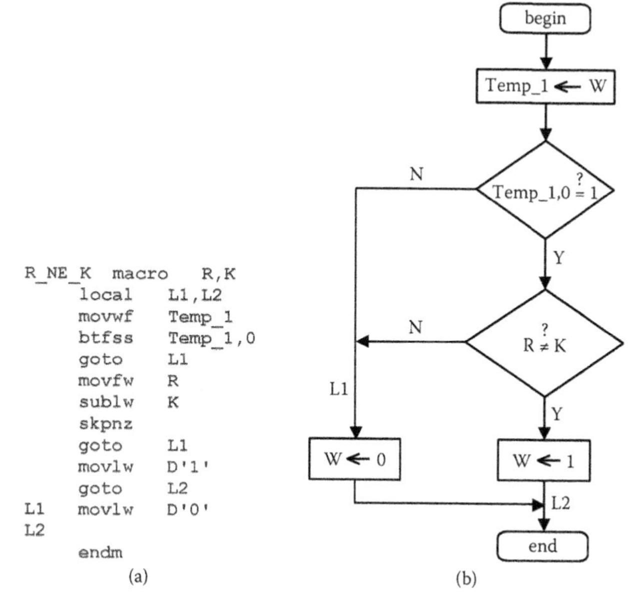

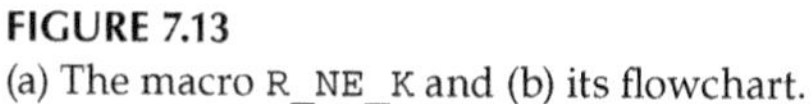

FIGURE 7.13
(a) The macro `R_NE_K` and (b) its flowchart.

Q (W) is forced to be 0. When EN = 1, if the content of R is not equal to the constant value K (R ≠ K), then the output Q (W) is forced to be 1. Otherwise, the output Q (W) is forced to be 0.

7.13 Examples for Comparison Macros

In this section, we will consider two examples, UZAM_plc_16i16o_ex12.asm and UZAM_plc_16i16o_ex13.asm, to show the usage of comparison macros. In order to test one of these examples, please take the related file UZAM_plc_16i16o_ex12.asm or UZAM_plc_16i16o_ex13.asm from the CD-ROM attached to this book, and then open the program by MPLAB IDE and compile it. After that, by using the PIC programmer software, take the compiled file UZAM_plc_16i16o_ex12.hex or UZAM_plc_16i16o_ex13.hex, and by your PIC programmer hardware, send it to the program memory of PIC16F648A microcontroller within the PIC16F648A-based PLC. To do this, switch the 4PDT in PROG position and the power switch in OFF position. After loading the file UZAM_plc_16i16o_ex12.hex or UZAM_plc_16i16o_ex13.hex, switch the 4PDT in RUN and the power switch in ON position. Please check the program's accuracy by cross-referencing it with the related macros.

Let us now consider these example programs: The first example program, UZAM_plc_16i16o_ex12.asm, is shown in Figure 7.14. It shows the usage of

```
;--------------- user program starts here --
        ld          LOGIC1        ;rung 1
        R1_GT_R2    I1,I0
        out         Q1.7

        ld          LOGIC1        ;rung 2
        R1_GE_R2    I1,I0
        out         Q1.4

        ld          LOGIC1        ;rung 3
        R1_EQ_R2    I1,I0
        out         Q1.1

        ld          LOGIC1        ;rung4
        R1_LT_R2    I1,I0
        out         Q0.6

        ld          LOGIC1        ;rung5
        R1_LE_R2    I1,I0
        out         Q0.3

        ld          LOGIC1        ;rung6
        R1_NE_R2    I1,I0
        out         Q0.0
;--------------- user program ends here ----
```

FIGURE 7.14
The user program of UZAM_plc_16i16o_ex12.asm.

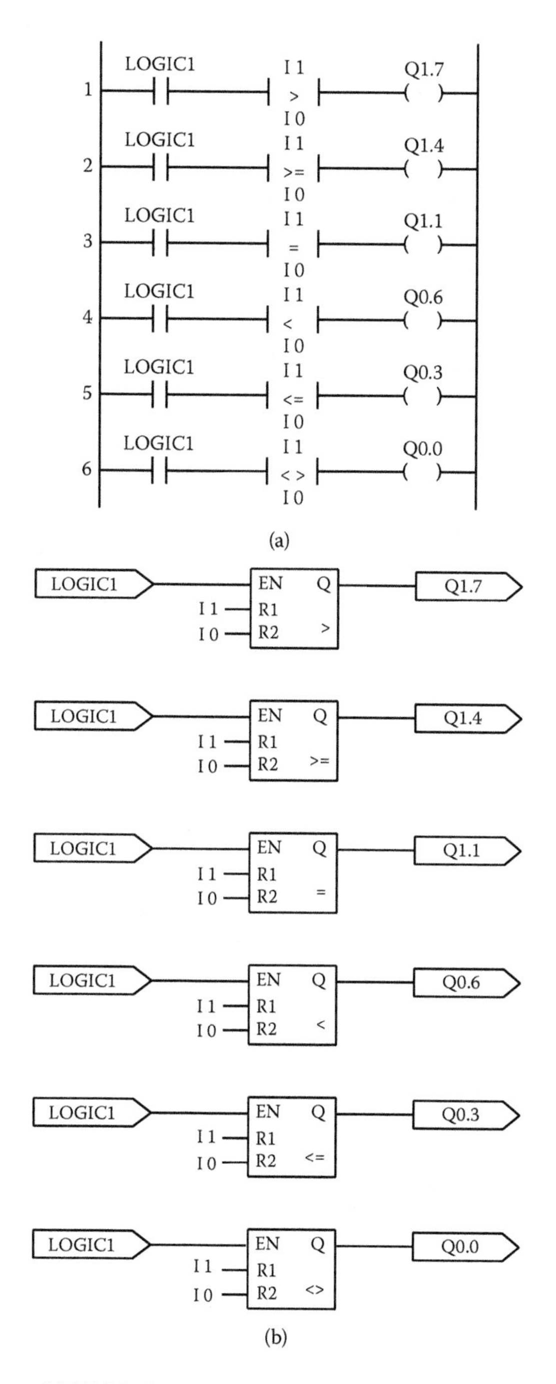

FIGURE 7.15
The user program of UZAM_plc_16i16o_ex12.asm: (a) ladder diagram and (b) schematic diagram.

```
;
;--------------- user program starts here --
        ld          LOGIC1      ;rung1
        R_GT_K      I1,0Fh
        out         Q1.7

        ld          LOGIC1      ;rung2
        R_GE_K      I1,0Fh
        out         Q1.4

        ld          LOGIC1      ;rung3
        R_EQ_K      I1,0Fh
        out         Q1.1

        ld          LOGIC1      ;rung4
        R_LT_K      I1,0Fh
        out         Q0.6

        ld          LOGIC1      ;rung5
        R_LE_K      I1,0Fh
        out         Q0.3

        ld          LOGIC1      ;rung6
        R_NE_K      I1,0Fh
        out         Q0.0
;--------------- user program ends here ----
;
```

FIGURE 7.16
The user program of UZAM_plc_16i16o_ex13.asm.

the macros in which the contents of two registers (R1 and R2) are compared. The ladder diagram and schematic diagram of the user program of UZAM_plc_16i16o_ex12.asm, shown in Figure 7.14, are depicted in Figure 7.15(a) and (b), respectively. In rungs 1 to 6, the content of I1 is compared with the content of I0 based on the following criteria, respectively: >, ≥, =, <, ≤, ≠. The result of each comparison is observed from the outputs Q1.7, Q1.4, Q1.1, Q0.6, Q0.3, and Q0.0, respectively. These outputs will be true or false based on the comparison being made and the input data entered from the inputs I1 and I0.

The second example program, UZAM_plc_16i16o_ex13.asm, is shown in Figure 7.16. It shows the usage of the macros in which the content of a register R is compared with a constant value K. The ladder diagram and schematic diagram of the user program of UZAM_plc_16i16o_ex13.asm, shown in Figure 7.16, are depicted in Figure 7.17(a) and (b), respectively. In rungs 1

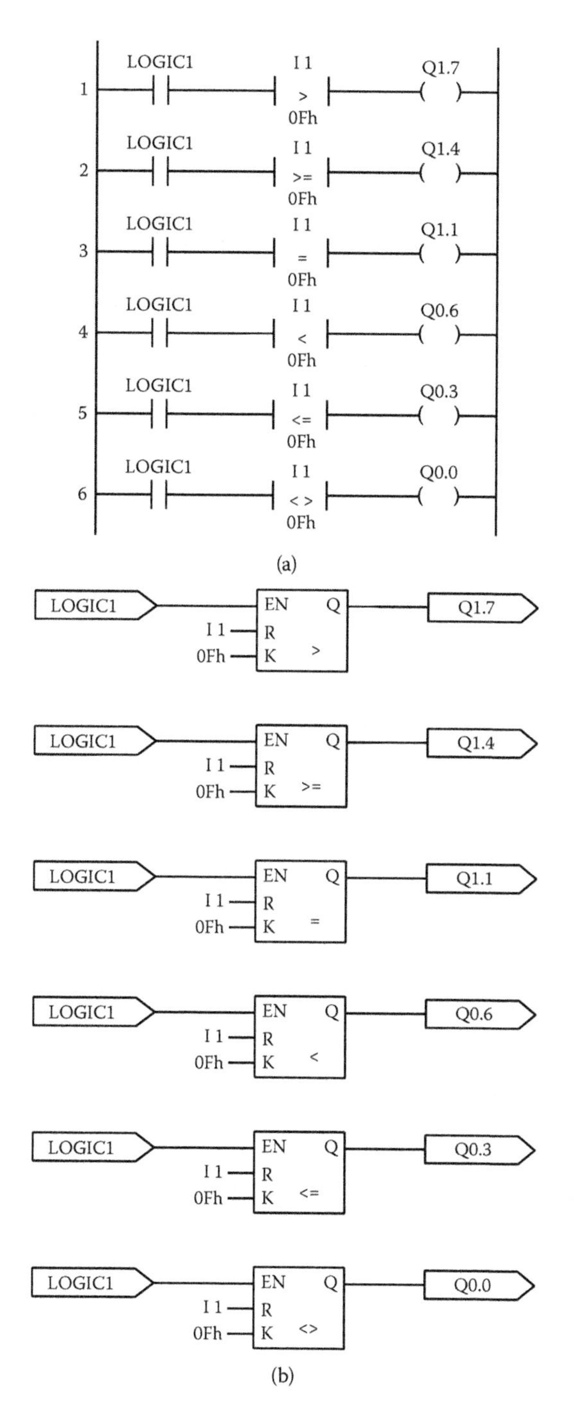

FIGURE 7.17
The user program of UZAM_plc_16i16o_ex13.asm: (a) ladder diagram and (b) schematic diagram.

to 6, the content of I1 is compared with the constant value 0Fh based on the following criteria, respectively: >, ≥, =, <, ≤, ≠. The result of each comparison is observed from the outputs Q1.7, Q1.4, Q1.1, Q0.6, Q0.3, and Q0.0, respectively. These outputs will be true or false based on the comparison being made and the input data entered from the input register I1.

8

Arithmetical Macros

Numerical data imply the ability to do arithmetical operations, and almost all PLCs provide some arithmetical operations, such as add, subtract, multiply, and divide. Arithmetical functions will retrieve one or more values, perform an operation, and store the result in memory. As an example, Figure 8.1 shows an *ADD* function that will retrieve and add two values from sources labeled *source A* and *source B* and will store the result in *destination C*. The list of arithmetical functions (macros) described for the PIC16F648A-based PLC is as follows. The increment and decrement functions are unary, so there is only one source.

ADD (source value 1, source value 2, destination): Add two source values and put the result in the destination.

SUB (source value 1, source value 2, destination): Subtract the second source value from the first one and put the result in the destination.

INC (source value, destination): Increment the source and put the result in the destination.

DEC (source value, destination): Decrement the source and put the result in the destination.

In this chapter, the following six arithmetical macros are described for the PIC16F648A-based PLC:

```
R1addR2
RaddK
R1subR2
RsubK
incR
decR
```

The file definitions.inc, included within the CD-ROM attached to this book, contains all arithmetical macros defined for the PIC16F648A-based PLC. Let us now consider these macros in detail.

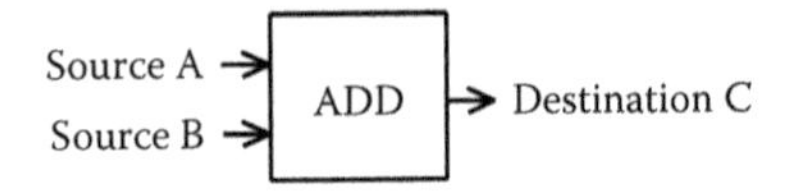

FIGURE 8.1
The ADD function.

TABLE 8.1
Algorithm and Symbol of the Macro `R1addR2`

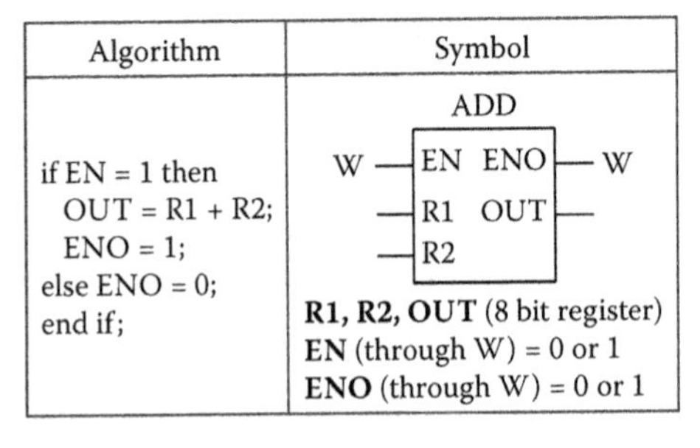

Algorithm	Symbol

if EN = 1 then
OUT = R1 + R2;
ENO = 1;
else ENO = 0;
end if;
	ADD
	W — EN ENO — W
	— R1 OUT —
	— R2
	R1, R2, OUT (8 bit register)
	EN (through W) = 0 or 1
	ENO (through W) = 0 or 1

8.1 Macro `R1addR2`

The algorithm and the symbol of the macro `R1addR2` are depicted in Table 8.1. Figure 8.2 shows the macro `R1addR2` and its flowchart. In this macro, EN is a Boolean input variable taken into the macro through W, and ENO is a Boolean output variable sent out from the macro through W. Output ENO

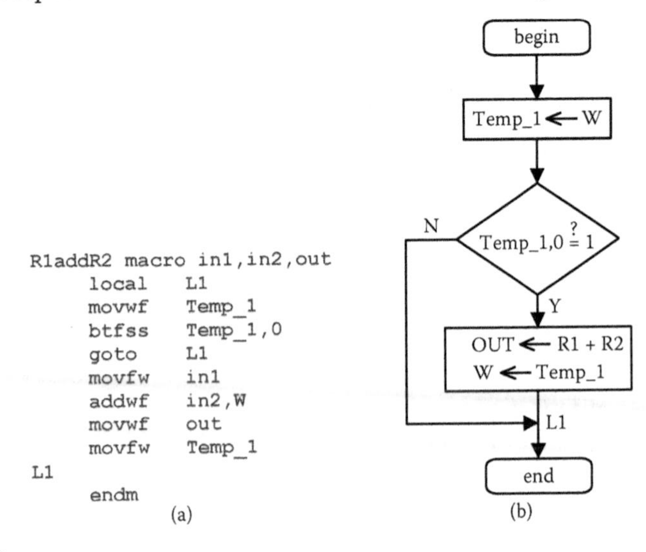

```
R1addR2 macro in1,in2,out
        local   L1
        movwf   Temp_1
        btfss   Temp_1,0
        goto    L1
        movfw   in1
        addwf   in2,W
        movwf   out
        movfw   Temp_1
L1
        endm
```

FIGURE 8.2
(a) The macro `R1addR2` and (b) its flowchart.

TABLE 8.2

Algorithm and Symbol of the Macro `RaddK`

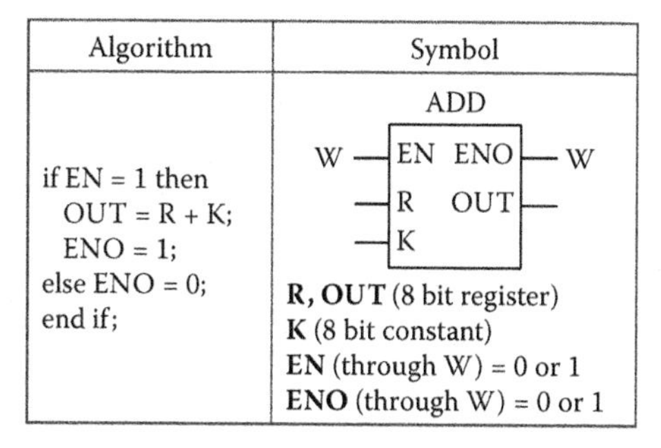

Algorithm	Symbol

if EN = 1 then
  OUT = R + K;
  ENO = 1;
else ENO = 0;
end if;
	ADD
	W — EN ENO — W
	— R OUT —
	— K
	R, OUT (8 bit register)
	K (8 bit constant)
	EN (through W) = 0 or 1
	ENO (through W) = 0 or 1

follows the input EN. This means that when EN = 0, ENO is forced to be 0, and when EN = 1, ENO is forced to be 1. This is especially useful if we want to carry out more than one operation based on a single input condition. R1 and R2 refer to 8-bit source variables from where the source values are taken into the macro, while OUT refers to an 8-bit destination variable to which the result of the macro is stored. When EN = 1, the macro `R1addR2` adds the contents of two 8-bit variables R1 and R2 and stores the result into the 8-bit output variable OUT (OUT = R1 + R2).

8.2 Macro `RaddK`

The algorithm and the symbol of the macro `RaddK` are depicted in Table 8.2. Figure 8.3 shows the macro `RaddK` and its flowchart. In this macro, EN is a Boolean input variable taken into the macro through W, and ENO is a Boolean output variable sent out from the macro through W. Output ENO follows the input EN. This means that when EN = 0, ENO is forced to be 0, and when EN = 1, ENO is forced to be 1. R and K are source values. R refers to an 8-bit source variable, while K represents an 8-bit constant value. OUT refers to an 8-bit destination variable to which the result of the macro is stored. When EN = 1, the macro `RaddK` adds the content of the 8-bit variable R and the 8-bit constant value K and stores the result into the 8-bit output variable OUT (OUT = R + K).

8.3 Macro `R1subR2`

The algorithm and the symbol of the macro `R1subR2` are depicted in Table 8.3. Figure 8.4 shows the macro `R1subR2` and its flowchart. In this macro, EN is a Boolean input variable taken into the macro through W,

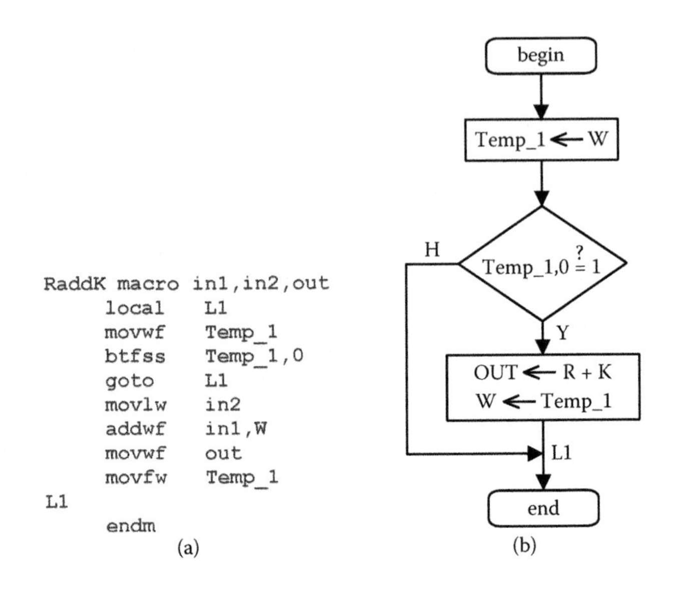

FIGURE 8.3
(a) The macro `RaddK` and (b) its flowchart.

and ENO is a Boolean output variable sent out from the macro through W. Output ENO follows the input EN. This means that when EN = 0, ENO is forced to be 0, and when EN = 1, ENO is forced to be 1. R1 and R2 refer to 8-bit source variables from where the source values are taken into the macro, while OUT refers to an 8-bit destination variable to which the result of the macro is stored. When EN = 1, the macro `R1subR2` subtracts the content of the 8-bit variable R2 from the content of the 8-bit variable R1 and stores the result into the 8-bit output variable OUT (OUT = R1 – R2).

TABLE 8.3
Algorithm and Symbol of the Macro `R1subR2`

Algorithm	Symbol

if EN = 1 then
OUT = R1 – R2;
ENO = 1;
else ENO = 0;
end if;
	SUB
	W — EN ENO — W
	— R1 OUT —
	— R2
	R1, R2, OUT (8 bit register)
	EN (through W) = 0 or 1
	ENO (through W) = 0 or 1

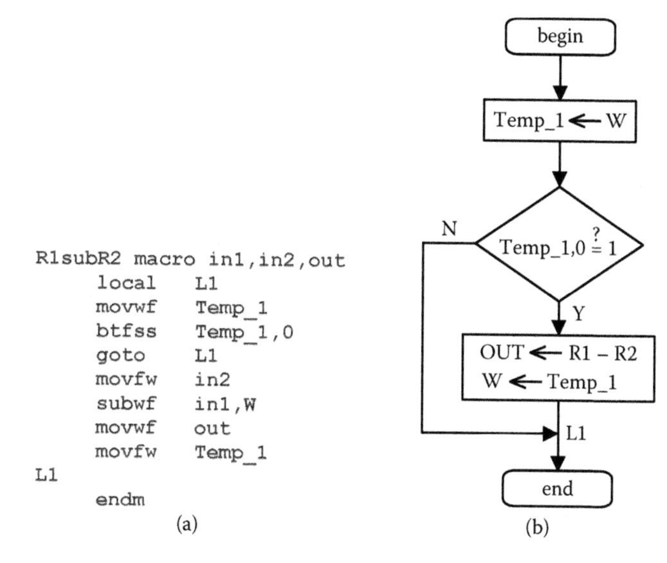

FIGURE 8.4
(a) The macro `R1subR2` and (b) its flowchart.

8.4 Macro `RsubK`

The algorithm and the symbol of the macro `RsubK` are depicted in Table 8.4. Figure 8.5 shows the macro `RsubK` and its flowchart. In this macro, EN is a Boolean input variable taken into the macro through W, and ENO is a Boolean output variable sent out from the macro through W. Output ENO follows the input EN. This means that when EN = 0, ENO is forced to be 0, and when EN = 1, ENO is forced to be 1. R and K are source values. R refers to an 8-bit source variable, while K represents an 8-bit constant value.

TABLE 8.4
Algorithm and Symbol of the Macro `RsubK`

Algorithm	Symbol

if EN = 1 then
OUT = R – K;
ENO = 1;
else ENO = 0;
end if;
	SUB
	W — EN ENO — W
	— R OUT —
	— K
	R, OUT (8 bit register)
	K (8 bit constant)
	EN (through W) = 0 or 1
	ENO (through W) = 0 or 1

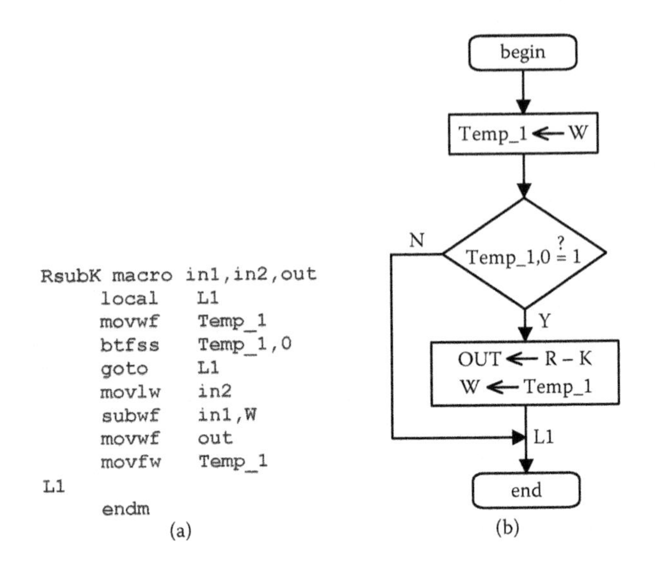

FIGURE 8.5
(a) The macro `RsubK` and (b) its flowchart.

OUT refers to an 8-bit destination variable to which the result of the macro is stored. When EN = 1, the macro `RsubK` subtracts the 8-bit constant value K from the content of the 8-bit variable R and stores the result into the 8-bit output variable OUT (OUT = R – K).

8.5 Macro `incR`

The algorithm and the symbol of the macro `incR` are depicted in Table 8.5. Figure 8.6 shows the macro `incR` and its flowchart. In this macro, EN is a Boolean input variable taken into the macro through W, and ENO is a Boolean output variable sent out from the macro through W. Output ENO

TABLE 8.5

Algorithm and Symbol of the Macro `incR`

Algorithm	Symbol

if EN = 1 then
OUT = IN + 1;
ENO = 1;
else ENO = 0;
end if;
	INC
	W — EN ENO — W
	— IN OUT —
	IN, OUT (8 bit register)
	EN (through W) = 0 or 1
	ENO (through W) = 0 or 1

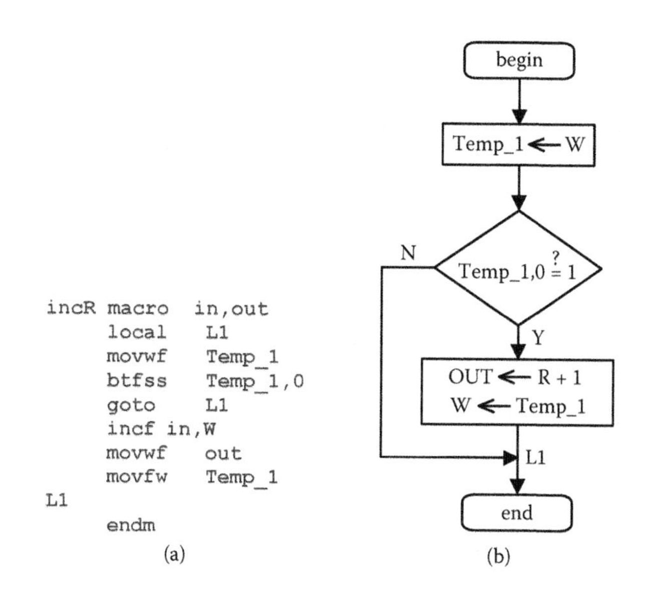

FIGURE 8.6
(a) The macro `incR` and (b) its flowchart.

follows the input EN. This means that when EN = 0, ENO is forced to be 0, and when EN = 1, ENO is forced to be 1. IN refers to an 8-bit source variable from where the source value is taken into the macro, while OUT refers to an 8-bit destination variable to which the result of the macro is stored. When EN = 1, the macro `incR` increments the content of the 8-bit variable IN and stores the result into the 8-bit output variable OUT (OUT = IN + 1).

8.6 Macro `decR`

The algorithm and the symbol of the macro `decR` are depicted in Table 8.6. Figure 8.7 shows the macro `decR` and its flowchart. In this macro, EN is a Boolean input variable taken into the macro through W, and ENO is a

TABLE 8.6

Algorithm and Symbol of the Macro `decR`

Algorithm	Symbol

if EN = 1 then
  OUT = IN – 1;
  ENO = 1;
else ENO = 0;
end if;
	DEC
	W — EN ENO — W
	— IN OUT —
	IN, OUT (8 bit register)
	EN (through W) = 0 or 1
	ENO (through W) = 0 or 1

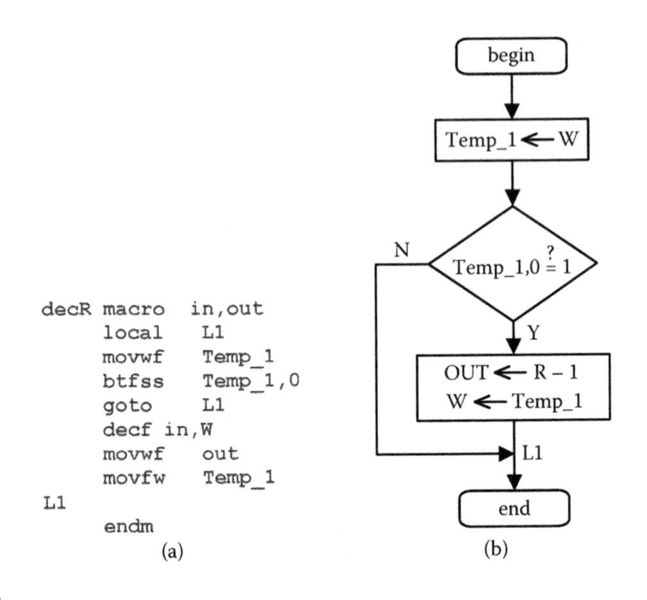

FIGURE 8.7
(a) The macro `decR` and (b) its flowchart.

Boolean output variable sent out from the macro through W. Output ENO follows the input EN. This means that when EN = 0, ENO is forced to be 0, and when EN = 1, ENO is forced to be 1. IN refers to an 8-bit source variable from where the source value is taken into the macro, while OUT refers to an 8-bit destination variable to which the result of the macro is stored. When EN = 1, the macro `decR` decrements the content of the 8-bit variable IN and stores the result into the 8-bit output variable OUT (OUT = IN – 1).

8.7 Examples for Arithmetical Macros

In this section, we will consider two examples, UZAM_plc_16i16o_ex14.asm and UZAM_plc_16i16o_ex15.asm, to show the usage of arithmetical macros. In order to test one of these examples, please take the related file UZAM_plc_16i16o_ex14.asm or UZAM_plc_16i16o_ex15.asm from the CD-ROM attached to this book, and then open the program by MPLAB IDE and compile it. After that, by using the PIC programmer software, take the compiled file UZAM_plc_16i16o_ex14.hex or UZAM_plc_16i16o_ex15.hex, and by your PIC programmer hardware, send it to the program memory of PIC16F648A microcontroller within the PIC16F648A-based PLC. To do this, switch the 4PDT in PROG position and the power switch in OFF position. After loading the file UZAM_plc_16i16o_ex14.hex or UZAM_plc_16i16o_ex15.hex, switch the 4PDT in RUN and the power switch in ON

```
.
;--------------- user program starts here --
        ld          FRSTSCN     ;rung 1
        or          I0.0
        load_R      00h,Q1

        ld          I0.1        ;rung 2
        R1addR2     I1,Q1,Q1

        ld          I0.2        ;rung 3
        r_edge      0
        R1addR2     I1,Q1,Q1

        ld          I0.3        ;rung 4
        R1subR2     Q1,I1,Q1

        ld          I0.4        ;rung 5
        r_edge      1
        R1subR2     Q1,I1,Q1

        ld          I0.5        ;rung 6
        RaddK       Q1,.2,Q1

        ld          I0.6        ;rung 7
        r_edge      2
        RaddK       Q1,.2,Q1

        ld          I0.7        ;rung 8
        r_edge      3
        RsubK       Q1,.3,Q1
;--------------- user program ends here ----
.
```

FIGURE 8.8
The user program of UZAM_plc_16i16o_ex14.asm.

position. Please check the program's accuracy by cross-referencing it with the related macros.

Let us now consider these example programs: The first example program UZAM_plc_16i16o_ex14.asm is shown in Figure 8.8. It shows the usage of the following arithmetical macros: `R1addR2`, `RaddK`, `R1subR2`, and `RsubK`. The ladder diagram of the user program of UZAM_plc_16i16o_ex14.asm, shown in Figure 8.8, is depicted in Figure 8.9.

In the first rung, Q1 is cleared, i.e., 8-bit constant value 00h is loaded into Q1, by using the macro `load_R`. This process is carried out once at the first program scan by using the FRSTSCN NO contact. Another condition to carry out the same process is the NO contact of the input I0.0. This means that when this program is run, during the normal PLC operation, if we force the input I0.0 to be true, then the above-mentioned process will take place.

In rungs 2 and 3, we see how the arithmetical macro `R1addR2` could be used. In rung 2, the addition process Q1 = I1 + Q1 is carried out, when I0.1 goes true. With this rung, if I0.1 goes and stays true, the content of I1 will be

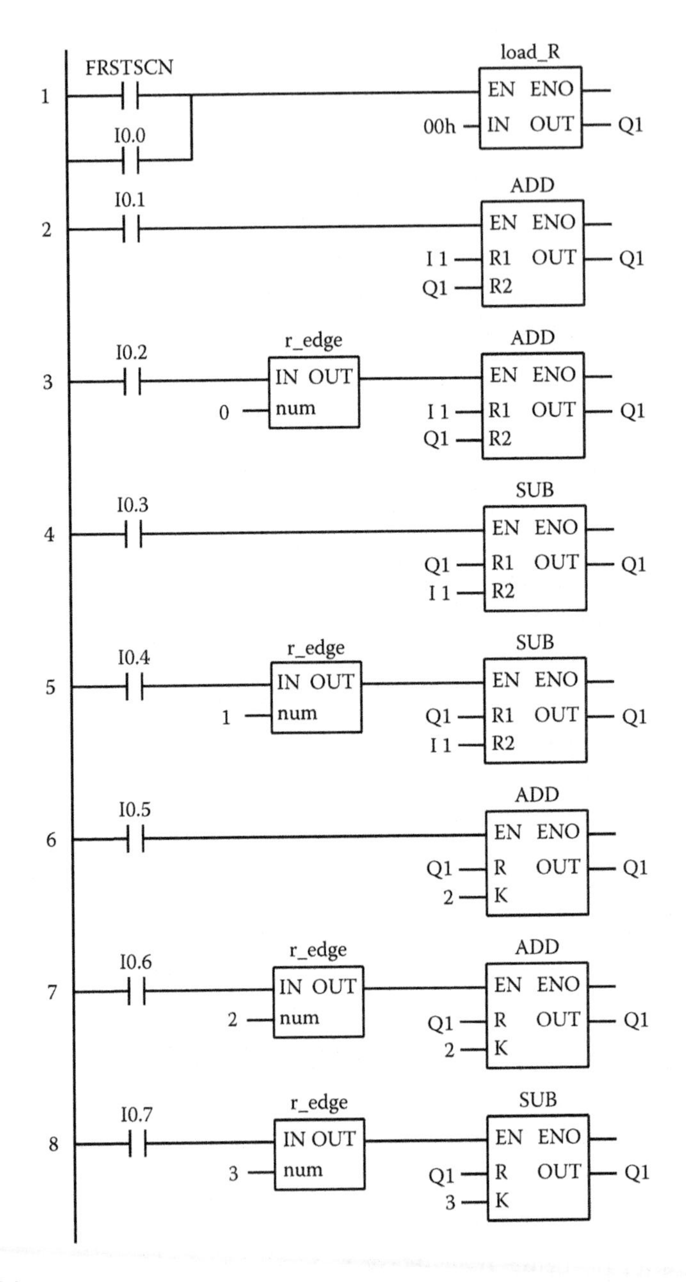

FIGURE 8.9
The ladder diagram of the user program of UZAM_plc_16i16o_ex14.asm.

added to the content of Q1 on every PLC scan. Rung 3 provides a little bit different usage of the arithmetical macro `R1addR2`. Here, we use a rising edge detector macro in order to detect the state change of input I0.2 from OFF to ON. So this time, the addition process Q1 = I1 + Q1 is carried out only at the rising edges of I0.2.

In rungs 4 and 5, we see how the arithmetical macro `R1subR2` could be used. In rung 4, the subtraction process Q1 = Q1 – I1 is carried out when I0.3 goes true. With this rung, if I0.3 goes and stays true, the content of I1 will be subtracted from the content of Q1, on every PLC scan. In rung 5, a rising edge detector macro is used in order to detect the state change of input I0.4 from OFF to ON. So this time, the subtraction process Q1 = Q1 – I1 is carried out only at the rising edges of I0.4.

In rungs 6 and 7, we see how the arithmetical macro `RaddK` could be used. In rung 6, the addition process Q1 = Q1 + 2 is carried out, when I0.5 goes true. With this rung, if I0.5 goes and stays true, the constant value 2 will be added to the content of Q1 on every PLC scan. In rung 7, a rising edge detector macro is used in order to detect the state change of input I0.6 from OFF to ON. So this time, the addition process Q1 = Q1 + 2 is carried out only at the rising edges of I0.6.

In the last rung, the subtraction process Q1 = Q1 – 3 is carried out at the rising edges of I0.7.

The second example program, UZAM_plc_16i16o_ex15.asm, is shown in Figure 8.10. It shows the usage of the following arithmetical macros: `incR` and

```
.
;--------------- user program starts here --
        ld          FRSTSCN     ;rung 1
        or          I0.0
        load_R      00h,Q1

        ld          I0.1        ;rung 2
        incR        Q1,Q1

        ld          I0.2        ;rung 3
        r_edge      0
        incR        Q1,Q1

        ld          I0.3        ;rung 4
        decR        Q1,Q1

        ld          I0.4        ;rung 5
        r_edge      1
        decR        Q1,Q1
;--------------- user program ends here ----
.
```

FIGURE 8.10
The user program of UZAM_plc_16i16o_ex15.asm.

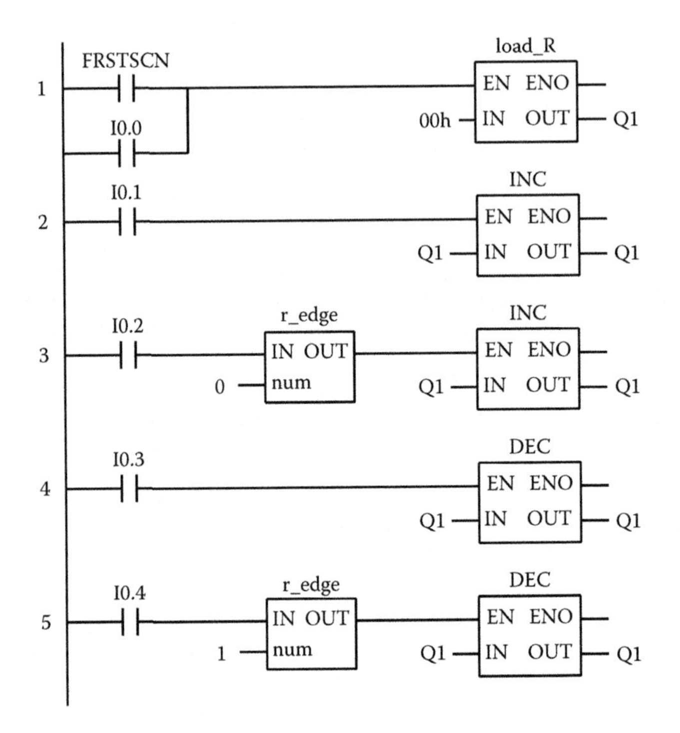

FIGURE 8.11
The ladder diagram of the user program of UZAM_plc_16i16o_ex15.asm.

`decR`. The ladder diagram of the user program of UZAM_plc_16i16o_ex15 .asm, shown in Figure 8.10, is depicted in Figure 8.11.

In the first rung, Q1 is cleared, i.e., 8-bit constant value 00h is loaded into Q1, by using the macro `load_R`. This process is carried out once at the first program scan by using the FRSTSCN NO contact. Another condition to carry out the same process is the NO contact of the input I0.0. This means that when this program is run, during the normal PLC operation, if we force the input I0.0 to be true, then the above-mentioned process will take place.

In rung 2, when I0.1 goes and stays true, Q1 is incremented on every PLC scan.

In rung 3, Q1 is incremented at each rising edge of I0.2.

In rung 4, when I0.3 goes and stays true, Q1 is decremented on every PLC scan.

In rung 5, Q1 is decremented at each rising edge of I0.4.

9

Logical Macros

A *logical* function performs AND, NAND, OR, NOR, exclusive OR (XOR), exclusive NOR (XNOR), logical operations on two registers (or one register plus one constant value), and NOT (invert) logical operations on one register. As an example, Figure 9.1 shows an *AND logical* function that will retrieve AND and two values from sources labeled *source A* and *source B* and will store the result in *destination C*. AND, NAND, OR, NOR, XOR, and XNOR logical functions have the form of Figure 9.1, with two source values and one destination register. In these, the logical function is applied to the two source values and the result is put in the destination register. However, the unary *invert* (INV) logical function has one source register and one destination register. It inverts all of the bits in the source register and puts the result in the destination register.

In this chapter, the following logical macros are described for the PIC16F648A-based PLC:

```
R1andR2
RandK
R1nandR2
RnandK
R1orR2
RorK
R1norR2
RnorK
R1xorR2
RxorK
R1xnorR2
RxnorK
inv_R
```

The file definitions.inc, included within the CD-ROM attached to this book, contains all logical macros defined for the PIC16F648A-based PLC. Let us now consider these macros in detail.

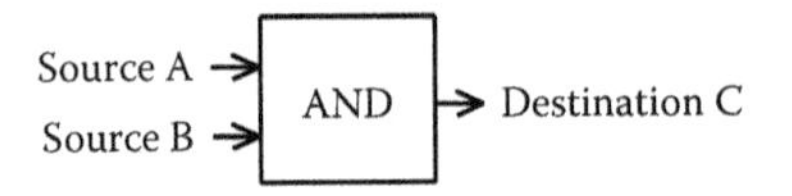

FIGURE 9.1
The AND function.

9.1 Macro `R1andR2`

The algorithm and the symbol of the macro `R1andR2` are depicted in Table 9.1. Figure 9.2 shows the macro `R1andR2` and its flowchart. In this macro, EN is a Boolean input variable taken into the macro through W, and ENO is a Boolean output variable sent out from the macro through W. Output ENO

TABLE 9.1

Algorithm and Symbol of the Macro `R1andR2`

Algorithm	Symbol

if EN = 1 then
OUT = R1 AND R2;
ENO = 1;
else ENO = 0;
end if;
	AND
	W — EN ENO — W
	— R1 OUT —
	— R2
	R1, R2, OUT (8 bit register)
	EN (through W) = 0 or 1
	ENO (through W) = 0 or 1

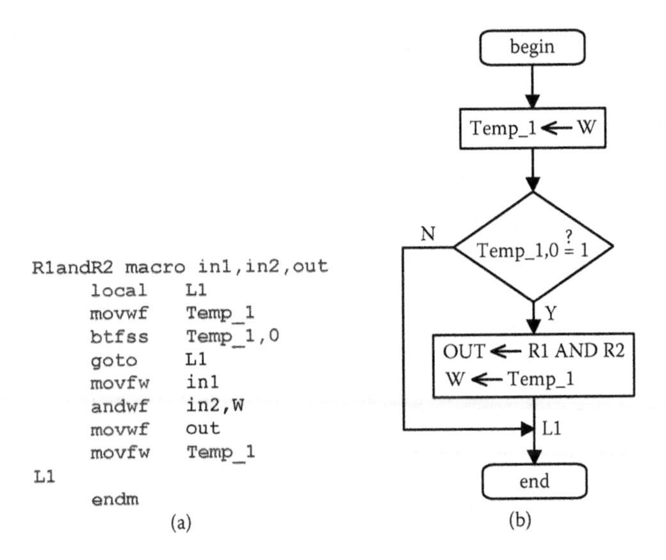

FIGURE 9.2
(a) The macro `R1andR2` and (b) its flowchart.

TABLE 9.2

Algorithm and Symbol of the Macro RandK

Algorithm	Symbol

if EN = 1 then
  OUT = R AND K;
  ENO = 1;
else ENO = 0;
end if;
	AND
	W — EN ENO — W
	— R OUT —
	— K
	R, OUT (8 bit register)
	K (8 bit constant)
	EN (through W) = 0 or 1
	ENO (through W) = 0 or 1

follows the input EN. This means that when EN = 0, ENO is forced to be 0, and when EN = 1, ENO is forced to be 1. This is especially useful if we want to carry out more than one operation based on a single input condition. R1 and R2 refer to 8-bit source variables from where the source values are taken into the macro, while OUT refers to an 8-bit destination variable to which the result of the macro is stored. When EN = 1, the macro `R1andR2` applies the logical AND function to the two 8-bit input variables R1 and R2 and stores the result in the 8-bit output variable OUT (OUT = R1 AND R2).

9.2 Macro `RandK`

The algorithm and the symbol of the macro `RandK` are depicted in Table 9.2. Figure 9.3 shows the macro `RandK` and its flowchart. In this macro, EN is a Boolean input variable taken into the macro through W, and ENO is a Boolean output variable sent out from the macro through W. Output ENO follows the input EN. This means that when EN = 0, ENO is forced to be 0, and when EN = 1, ENO is forced to be 1. R and K are source values. R refers to an 8-bit source variable, while K represents an 8-bit constant value. OUT refers to an 8-bit destination variable to which the result of the macro is stored. When EN = 1 the macro `RandK` applies the logical AND function to the 8-bit input variable R and the 8-bit constant value K and stores the result in the 8-bit output variable OUT (OUT = R AND K).

9.3 Macro `R1nandR2`

The algorithm and the symbol of the macro `R1nandR2` are depicted in Table 9.3. Figure 9.4 shows the macro `R1nandR2` and its flowchart. In this macro, EN is a Boolean input variable taken into the macro through W,

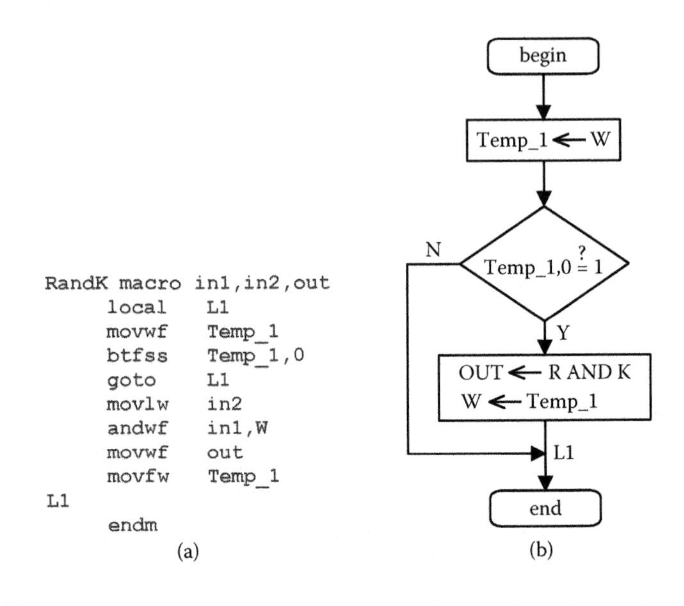

FIGURE 9.3
(a) The macro `RandK` and (b) its flowchart.

and ENO is a Boolean output variable sent out from the macro through W. Output ENO follows the input EN. This means that when EN = 0, ENO is forced to be 0, and when EN = 1, ENO is forced to be 1. R1 and R2 refer to 8-bit source variables from where the source values are taken into the macro, while OUT refers to an 8-bit destination variable to which the result of the macro is stored. When EN = 1, the macro `R1nandR2` applies the logical NAND function to the two 8-bit input variables R1 and R2 and stores the result in the 8-bit output variable OUT (OUT = R1 NAND R2).

TABLE 9.3

Algorithm and Symbol of the Macro `R1nandR2`

Algorithm	Symbol

if EN = 1 then
OUT = R1 NAND R2;
ENO = 1;
else ENO = 0;
end if;
	NAND
	W — EN ENO — W
	— R1 OUT —
	— R2
	R1, R2, OUT (8 bit register)
	EN (through W) = 0 or 1
	ENO (through W) = 0 or 1

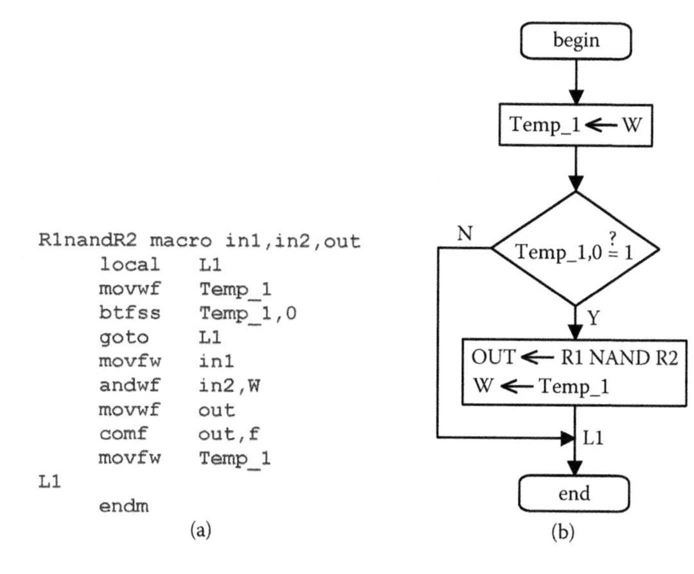

FIGURE 9.4
(a) The macro `R1nandR2` and (b) its flowchart.

9.4 Macro `RnandK`

The algorithm and the symbol of the macro `RnandK` are depicted in Table 9.4. Figure 9.5 shows the macro `RnandK` and its flowchart. In this macro, EN is a Boolean input variable taken into the macro through W, and ENO is a Boolean output variable sent out from the macro through W. Output ENO follows the input EN. This means that when EN = 0, ENO is forced to be 0, and when EN = 1, ENO is forced to be 1. R and K are source values. R refers to an 8-bit source variable, while K represents an 8-bit constant value.

TABLE 9.4

Algorithm and Symbol of the Macro `RnandK`

Algorithm	Symbol

if EN = 1 then
OUT = R NAND K;
ENO = 1;
else ENO = 0;
end if;
	NAND
	W — EN ENO — W
	— R OUT —
	— K
	R, OUT (8 bit register)
	K (8 bit constant)
	EN (through W) = 0 or 1
	ENO (through W) = 0 or 1

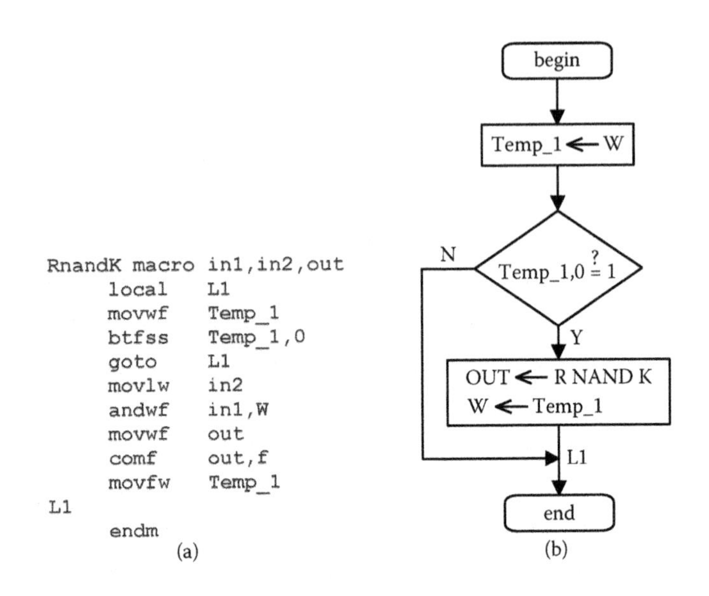

FIGURE 9.5
(a) The macro `RnandK` and (b) its flowchart.

OUT refers to an 8-bit destination variable to which the result of the macro is stored. When EN = 1 the macro `RnandK` applies the logical NAND function to the 8-bit input variable R and the 8-bit constant value K and stores the result in the 8-bit output variable OUT (OUT = R NAND K).

9.5 Macro `R1orR2`

The algorithm and the symbol of the macro `R1orR2` are depicted in Table 9.5. Figure 9.6 shows the macro `R1orR2` and its flowchart. In this macro, EN is a Boolean input variable taken into the macro through W, and ENO is a

TABLE 9.5

Algorithm and Symbol of the Macro `R1orR2`

Algorithm	Symbol

if EN = 1 then
OUT = R1 OR R2;
ENO = 1;
else ENO = 0;
end if;
	OR
	W — EN ENO — W
	— R1 OUT —
	— R2
	R1, R2, OUT (8 bit register)
	EN (through W) = 0 or 1
	ENO (through W) = 0 or 1

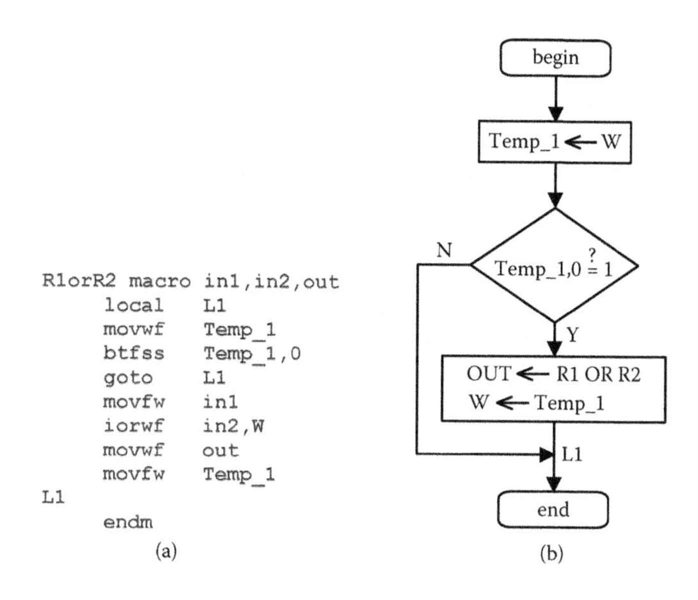

FIGURE 9.6
(a) The macro `R1orR2` and (b) its flowchart.

Boolean output variable sent out from the macro through W. Output ENO follows the input EN. This means that when EN = 0, ENO is forced to be 0, and when EN = 1, ENO is forced to be 1. R1 and R2 refer to 8-bit source variables from where the source values are taken into the macro, while OUT refers to an 8-bit destination variable to which the result of the macro is stored. When EN = 1, the macro `R1orR2` applies the logical OR function to the two 8-bit input variables R1 and R2 and stores the result in the 8-bit output variable OUT (OUT = R1 OR R2).

9.6 Macro `RorK`

The algorithm and the symbol of the macro `RorK` are depicted in Table 9.6. Figure 9.7 shows the macro `RorK` and its flowchart. In this macro, EN is a Boolean input variable taken into the macro through W, and ENO is a Boolean output variable sent out from the macro through W. Output ENO follows the input EN. This means that when EN = 0, ENO is forced to be 0, and when EN = 1, ENO is forced to be 1. R and K are source values. R refers to an 8-bit source variable, while K represents an 8-bit constant value. OUT refers to an 8-bit destination variable to which the result of the macro is stored. When EN = 1 the macro `RorK` applies the logical OR function to the 8-bit input variable R and the 8-bit constant value K and stores the result in the 8-bit output variable OUT (OUT = R OR K).

TABLE 9.6

Algorithm and Symbol of the Macro `RorK`

Algorithm	Symbol

if EN = 1 then
OUT = R OR K;
ENO = 1;
else ENO = 0;
end if;
	OR
	W — EN ENO — W
	— R OUT —
	— K
	R, OUT (8 bit register)
	K (8 bit constant)
	EN (through W) = 0 or 1
	ENO (through W) = 0 or 1

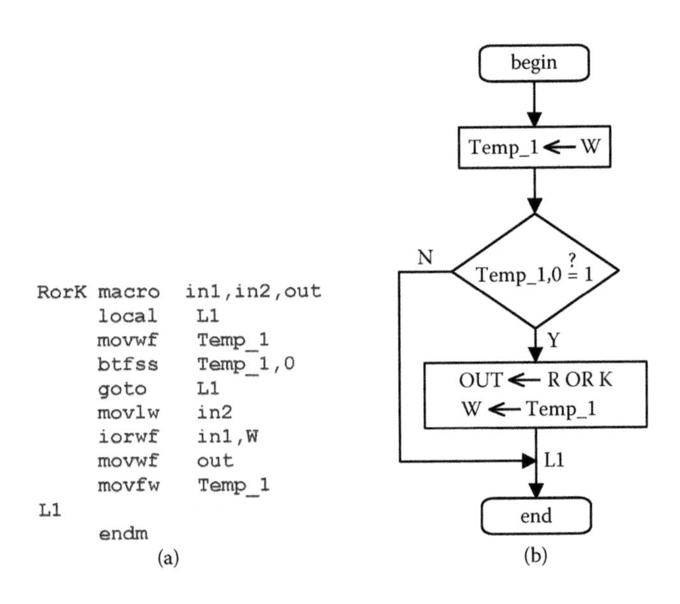

FIGURE 9.7
(a) The macro `RorK` and (b) its flowchart.

9.7 Macro `R1norR2`

The algorithm and the symbol of the macro `R1norR2` are depicted in Table 9.7. Figure 9.8 shows the macro `R1norR2` and its flowchart. In this macro, EN is a Boolean input variable taken into the macro through W, and ENO is a Boolean output variable sent out from the macro through W. Output ENO follows the input EN. This means that when EN = 0, ENO is forced to be 0, and when EN = 1, ENO is forced to be 1. R1 and R2 refer to 8-bit source variables from where the source values are taken into the macro, while OUT

TABLE 9.7

Algorithm and Symbol of the Macro `R1norR2`

Algorithm	Symbol

if EN = 1 then
OUT = R1 NOR R2;
ENO = 1;
else ENO = 0;
end if;
	NOR
	W — EN ENO — W
	— R1 OUT —
	— R2
	R1, R2, OUT (8 bit register)
	EN (through W) = 0 or 1
	ENO (through W) = 0 or 1

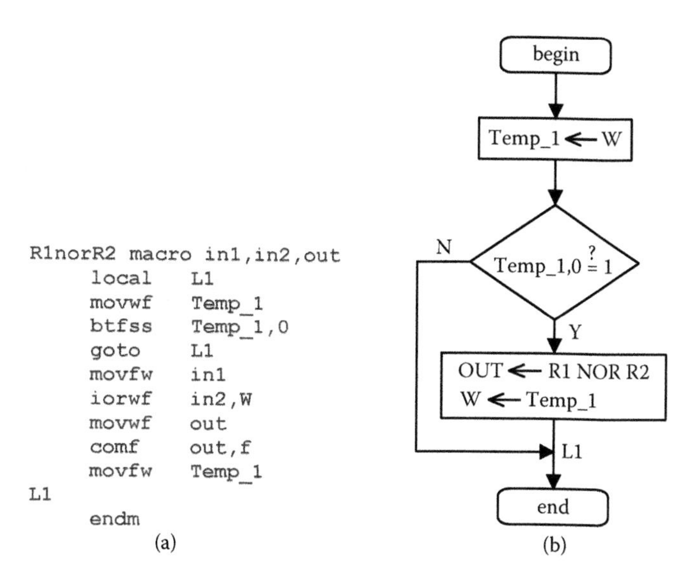

FIGURE 9.8
(a) The macro `R1norR2` and (b) its flowchart.

refers to an 8-bit destination variable to which the result of the macro is stored. When EN = 1, the macro `R1norR2` applies the logical NOR function to the two 8-bit input variables R1 and R2 and stores the result in the 8-bit output variable OUT (OUT = R1 NOR R2).

9.8 Macro `RnorK`

The algorithm and the symbol of the macro `RnorK` are depicted in Table 9.8. Figure 9.9 shows the macro `RnorK` and its flowchart. In this macro, EN is a Boolean input variable taken into the macro through W, and ENO is a

TABLE 9.8

Algorithm and Symbol of the Macro `RnorK`

Algorithm	Symbol

if EN = 1 then
OUT = R NOR K;
ENO = 1;
else ENO = 0;
end if;
	NOR
	W — EN ENO — W
	— R OUT —
	— K
	R, OUT (8 bit register)
	K (8 bit constant)
	EN (through W) = 0 or 1
	ENO (through W) = 0 or 1

Boolean output variable sent out from the macro through W. Output ENO follows the input EN. This means that when EN = 0, ENO is forced to be 0, and when EN = 1, ENO is forced to be 1. R and K are source values. R refers to an 8-bit source variable, while K represents an 8-bit constant value. OUT refers to an 8-bit destination variable to which the result of the macro is stored. When EN = 1, the macro `RnorK` applies the logical NOR function to the 8-bit input variable R and the 8-bit constant value K and stores the result in the 8-bit output variable (OUT = R NOR K).

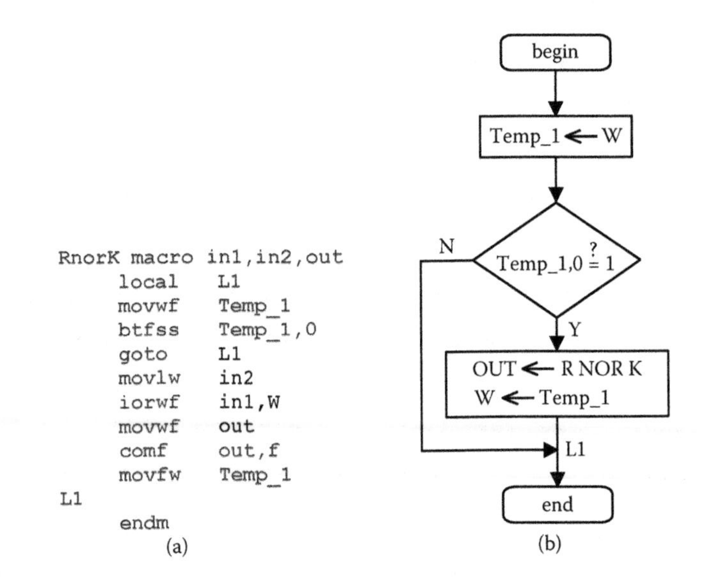
```
RnorK macro in1,in2,out
      local   L1
      movwf   Temp_1
      btfss   Temp_1,0
      goto    L1
      movlw   in2
      iorwf   in1,W
      movwf   out
      comf    out,f
      movfw   Temp_1
L1
      endm
```


FIGURE 9.9
(a) The macro `RnorK` and (b) its flowchart.

9.9 Macro R1xorR2

The algorithm and the symbol of the macro R1xorR2 are depicted in Table 9.9. Figure 9.10 shows the macro R1xorR2 and its flowchart. In this macro, EN is a Boolean input variable taken into the macro through W, and ENO is a Boolean output variable sent out from the macro through W. Output ENO follows the input EN. This means that when EN = 0, ENO is forced to be 0, and when EN = 1, ENO is forced to be 1. R1 and R2 refer to 8-bit source variables from where the source values are taken into the macro, while OUT refers to an 8-bit destination variable to which the result of the

TABLE 9.9

Algorithm and Symbol of the Macro R1xorR2

Algorithm	Symbol

if EN = 1 then
OUT = R1 EXOR R2;
ENO = 1;
else ENO = 0;
end if;
	XOR
	W — EN ENO — W
	— R1 OUT —
	— R2
	R1, R2, OUT (8 bit register)
	EN (through W) = 0 or 1
	ENO (through W) = 0 or 1

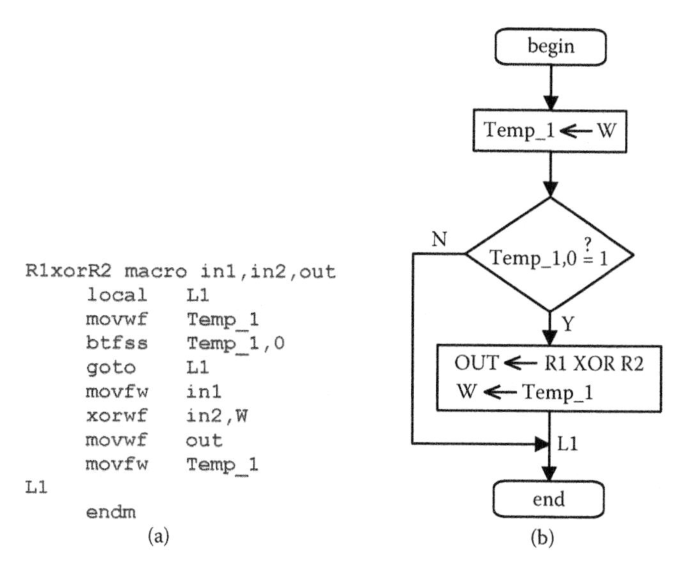

FIGURE 9.10
(a) The macro R1xorR2 and (b) its flowchart.

macro is stored. When EN = 1, the macro `R1xorR2` applies the logical EXOR function to the two 8-bit input variables R1 and R2 and stores the result in the 8-bit output variable OUT (OUT = R1 EXOR R2).

9.10 Macro `RxorK`

The algorithm and the symbol of the macro `RxorK` are depicted in Table 9.10. Figure 9.11 shows the macro `RxorK` and its flowchart. In this macro, EN is a Boolean input variable taken into the macro through W, and ENO is a

TABLE 9.10

Algorithm and Symbol of the Macro `RxorK`

Algorithm	Symbol

if EN = 1 then
OUT = R EXOR K;
ENO = 1;
else ENO = 0;
end if;
	XOR
	W — EN ENO — W
	— R OUT —
	— K
	R, OUT (8 bit register)
	K (8 bit constant)
	EN (through W) = 0 or 1
	ENO (through W) = 0 or 1

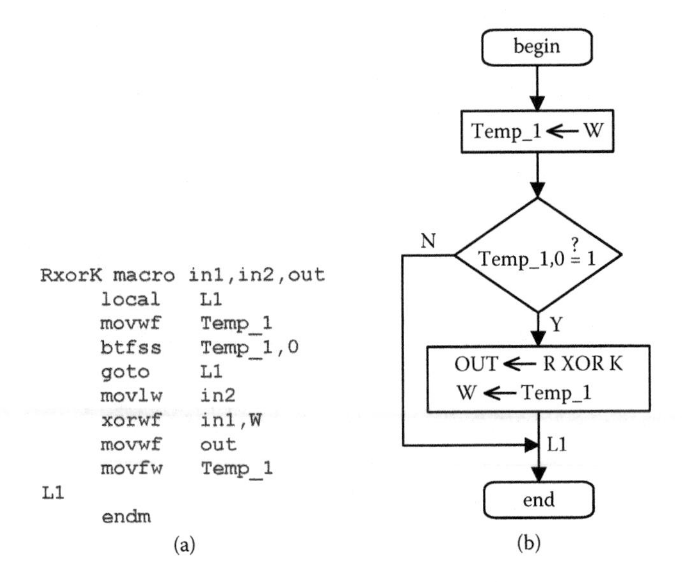

FIGURE 9.11
(a) The macro `RxorK` and (b) its flowchart.

TABLE 9.11

Algorithm and Symbol of the Macro `R1xnorR2`

Algorithm	Symbol

if EN = 1 then
OUT = R1 EXNOR R2;
ENO = 1;
else ENO = 0;
end if;
	XNOR
	W — EN ENO — W
	— R1 OUT —
	— R2
	R1, R2, OUT (8 bit register)
	EN (through W) = 0 or 1
	ENO (through W) = 0 or 1

Boolean output variable sent out from the macro through W. Output ENO follows the input EN. This means that when EN = 0, ENO is forced to be 0, and when EN = 1, ENO is forced to be 1. R and K are source values. R refers to an 8-bit source variable, while K represents an 8-bit constant value. OUT refers to an 8-bit destination variable to which the result of the macro is stored. When EN = 1, the macro `RxorK` applies the logical EXOR function to the 8-bit input variable R and the 8-bit constant value K and stores the result in the 8-bit output variable OUT (OUT = R EXOR K).

9.11 Macro `R1xnorR2`

The algorithm and the symbol of the macro `R1xnorR2` are depicted in Table 9.11. Figure 9.12 shows the macro `R1xnorR2` and its flowchart. In this macro, EN is a Boolean input variable taken into the macro through W, and ENO is a Boolean output variable sent out from the macro through W. Output ENO follows the input EN. This means that when EN = 0, ENO is forced to be 0, and when EN = 1, ENO is forced to be 1. R1 and R2 refer to 8-bit source variables from where the source values are taken into the macro, while OUT refers to an 8-bit destination variable to which the result of the macro is stored. When EN = 1, the macro `R1xnorR2` applies the logical EXNOR function to the two 8-bit input variables R1 and R2 and stores the result in the 8-bit output variable OUT (OUT = R1 EXNOR R2).

9.12 Macro `RxnorK`

The algorithm and the symbol of the macro `RxnorK` are depicted in Table 9.12. Figure 9.13 shows the macro `RxnorK` and its flowchart. In this macro, EN is a Boolean input variable taken into the macro through W,

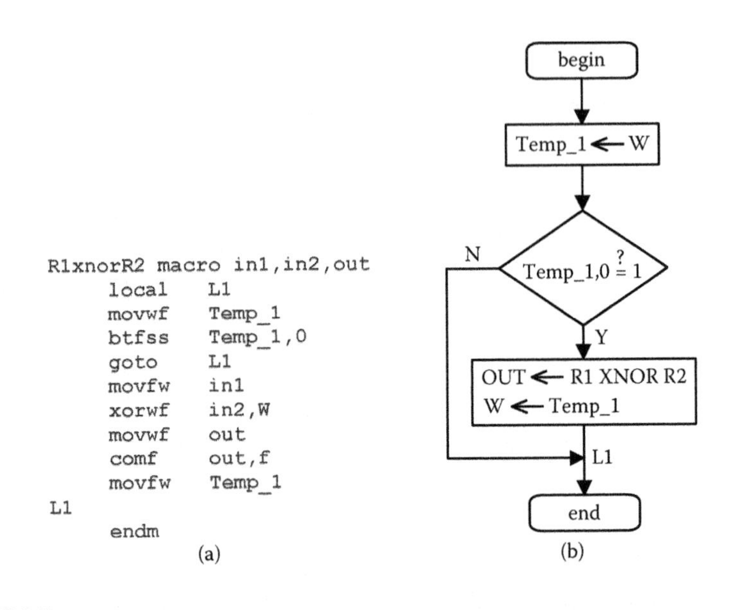

FIGURE 9.12
(a) The macro `R1xnorR2` and (b) its flowchart.

and ENO is a Boolean output variable sent out from the macro through W. Output ENO follows the input EN. This means that when EN = 0, ENO is forced to be 0, and when EN = 1, ENO is forced to be 1. R and K are source values. R refers to an 8-bit source variable, while K represents an 8-bit constant value. OUT refers to an 8-bit destination variable to which the result of the macro is stored. When EN = 1, the macro `RxnorK` applies the logical EXNOR function to the 8-bit input variable R and the 8-bit constant value K and stores the result in the 8-bit output variable OUT (OUT = R EXNOR K).

TABLE 9.12

Algorithm and Symbol of the Macro `RxnorK`

Algorithm	Symbol

if EN = 1 then
OUT = R EXNOR K;
ENO = 1;
else ENO = 0;
end if;
	XNOR
	W — EN ENO — W
	— R OUT —
	— K
	R, OUT (8 bit register)
	K (8 bit constant)
	EN (through W) = 0 or 1
	ENO (through W) = 0 or 1

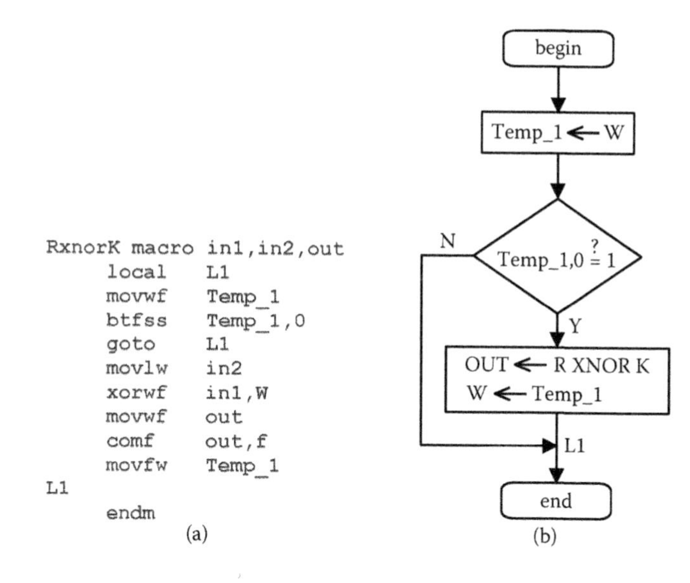

FIGURE 9.13
(a) The macro `RxnorK` and (b) its flowchart.

9.13 Macro `inv_R`

The algorithm and the symbol of the macro `inv_R` are depicted in Table 9.13. Figure 9.14 shows the macro `inv_R` and its flowchart. In this macro, EN is a Boolean input variable taken into the macro through W, and ENO is a Boolean output variable sent out from the macro through W. Output ENO follows the input EN. This means that when EN = 0, ENO is forced to be 0, and when EN = 1, ENO is forced to be 1. IN refers to an 8-bit source variable from where the source value is taken into the macro, while OUT refers to an 8-bit destination variable to which the result of the macro is stored. When EN = 1, the

TABLE 9.13
Algorithm and Symbol of the Macro `inv_R`

Algorithm	Symbol

if EN = 1 then
  OUT = invert IN;
  ENO = 1;
else ENO = 0;
end if;
	inv_R
	W — EN ENO — W
	— IN OUT —
	IN, OUT (8 bit register)
	EN (through W) = 0 or 1
	ENO (through W) = 0 or 1

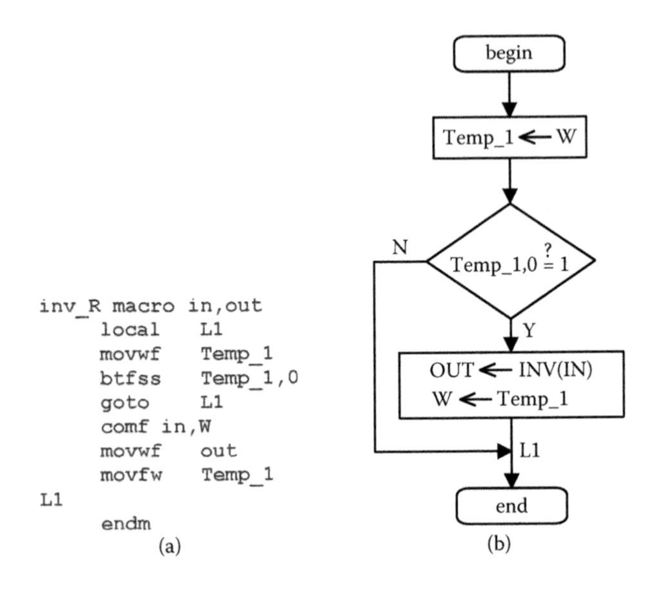

FIGURE 9.14
(a) The macro `inv_R` and (b) its flowchart.

macro `inv_R` inverts all of the bits in the 8-bit source register IN and stores the result in the 8-bit destination register OUT (OUT = invert IN).

9.14 Example for Logical Macros

In this section, we will consider an example, UZAM_plc_16i16o_ex16.asm, to show the usage of logical macros. In order to test the example, please take the file UZAM_plc_16i16o_ex16.asm from the CD-ROM attached to this book, and then open the program by MPLAB IDE and compile it. After that, by using the PIC programmer software, take the compiled file UZAM_plc_16i16o_ex16.hex, and by your PIC programmer hardware send it to the program memory of the PIC16F648A microcontroller within the PIC16F648A-based PLC. To do this, switch the 4PDT in PROG position and the power switch in OFF position. After loading the file UZAM_plc_16i16o_ex16.hex, switch the 4PDT in RUN and the power switch in ON position. Please check the program's accuracy by cross-referencing it with the related macros.

Let us now consider this example program: The example program, UZAM_plc_16i16o_ex16.asm, is shown in Figure 9.15. It shows the usage of all logical macros. The ladder diagram of the user program of UZAM_plc_16i16o_ex16.asm, shown in Figure 9.15, is depicted in Figure 9.16.

In the first rung, both Q1 and Q0 are cleared, i.e., 8-bit value 00h is loaded into both Q0 and Q1, by using the macro `load_R`. This process is carried out once at the first program scan by using the FRSTSCN NO contact.

```
.
;--------------- user program starts here --
        ld          FRSTSCN     ;rung 1
        load_R      00h,Q1
        load_R      00h,Q0


        ld          I0.0        ;rung 2
        and_not     I0.1
        and_not     I0.2
        and_not     I0.3
        load_R      03h,Q0


        ld          I0.1        ;rung 3
        and_not     I0.0
        and_not     I0.2
        and_not     I0.3
        load_R      05h,Q0

        ld          I0.2        ;rung 4
        and_not     I0.0
        and_not     I0.1
        and_not     I0.3
        load_R      0Fh,Q0

        ld          I0.3        ;rung 5
        and_not     I0.0
        and_not     I0.1
        and_not     I0.2
        load_R      0F0h,Q0

        ld_not      I0.7        ;rung 6
        and_not     I0.6
        and_not     I0.5
        and         I0.4
        out         M0.1

        ld_not      I0.7        ;rung 7
        and_not     I0.6
        and         I0.5
        and_not     I0.4
        out         M0.2

        ld_not      I0.7        ;rung 8
        and_not     I0.6
        and         I0.5
        and         I0.4
        out         M0.3

        ld_not      I0.7        ;rung 9
        and         I0.6
        and_not     I0.5
        and_not     I0.4
        out         M0.4

        ld_not      I0.7        ;rung 10
        and         I0.6
        and_not     I0.5
        and         I0.4
        out         M0.5
```

FIGURE 9.15
The user program of UZAM_plc_16i16o_ex16.asm. (*Continued*)

```
ld_not      I0.7        ;rung 11
and         I0.6
and         I0.5
and_not     I0.4
out         M0.6

ld_not      I0.7        ;rung 12
and         I0.6
and         I0.5
and         I0.4
out         M0.7

ld          I0.7        ;rung 13
and_not     I0.6
and_not     I0.5
and_not     I0.4
out         M1.0

ld          I0.7        ;rung 14
and_not     I0.6
and_not     I0.5
and         I0.4
out         M1.1

ld          I0.7        ;rung 15
and_not     I0.6
and         I0.5
and_not     I0.4
out         M1.2

ld          I0.7        ;rung 16
and_not     I0.6
and         I0.5
and         I0.4
out         M1.3

ld          I0.7        ;rung 17
and         I0.6
and_not     I0.5
and_not     I0.4
out         M1.4

ld          I0.7        ;rung 18
and         I0.6
and_not     I0.5
and         I0.4
out         M1.5

ld          I0.7        ;rung 19
and         I0.6
and         I0.5
and_not     I0.4
out         M1.6
```

FIGURE 9.15 (*Continued*)
The user program of UZAM_plc_16i16o_ex16.asm. (*Continued*)

```
        ld          M0.1        ;rung 20
        inv_R       I1,Q1

        ld          M0.2        ;rung 21
        R1andR2     I1,Q0,Q1

        ld          M0.3        ;rung 22
        R1andR2     I1,Q0,M3
        inv_R       M3,Q1

        ld          M0.4        ;rung 23
        RandK       I1,50h,Q1

        ld          M0.5        ;rung 24
        R1nandR2    I1,Q0,Q1

        ld          M0.6        ;rung 25
        RnandK      I1,50h,Q1

        ld          M0.7        ;rung 26
        R1orR2      I1,Q0,Q1

        ld          M1.0        ;rung 27
        RorK        I1,50h,Q1

        ld          M1.1        ;rung 28
        R1norR2     I1,Q0,Q1

        ld          M1.2        ;rung 29
        RnorK       I1,50h,Q1

        ld          M1.3        ;rung 30
        R1xorR2     I1,Q0,Q1

        ld          M1.4        ;rung 31
        RxorK       I1,50h,Q1

        ld          M1.5        ;rung 32
        R1xnorR2    I1,Q0,Q1

        ld          M1.6        ;rung 33
        RxnorK      I1,50h,Q1
;--------------- user program ends here --
```

FIGURE 9.15 (*Continued*)
The user program of UZAM_plc_16i16o_ex16.asm.

In each rung between 2 and 5, an 8-bit value, namely, 03h, 05h, 0Fh, and F0h, is loaded into Q0 based on the inputs I0.3, I0.2, I0.1, and I0.0, by using the macro `load_R`, as shown in Table 9.14. If I0.3,I0.2,I0.1,I0.0 = 0001 (0010, 0100, and 1000, respectively), then Q0 = 03h (05h, 0Fh, and F0h, respectively).

In the 14 rungs between 6 and 19, a 4-to-16 decoder is implemented, whose inputs are I0.7, I0.6, I0.5, and I0.4, and whose outputs are M0.1, M0.2, …, M0.7, M1.0, M1.1, …, M1.6. Note that only 14 combinations are utilized,

TABLE 9.14

Selection of 8-Bit Values to Be Deposited in Q0 Based on the Inputs I0.0, I0.1, I0.2, and I0.3

I0.0	I0.1	I0.2	I0.3	8-Bit Value Selected to Be Deposited in Q0
1	0	0	0	Q0 = 03h (0 0 0 0 0 0 1 1)
0	1	0	0	Q0 = 05h (0 0 0 0 0 1 0 1)
0	0	1	0	Q0 = 0Fh (0 0 0 0 1 1 1 1)
0	0	0	1	Q0 = F0h (1 1 1 1 0 0 0 0)

while the following combinations for inputs (I0.7, I0.6, I0.5, I0.4), 0000 and 1111, are not implemented. Therefore, for these combinations of the inputs I0.7, I0.6, I0.5, and I0.4, the program will not produce any output. This arrangement is made to choose 14 different markers based on the input data given through the inputs I0.7, I0.6, I0.5, and I0.4. Table 9.15 shows the

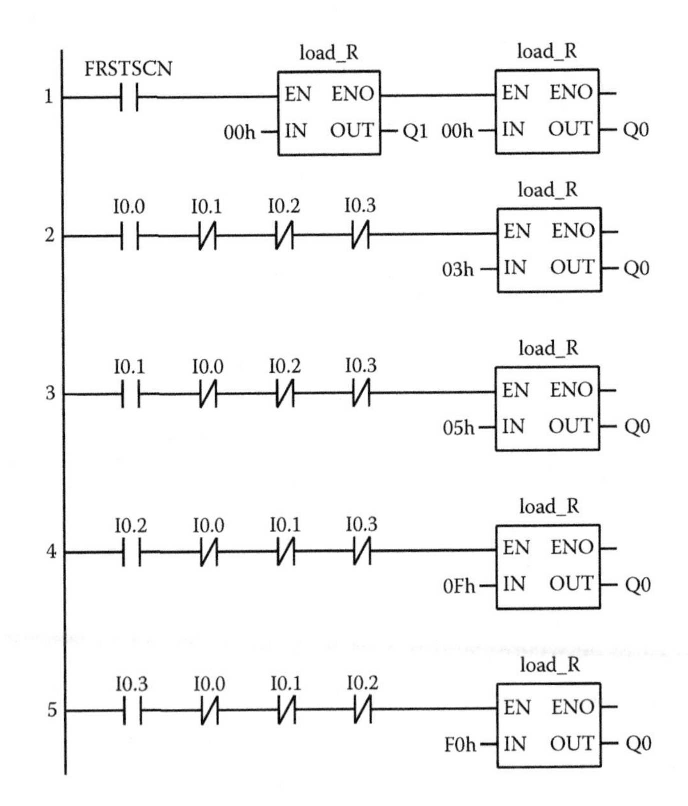

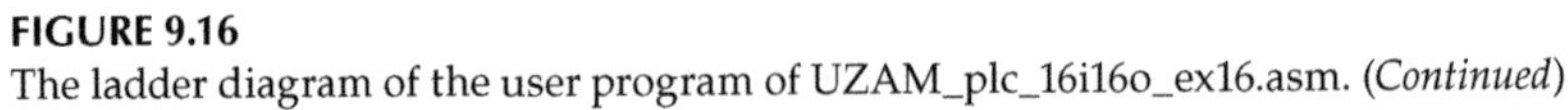

FIGURE 9.16
The ladder diagram of the user program of UZAM_plc_16i16o_ex16.asm. (*Continued*)

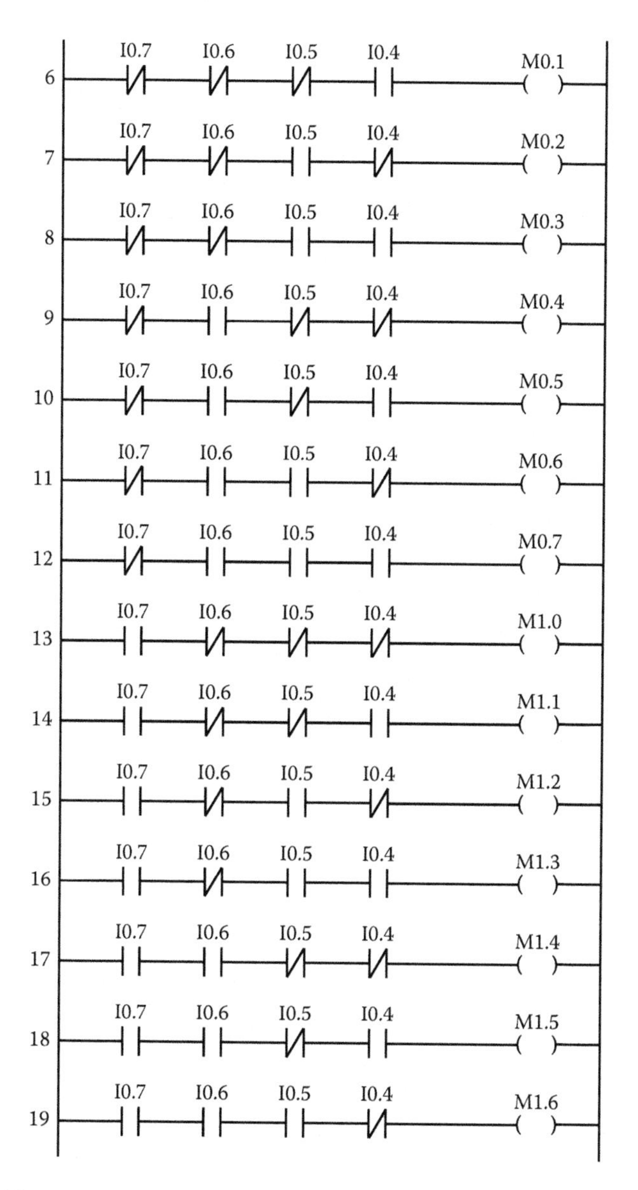

FIGURE 9.16 *(Continued)*
The ladder diagram of the user program of UZAM_plc_16i16o_ex16.asm. *(Continued)*

truth table based on the input data entered through I0.7, I0.6, I0.5, and I0.4, and the 14 markers chosen.

In the 14 PLC rungs between 20 and 33, we define different logical operations according to the decoder outputs represented by the marker bits M0.1, M0.2, …, M0.7, M1.0, M1.1, …, M1.6. In each of these 14 rungs, a logical process

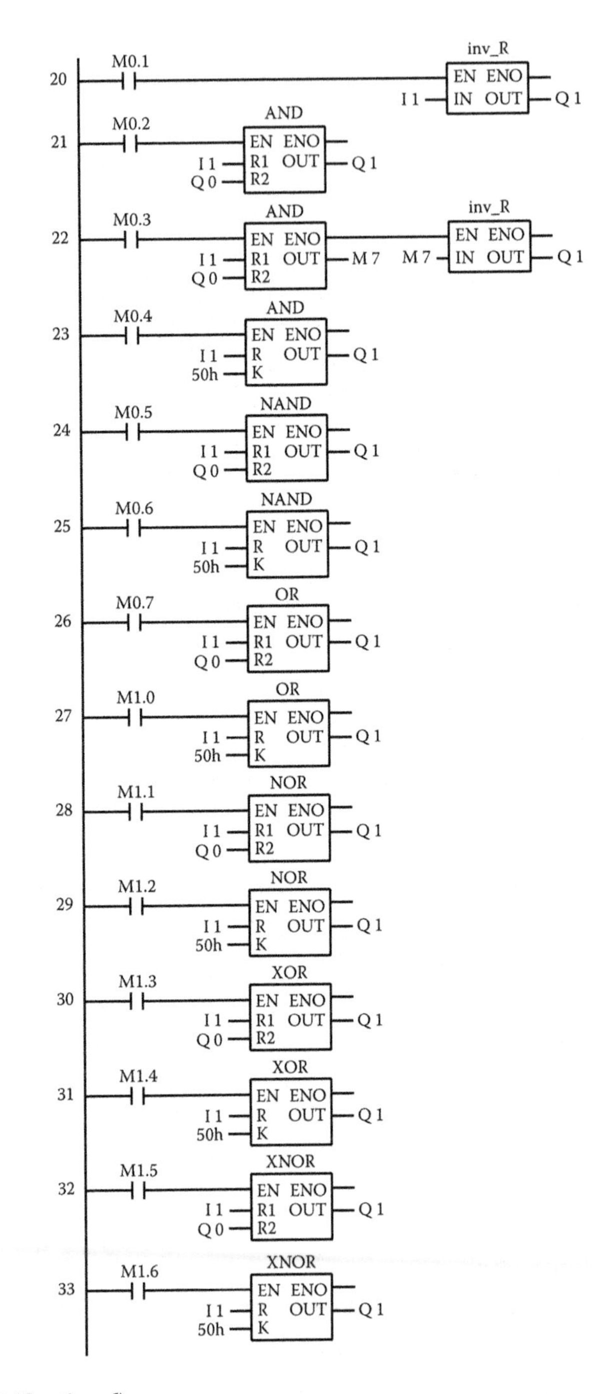

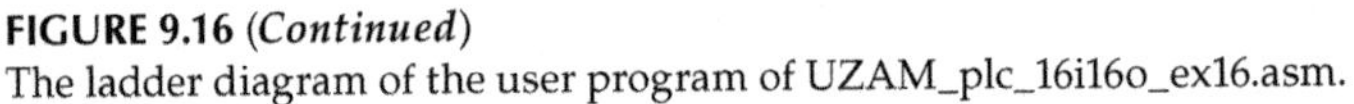

FIGURE 9.16 *(Continued)*
The ladder diagram of the user program of UZAM_plc_16i16o_ex16.asm.

TABLE 9.15

Selection of Markers Based on the Inputs I0.7, I0.6, I0.5, and I0.4

I0.7	I0.6	I0.5	I0.4	Marker
0	0	0	1	M0.1
0	0	1	0	M0.2
0	0	1	1	M0.3
0	1	0	0	M0.4
0	1	0	1	M0.5
0	1	1	0	M0.6
0	1	1	1	M0.7
1	0	0	0	M1.0
1	0	0	1	M1.1
1	0	1	0	M1.2
1	0	1	1	M1.3
1	1	0	0	M1.4
1	1	0	1	M1.5
1	1	1	0	M1.6

is carried out, as shown in Table 9.16. For example, if M0.7 = 1, then the following operation is done: Q1 = I1 OR Q0. This means that the macro `R1orR2` applies the logical OR function to the two 8-bit input variables I1 and Q0 and stores the result to the 8-bit output variable Q1. It should be obvious that since only one of the markers (M0.1, M0.2, …, M0.7, M1.0, M1.1, …, M1.6) is active at any time, only one of the processes shown in Table 9.16 can be carried out at a time.

TABLE 9.16

Selection of Logical Processes Based on Markers

Marker	Logical Process Selected
M0.1	Q1 = **INV** I1
M0.2	Q1 = I1 **AND** Q0
M0.3	Q1 = I1 **NAND** Q0 = **INV** M7 (M7 = I1 **AND** Q0)
M0.4	Q1 = I1 **AND** 50h
M0.5	Q1 = I1 **NAND** Q0
M0.6	Q1 = I1 **NAND** 50h
M0.7	Q1 = I1 **OR** Q0
M1.0	Q1 = I1 **OR** 50h
M1.1	Q1 = I1 **NOR** Q0
M1.2	Q1 = I1 **NOR** 50h
M1.3	Q1 = I1 **XOR** Q0
M1.4	Q1 = I1 **XOR** 50h
M1.5	Q1 = I1 **XNOR** Q0
M1.6	Q1 = I1 **XNOR** 50h

10

Shift and Rotate Macros

A *shift* (SHIFT) function moves the bits in a register to the right or to the left. As an example, Figure 10.1 shows a *shift right* function that retrieves the input data from the *source register A* and shifts the bits of the *source register A* toward the right as many numbers as specified by the *number of shift*, while the serial data are taken from the left through the Boolean input variable `shift in bit`. The result of the shift operation is stored in a *destination register B*. In this case, the least significant bit (LSB) is shifted out as many numbers as specified by the number of shift. A *shift left* function is identical, except that the `shift in bit`, taken from the right, is moved in the opposite direction toward left, shifting out the most significant bit (MSB) as many numbers as specified by the number of shift. A *rotate* (ROTATE) function, like a shift function, shifts data to the right or left, but instead of losing the `shift out bit`, this bit becomes the `shift in bit` at the other end of the register (rotated bit). The *number of rotation* defines how many bits will be rotated to the right or left. Similar to the shift function, the result of the rotate operation is stored in the destination register B.

In this chapter, the following shift and rotate macros are described for the PIC16F648A-based PLC:

```
shift_R
shift_L
rotate_R
rotate_L
Swap
```

The file definitions.inc, included within the CD-ROM attached to this book, contains all shift and rotate macros defined for the PIC16F648A-based PLC. Let us now consider these macros in detail.

10.1 Macro `shift_R`

The algorithm and the symbol of the macro `shift_R` are depicted in Table 10.1. Figure 10.2 shows the macro `shift_R` and its flowchart. In this macro, EN is a Boolean input variable taken into the macro through W, and

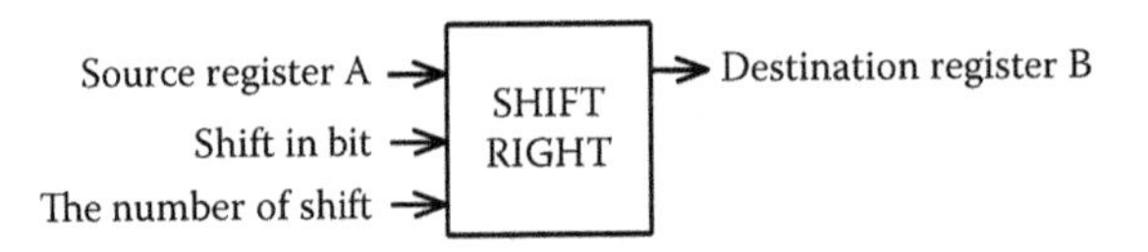

FIGURE 10.1
The shift right function.

ENO is a Boolean output variable sent out from the macro through W. Output ENO follows the input EN. This means that when EN = 0, ENO is forced to be 0, and when EN = 1, ENO is forced to be 1. This is especially useful if we want to carry out more than one operation based on a single input condition. RIN refers to an 8-bit source variable from where the source value is taken into the macro, while ROUT refers to an 8-bit destination variable to which the result of the macro is stored. N represents the number of shift, which can be any number in 1, 2, …, 8. SIN is the Boolean input variable `shift in bit`. When EN = 1, the macro `shift_R` retrieves the 8-bit input data from RIN and shifts the bits of RIN toward right as many numbers as specified by N, while the serial data are taken from left through SIN. The result of the shift right operation is stored in the 8-bit output register ROUT.

10.2 Macro `shift_L`

The algorithm and the symbol of the macro `shift_L` are depicted in Table 10.2. Figure 10.3 shows the macro `shift_L` and its flowchart. In this macro, EN is a Boolean input variable taken into the macro through W,

TABLE 10.1

Algorithm and Symbol of the Macro `shift_R`

Algorithm	Symbol

if EN = 1 then
ROUT = N times shift right(RIN)
and take the serial `data_in` from SIN;
ENO = 1;
else ENO = 0;
end if;
	SHIFT_R
	W — EN ENO — W
	— SIN
	— RIN ROUT —
	— N
	RIN, ROUT (8 bit register)
	SIN (reg,bit) = 0 or 1
	N (number of shift) = 1,2, ..., 8
	EN (through W) = 0 or 1
	ENO (through W) = 0 or 1

```
shift_R macro n,reg,bit,Rin,Rout
        local   L1,L2
        movwf   Temp_1
        btfss   Temp_1,0
        goto    L1
        movlw   n
        xorlw   00h
        skpnz
        goto    L1
        movlw   .9
        sublw   n
        skpnc
        goto    L1
        movfw   Rin
        movwf   Rout
        movlw   n
        movwf   Temp_1
L2      bcf     STATUS,C
        btfsc   reg,bit
        bsf     STATUS,C
        rrf     Rout,f
        decfsz  Temp_1,f
        goto    L2
        bsf     Temp_1,0
        movfw   Temp_1
L1
        endm
```

(a)

FIGURE 10.2
(a) The macro `shift_R` and (b) its flowchart. (*Continued*)

and ENO is a Boolean output variable sent out from the macro through W. Output ENO follows the input EN. This means that when EN = 0, ENO is forced to be 0, and when EN = 1, ENO is forced to be 1. RIN refers to an 8-bit source variable from where the source value is taken into the macro, while ROUT refers to an 8-bit destination variable to which the result of the macro is stored. N represents the number of shift, which can be any number in 1, 2, …, 8. SIN is the Boolean input variable `shift in bit`. When EN = 1, the macro `shift_L` retrieves the 8-bit input data from RIN and shifts the bits of RIN toward left as many numbers as specified by N, while the serial data are taken from right through SIN. The result of the shift left operation is stored in the 8-bit output register ROUT.

10.3 Macro `rotate_R`

The algorithm and the symbol of the macro `rotate_R` are depicted in Table 10.3. Figure 10.4 shows the macro `rotate_R` and its flowchart. In this macro, EN is a Boolean input variable taken into the macro through W,

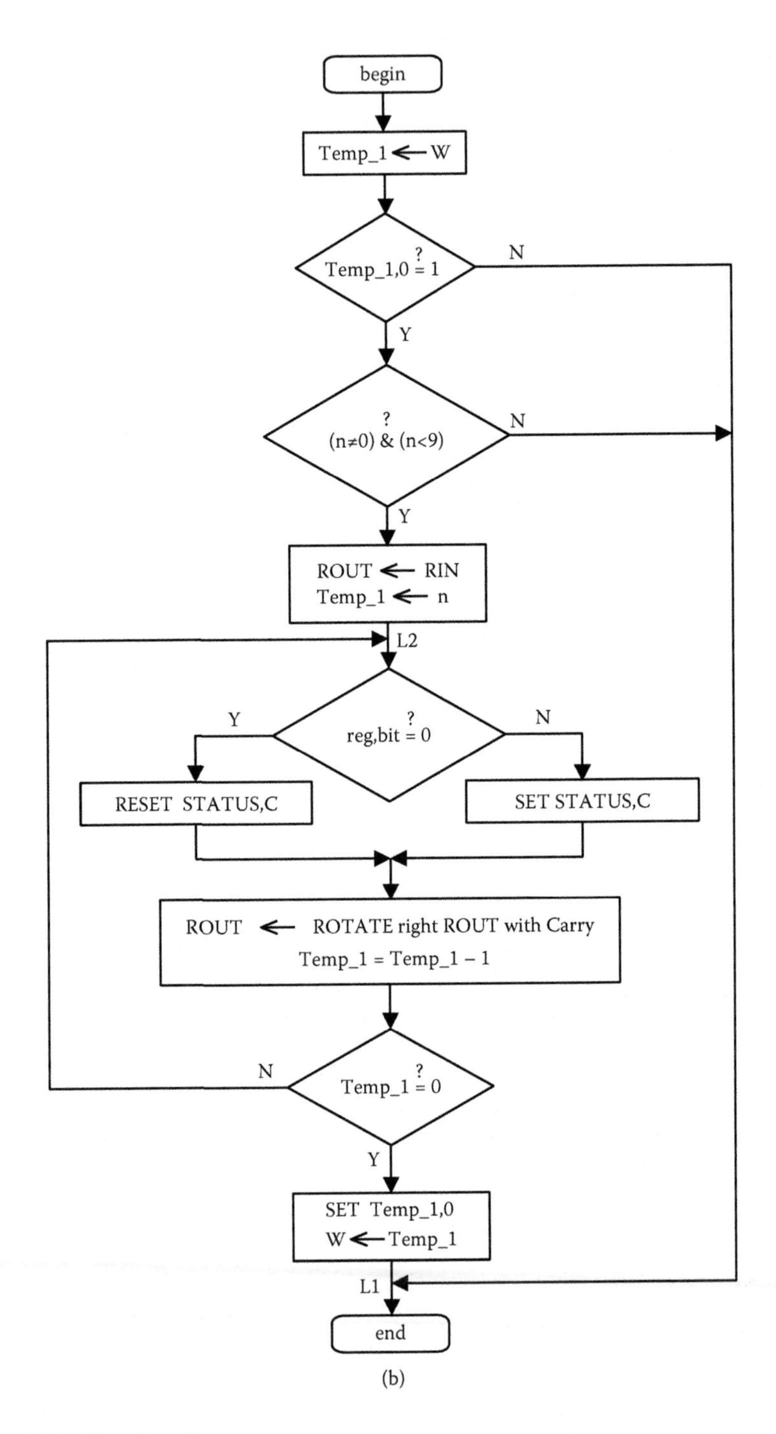

FIGURE 10.2 (*Continued*)
(a) The macro `shift_R` and (b) its flowchart.

TABLE 10.2

Algorithm and Symbol of the Macro `shift_L`

Algorithm	Symbol

if EN = 1 then
ROUT = N times shift left(RIN)
and take the serial `data_in` from SIN;
ENO = 1;
else ENO = 0;
end if;
	SHIFT_L
	W — EN ENO — W
	— SIN
	— RIN ROUT —
	— N
	RIN, ROUT (8 bit register)
	SIN (reg,bit) = 0 or 1
	N (number of shift) = 1,2, ..., 8
	EN (through W) = 0 or 1
	EN0 (through W) = 0 or 1

and ENO is a Boolean output variable sent out from the macro through W. Output ENO follows the input EN. This means that when EN = 0, ENO is forced to be 0, and when EN = 1, ENO is forced to be 1. RIN refers to an 8-bit source variable from where the source value is taken into the macro, while ROUT refers to an 8-bit destination variable to which the result of the macro

```
shift_L macro n,reg,bit,Rin,Rout
     local   L1,L2
     movwf   Temp_1
     btfss   Temp_1,0
     goto    L1
     movlw   n
     xorlw   00h
     skpnz
     goto    L1
     movlw   .9
     sublw   n
     skpnc
     goto    L1
     movfw   Rin
     movwf   Rout
     movlw   n
     movwf   Temp_1
L2   bcf     STATUS,C
     btfsc   reg,bit
     bsf     STATUS,C
     rlf     Rout,f
     decfsz  Temp_1,f
     goto    L2
     bsf     Temp_1,0
     movfw   Temp_1
L1
     endm
```

(a)

FIGURE 10.3
(a) The macro `shift_L` and (b) its flowchart. (*Continued*)

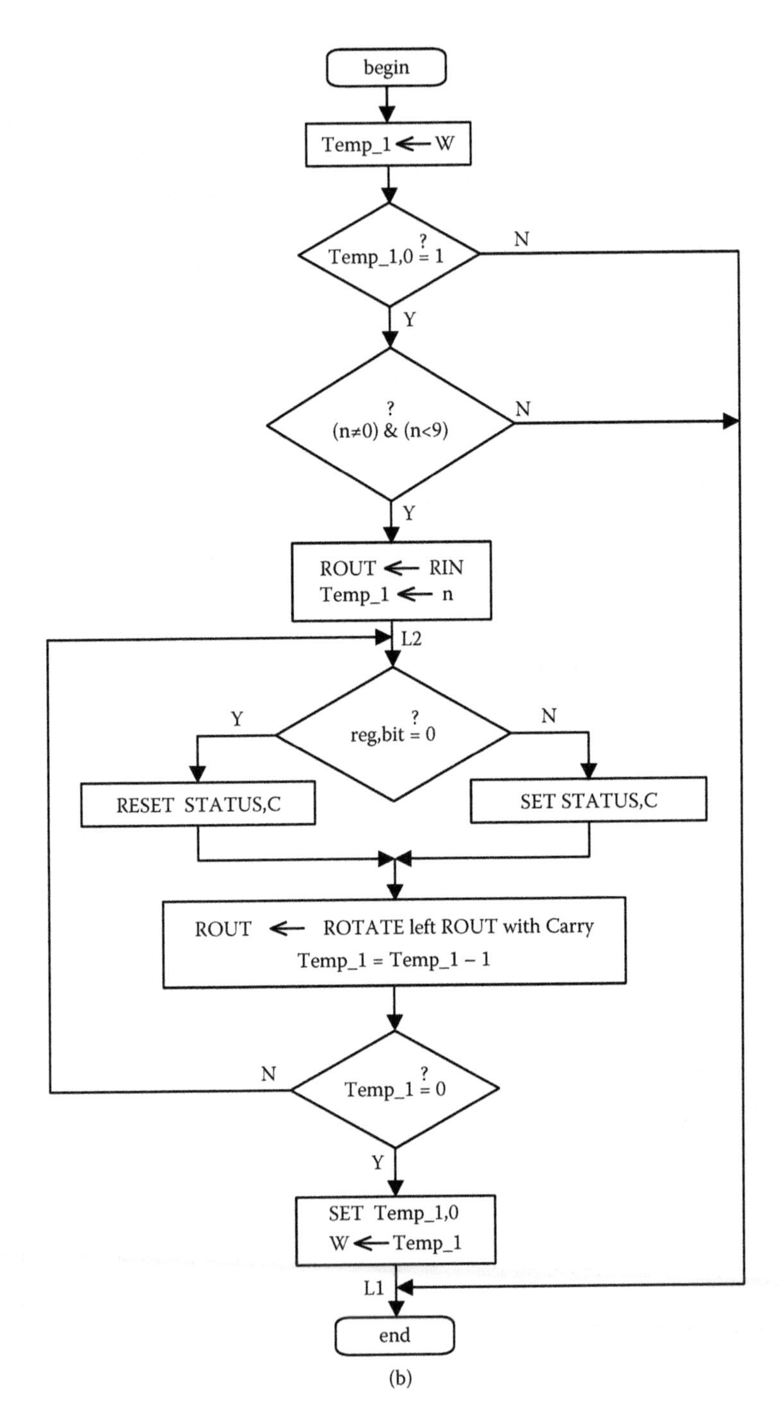

FIGURE 10.3 (*Continued*)
(a) The macro `shift_L` and (b) its flowchart.

TABLE 10.3

Algorithm and Symbol of the Macro `rotate_R`

Algorithm	Symbol

if EN = 1 then
ROUT = N times rotate right(RIN);
ENO = 1;
else ENO = 0;
end if;
	ROTATE_R
	W — EN ENO — W
	— RIN ROUT —
	— N
	RIN, ROUT (8 bit register)
	N (number of rotation) = 1,2, ..., 7
	EN (through W) = 0 or 1
	EN0 (through W) = 0 or 1

is stored. N represents the number of rotation, which can be any number in 1, 2, …, 7. When EN = 1, the macro `rotate_R` retrieves the 8-bit input data from RIN and rotates the bits of RIN toward right as many numbers as specified by N. The result of the rotate right operation is stored in the 8-bit output register ROUT.

```
rotate_R  macro n,Rin,Rout
          local    L1,L2
          movwf    Temp_1
          btfss    Temp_1,0
          goto     L1
          movlw    n
          xorlw    00h
          skpnz
          goto     L1
          movlw    .8
          sublw    n
          skpnc
          goto     L1
          movfw    Rin
          movwf    Rout
          movlw    n
          movwf    Temp_1
L2        bcf      STATUS,C
          btfsc    Rout,0
          bsf      STATUS,C
          rrf      Rout,f
          decfsz   Temp_1,f
          goto     L2
          bsf      Temp_1,0
          movfw    Temp_1
L1
          endm
```

(a)

FIGURE 10.4
(a) The macro `rotate_R` and (b) its flowchart. (*Continued*)

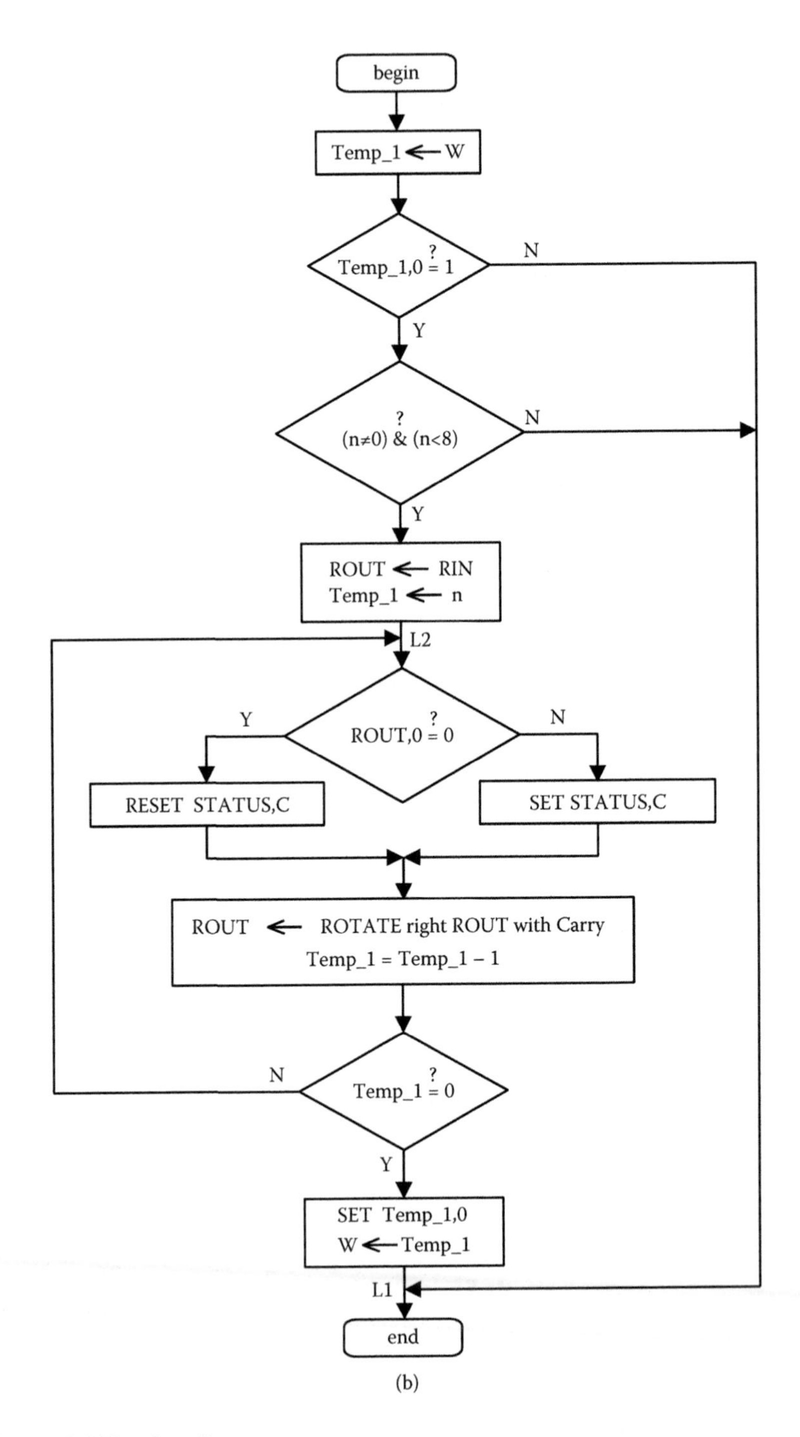

FIGURE 10.4 (*Continued*)

(a) The macro `rotate_R` and (b) its flowchart.

TABLE 10.4

Algorithm and Symbol of the Macro `rotate_L`

Algorithm	Symbol

if EN = 1 then
ROUT = N times rotate left(RIN);
ENO = 1;
else ENO = 0;
end if;
	ROTATE_L
	W — EN ENO — W
	— RIN ROUT —
	— N
	RIN, ROUT (8 bit register)
	N (number of rotation) = 1,2, ..., 7
	EN (through W) = 0 or 1
	ENO (through W) = 0 or 1

10.4 Macro `rotate_L`

The algorithm and the symbol of the macro `rotate_L` are depicted in Table 10.4. Figure 10.5 shows the macro `rotate_L` and its flowchart. In this macro, EN is a Boolean input variable taken into the macro through W, and ENO is a Boolean output variable sent out from the macro through W. Output ENO follows the input EN. This means that when EN = 0, ENO is forced to

```
rotate_L macro n,Rin,Rout
     local   L1,L2
     movwf   Temp_1
     btfss   Temp_1,0
     goto    L1
     movlw   n
     xorlw   00h
     skpnz
     goto    L1
     movlw   .8
     sublw   n
     skpnc
     goto    L1
     movfw   Rin
     movwf   Rout
     movlw   n
     movwf   Temp_1
L2   bcf     STATUS,C
     btfsc   Rout,7
     bsf     STATUS,C
     rlf     Rout,f
     decfsz  Temp_1,f
     goto    L2
     bsf     Temp_1,0
     movfw   Temp_1
L1
     endm
```

(a)

FIGURE 10.5
(a) The macro `rotate_L` and (b) its flowchart. (*Continued*)

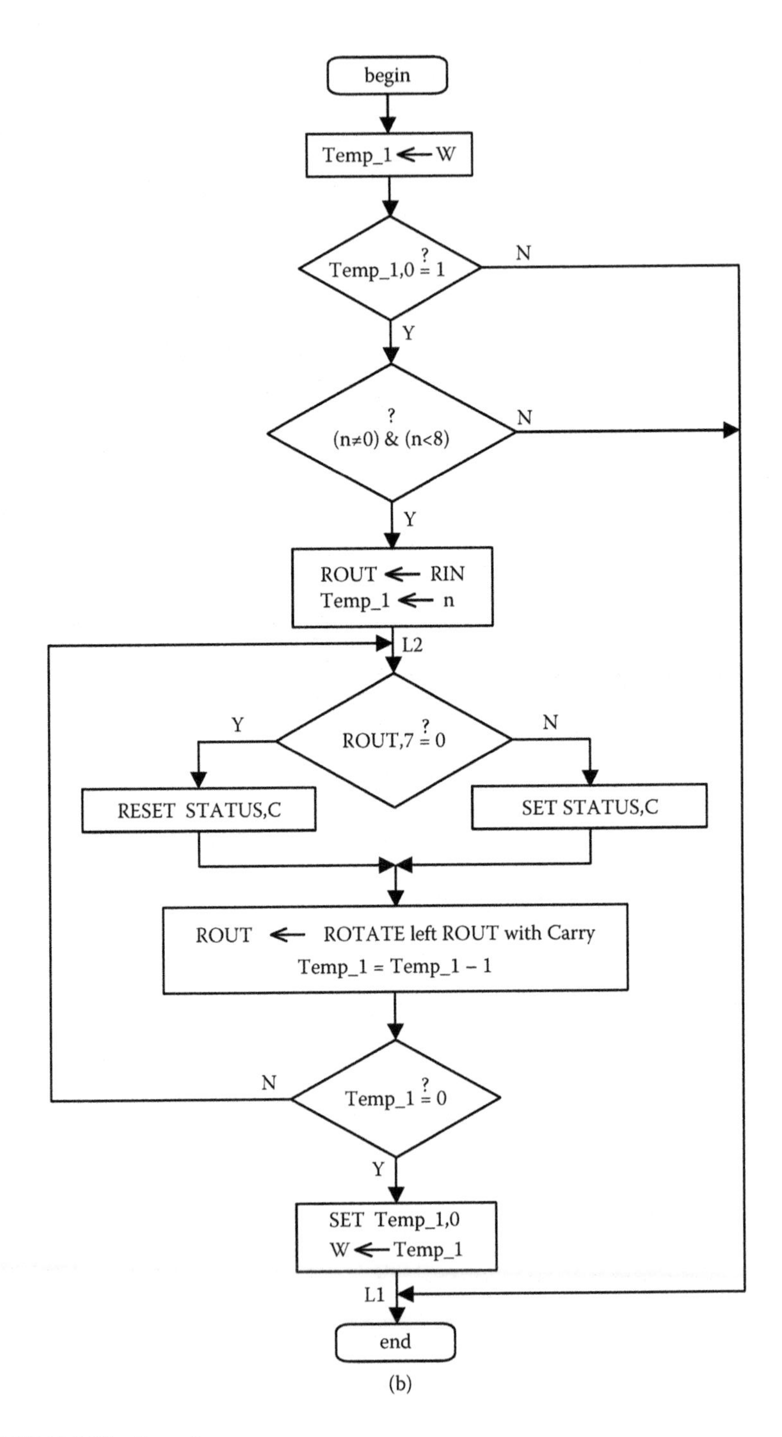

(b)

FIGURE 10.5 (*Continued*)
(a) The macro `rotate_L` and (b) its flowchart.

TABLE 10.5

Algorithm and Symbol of the Macro `Swap`

Algorithm	Symbol

if EN = 1 then
OUT = SWAP(IN);
ENO = 1;
else ENO = 0;
end if;
	SWAP
	W — EN ENO — W
	— IN OUT —
	IN, OUT (8 bit register)
	EN (through W) = 0 or 1
	ENO (through W) = 0 or 1

be 0, and when EN = 1, ENO is forced to be 1. RIN refers to an 8-bit source variable from where the source value is taken into the macro, while ROUT refers to an 8-bit destination variable to which the result of the macro is stored. N represents the number of rotation, which can be any number in 1, 2, …, 7. When EN = 1, the macro `rotate_L` retrieves the 8-bit input data from RIN and rotates the bits of RIN toward left as many numbers as specified by N. The result of the rotate left operation is stored in the 8-bit output register ROUT.

10.5 Macro `Swap`

The algorithm and the symbol of the macro `Swap` are depicted in Table 10.5. Figure 10.6 shows the macro `Swap` and its flowchart. In this macro, EN is a Boolean input variable taken into the macro through W,

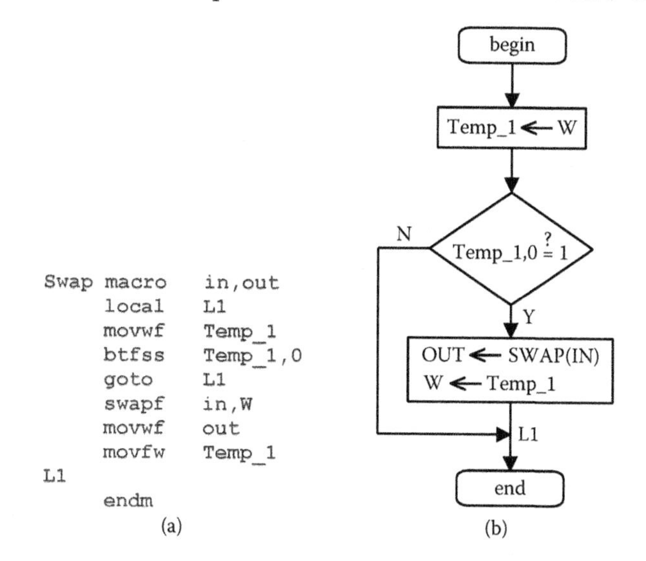

FIGURE 10.6
(a) The macro `Swap` and (b) its flowchart.

and ENO is a Boolean output variable sent out from the macro through W. Output ENO follows the input EN. This means that when EN = 0, ENO is forced to be 0, and when EN = 1, ENO is forced to be 1. IN refers to an 8-bit source variable from where the source value is taken into the macro, while OUT refers to an 8-bit destination variable to which the result of the macro is stored. When EN = 1, the macro `Swap` retrieves the 8-bit input data from IN and swaps (exchanges the upper and lower nibbles—4 bits) the nibbles of IN. The result of the swap operation is stored in the 8-bit output register OUT.

10.6 Examples for Shift and Rotate Macros

In this section, we will consider two examples, UZAM_plc_16i16o_ex17.asm and UZAM_plc_16i16o_ex18.asm, to show the usage of shift and rotate macros. In order to test one of these examples, please take the related file UZAM_plc_16i16o_ex17.asm or UZAM_plc_16i16o_ex18.asm from the CD-ROM attached to this book, and then open the program by MPLAB IDE and compile it. After that, by using the PIC programmer software, take the compiled file UZAM_plc_16i16o_ex17.hex or UZAM_plc_16i16o_ex18.hex, and by your PIC programmer hardware, send it to the program memory of PIC16F648A microcontroller within the PIC16F648A-based PLC. To do this, switch the 4PDT in PROG position and the power switch in OFF position. After loading the file UZAM_plc_16i16o_ex17.hex or UZAM_plc_16i16o_ex18.hex, switch the 4PDT in RUN and the power switch in ON position. Please check the program's accuracy by cross-referencing it with the related macros. When studying these two examples, note that the register Q0 (respectively, Q1, I0, and I1) is made up of 8 bits: Q0.7, Q0.6, …, Q0.0 (respectively, Q1.7, Q1.6, …, I1.0; I0.7, I0.6, …, I0.0; and I1.7, I1.6, …, I1.0), and that Q0.7 (respectively, Q1.7, I0.7, and I1.7) is the most significant bit (MSB), while Q0.0 (respectively, Q1.0, I0.0, and I1.0) is the least significant bit (LSB).

Let us now consider these example programs: The first example program, UZAM_plc_16i16o_ex17.asm, is shown in Figure 10.7. It shows the usage of two shift macros `shift_R` and `shift_L`. The ladder diagram of the user program of UZAM_plc_16i16o_ex17.asm, shown in Figure 10.7, is depicted in Figure 10.8.

In the first rung, 8-bit numerical data 3Ch are loaded to Q1, by using the macro `load_R`. This process is carried out once at the first program scan by using the FRSTSCN NO contact.

In the eight rungs between 2 and 9, a 3-to-8 decoder is implemented, whose inputs are I0.2, I0.1, and I0.0, and whose outputs are M0.0, M0.1, …, M0.7. This arrangement is made to choose the number of shift for the selected shift right or shift left operation based on the input data given through the input bits

```
.
;--------------- user program starts here --
        ld          FRSTSCN      ;rung 1
        load_R      3Ch,Q1

        ld_not      I0.2         ;rung 2
        and_not     I0.1
        and_not     I0.0
        out         M0.0

        ld_not      I0.2         ;rung 3
        and_not     I0.1
        and         I0.0
        out         M0.1

        ld_not      I0.2         ;rung 4
        and         I0.1
        and_not     I0.0
        out         M0.2

        ld_not      I0.2         ;rung 5
        and         I0.1
        and         I0.0
        out         M0.3

        ld          I0.2         ;rung 6
        and_not     I0.1
        and_not     I0.0
        out         M0.4

        ld          I0.2         ;rung 7
        and_not     I0.1
        and         I0.0
        out         M0.5

        ld          I0.2         ;rung 8
        and         I0.1
        and_not     I0.0
        out         M0.6

        ld          I0.2         ;rung 9
        and         I0.1
        and         I0.0
        out         M0.7
```

FIGURE 10.7
The user program of UZAM_plc_16i16o_ex17.asm. (*Continued*)

I0.2, I0.1, and I0.0. When these bits are 001, 010, 100, 100, 101, 110, 111, and 000, we define the number of shift for the selected shift right or shift left operation as 1, 2, 3, 4, 5, 6, 7, and 8 respectively.

In the eight rungs between 10 and 17, we define eight different shift right operations according to the 3-to-8 decoder outputs represented by the marker bits M0.0, M0.1, …, M0.7. Shift right operations defined in these rungs are applied to the 8-bit input variable Q1. The result of the shift right operations defined in these rungs will be stored in Q0. The `shift in bit` for these shift right operations defined in these rungs is I1.7. The only difference

```
ld          I0.3            ;rung 10
and_not     I0.4
and         M0.1
r_edge      0
shift_R     1,I1.7,Q1,Q0

ld          I0.3            ;rung 11
and_not     I0.4
and         M0.2
r_edge      0
shift_R     2,I1.7,Q1,Q0

ld          I0.3            ;rung 12
and_not     I0.4
and         M0.3
r_edge      0
shift_R     3,I1.7,Q1,Q0

ld          I0.3            ;rung 13
and_not     I0.4
and         M0.4
r_edge      0
shift_R     4,I1.7,Q1,Q0

ld          I0.3            ;rung 14
and_not     I0.4
and         M0.5
r_edge      0
shift_R     5,I1.7,Q1,Q0

ld          I0.3            ;rung 15
and_not     I0.4
and         M0.6
r_edge      0
shift_R     6,I1.7,Q1,Q0

ld          I0.3            ;rung 16
and_not     I0.4
and         M0.7
r_edge      0
shift_R     7,I1.7,Q1,Q0

ld          I0.3            ;rung 17
and_not     I0.4
and         M0.0
r_edge      0
shift_R     8,I1.7,Q1,Q0
```

FIGURE 10.7 (*Continued*)
The user program of UZAM_plc_16i16o_ex17.asm. (*Continued*)

```
        ld          I0.4        ;rung 18
        and_not     I0.3
        and         M0.1
        r_edge      0
        shift_L     1,I1.0,Q1,Q0

        ld          I0.4        ;rung 19
        and_not     I0.3
        and         M0.2
        r_edge      0
        shift_L     2,I1.0,Q1,Q0

        ld          I0.4        ;rung 20
        and_not     I0.3
        and         M0.3
        r_edge      0
        shift_L     3,I1.0,Q1,Q0

        ld          I0.4        ;rung 21
        and_not     I0.3
        and         M0.4
        r_edge      0
        shift_L     4,I1.0,Q1,Q0

        ld          I0.4        ;rung 22
        and_not     I0.3
        and         M0.5
        r_edge      0
        shift_L     5,I1.0,Q1,Q0

        ld          I0.4        ;rung 23
        and_not     I0.3
        and         M0.6
        r_edge      0
        shift_L     6,I1.0,Q1,Q0

        ld          I0.4        ;rung 24
        and_not     I0.3
        and         M0.7
        r_edge      0
        shift_L     7,I1.0,Q1,Q0

        ld          I0.4        ;rung 25
        and_not     I0.3
        and         M0.0
        r_edge      0
        shift_L     8,I1.0,Q1,Q0
;--------------- user program ends here --
;
```

FIGURE 10.7 (*Continued*)
The user program of UZAM_plc_16i16o_ex17.asm.

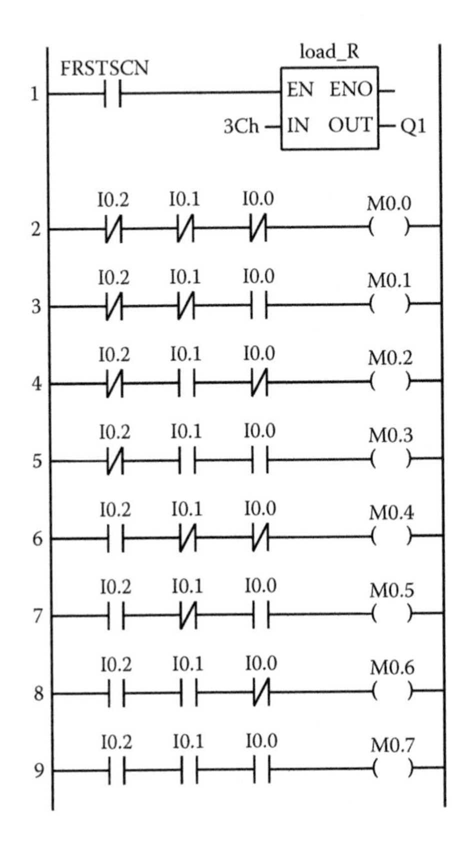

FIGURE 10.8
The ladder diagram of the user program of UZAM_plc_16i16o_ex17.asm. (*Continued*)

for these eight shift right operations is the number of shift. It can be seen that for each rung one rising edge detector is used. This is to make sure that when the related shift right operation is chosen, it will be carried out only once. In order to choose one of these eight shift right operations the input bits I0.4 and I0.3 must be as follows: I0.4 = 0, I0.3 = 1.

In the eight rungs between 18 and 25, we define eight different shift left operations according to the 3-to-8 decoder outputs represented by the marker bits M0.0, M0.1, …, M0.7. Shift left operations defined in these rungs are applied to the 8-bit input variable Q1. The result of the shift left operations defined in these rungs will be stored in Q0. The `shift in bit` for these shift left operations defined in these rungs is I1.0. The only difference for these eight shift left operations is the number of shift. It can be seen that for each rung one rising edge detector is used. This is to make sure that when the related shift left operation is chosen, it will be carried out only once. In order to choose one of these eight shift left operations, the input bits I0.4 and I0.3 must be set as follows: I0.4 = 1, I0.3 = 0.

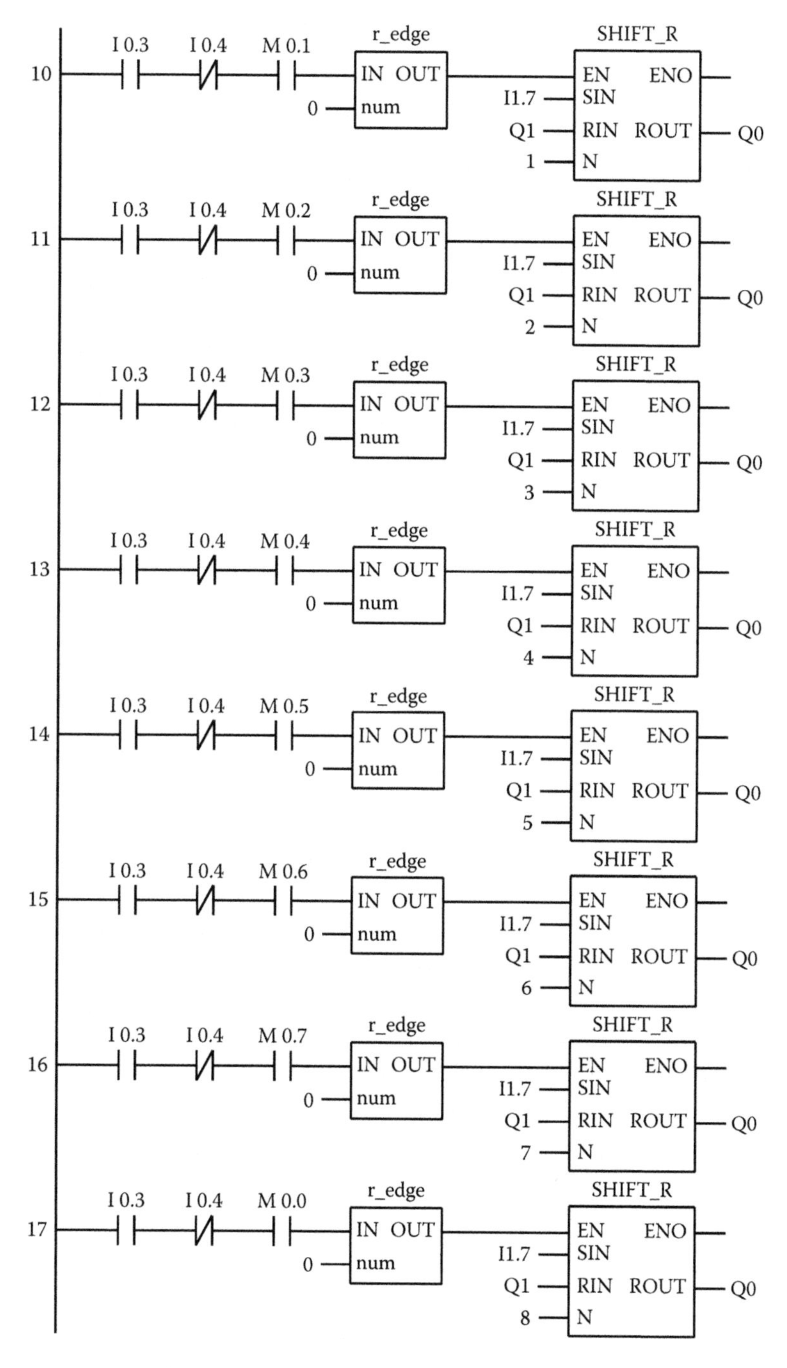

FIGURE 10.8 (*Continued*)
The ladder diagram of the user program of UZAM_plc_16i16o_ex17.asm. (*Continued*)

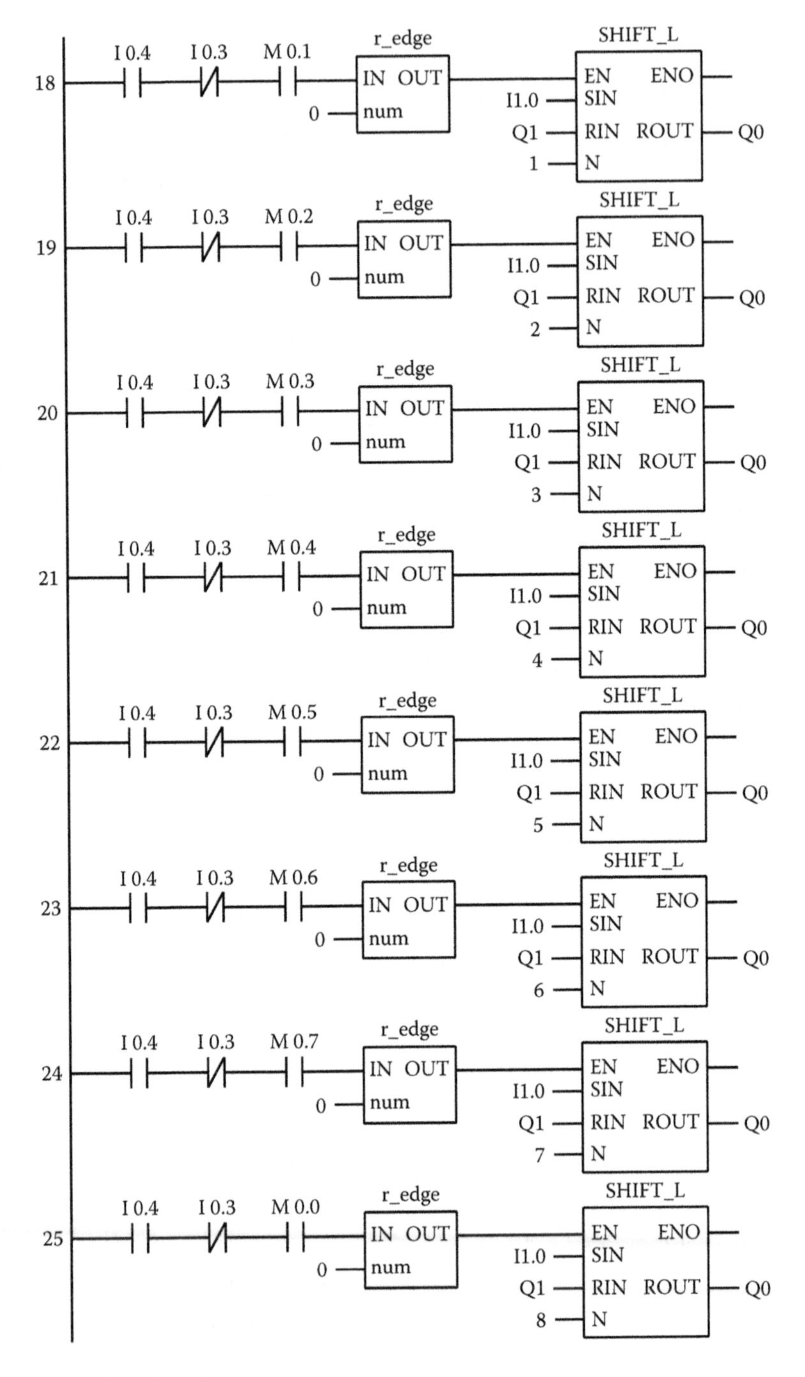

FIGURE 10.8 *(Continued)*
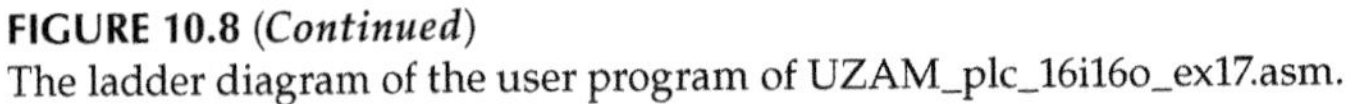
The ladder diagram of the user program of UZAM_plc_16i16o_ex17.asm.

TABLE 10.6

Truth Table of the User Program of UZAM_plc_16i16o_ex17.asm

I0.4	I0.3	I0.2	I0.1	I0.0	Selected Process
0	0	×	×	×	No process is selected
1	1	×	×	×	No process is selected
0	1	0	0	0	Shift right Q1 once; shift in bit = I1.7
0	1	0	0	1	Shift right Q1 twice; shift in bit = I1.7
0	1	0	1	0	Shift right Q1 3 times; shift in bit = I1.7
0	1	0	1	1	Shift right Q1 4 times; shift in bit = I1.7
0	1	1	0	0	Shift right Q1 5 times; shift in bit = I1.7
0	1	1	0	1	Shift right Q1 6 times; shift in bit = I1.7
0	1	1	1	0	Shift right Q1 7 times; shift in bit = I1.7
0	1	1	1	1	Shift right Q1 8 times; shift in bit = I1.7
1	0	0	0	0	Shift left Q1 once; shift in bit = I1.0
1	0	0	0	1	Shift left Q1 twice; shift in bit = I1.0
1	0	0	1	0	Shift left Q1 3 times; shift in bit = I1.0
1	0	0	1	1	Shift left Q1 4 times; shift in bit = I1.0
1	0	1	0	0	Shift left Q1 5 times; shift in bit = I1.0
1	0	1	0	1	Shift left Q1 6 times; shift in bit = I1.0
1	0	1	1	0	Shift left Q1 7 times; shift in bit = I1.0
1	0	1	1	1	Shift left Q1 8 times; shift in bit = I1.0

×: Don't care. Note that the result of the shift operations will be stored in Q0.

Table 10.6 shows the truth table of the user program of UZAM_plc_16i16o_ex17.asm.

The second example program, UZAM_plc_16i16o_ex18.asm, is shown in Figure 10.9. It shows usage of the following macros: `rotate_R`, `rotate_L`, and `Swap`. The ladder diagram of the user program of UZAM_plc_16i16o_ex18.asm, shown in Figure 10.9, is depicted in Figure 10.10.

In the first rung, 8-bit numerical data F0h are loaded to the 8-bit variable Q1, by using the macro `load_R`. This process is carried out once at the first program scan by using the FRSTSCN NO contact.

In the second rung, if the 8-bit input register I0 is set to 80h, then I1 is loaded to Q1, by using the macro `load_R`.

In the seven rungs between 3 and 9, a 3-to-8 decoder is implemented, whose inputs are I0.2, I0.1, and I0.0, and whose outputs are M0.1, M0.2, …, M0.7. Note that the first combination of 3-to-8 decoder, namely, (I0.2, I0.1, I0.0) = 000, is not implemented. This arrangement is made to choose the number of rotation for the selected rotate right or rotate left operation based on the input data given through the input bits I0.2, I0.1, and I0.0. When these bits are 001, 010, 100, 100, 101, 110, and 111, we define the number of rotation for the selected rotate right or rotate left operation as 1, 2, 3, 4, 5, 6, and 7, respectively.

In the seven rungs between 10 and 16, we define seven different rotate right operations according to the 3-to-8 decoder outputs represented by the

```
;--------------- user program starts here --
        ld          FRSTSCN       ;rung 1
        load_R      0F0h,Q1

        ld          I0.7          ;rung 2
        and_not     I0.6
        and_not     I0.5
        and_not     I0.4
        and_not     I0.3
        and_not     I0.2
        and_not     I0.1
        and_not     I0.0
        r_edge      0
        move_R      I1,Q1

        ld_not      I0.2          ;rung 3
        and_not     I0.1
        and         I0.0
        out         M0.1

        ld_not      I0.2          ;rung 4
        and         I0.1
        and_not     I0.0
        out         M0.2

        ld_not      I0.2          ;rung 5
        and         I0.1
        and         I0.0
        out         M0.3

        ld          I0.2          ;rung 6
        and_not     I0.1
        and_not     I0.0
        out         M0.4

        ld          I0.2          ;rung 7
        and_not     I0.1
        and         I0.0
        out         M0.5

        ld          I0.2          ;rung 8
        and         I0.1
        and_not     I0.0
        out         M0.6

        ld          I0.2          ;rung 9
        and         I0.1
        and         I0.0
        out         M0.7
```

FIGURE 10.9
The user program of UZAM_plc_16i16o_ex18.asm. (*Continued*)

```
ld          I0.3        ;rung 10
and_not     I0.4
and         M0.1
r_edge      1
rotate_R    1,Q1,Q0

ld          I0.3        ;rung 11
and_not     I0.4
and         M0.2
r_edge      2
rotate_R    2,Q1,Q0

ld          I0.3        ;rung 12
and_not     I0.4
and         M0.3
r_edge      3
rotate_R    3,Q1,Q0

ld          I0.3        ;rung 13
and_not     I0.4
and         M0.4
r_edge      4
rotate_R    4,Q1,Q0

ld          I0.3        ;rung 14
and_not     I0.4
and         M0.5
r_edge      5
rotate_R    5,Q1,Q0

ld          I0.3        ;rung 15
and_not     I0.4
and         M0.6
r_edge      6
rotate_R    6,Q1,Q0

ld          I0.3        ;rung 16
and_not     I0.4
and         M0.7
r_edge      7
rotate_R    7,Q1,Q0
```

FIGURE 10.9 (*Continued*)
The user program of UZAM_plc_16i16o_ex18.asm. (*Continued*)

```
        ld          I0.4          ;rung 17
        and_not     I0.3
        and         M0.1
        r_edge      1
        rotate_L    1,Q1,Q0

        ld          I0.4          ;rung 18
        and_not     I0.3
        and         M0.2
        r_edge      2
        rotate_L    2,Q1,Q0

        ld          I0.4          ;rung 19
        and_not     I0.3
        and         M0.3
        r_edge      3
        rotate_L    3,Q1,Q0

        ld          I0.4          ;rung 20
        and_not     I0.3
        and         M0.4
        r_edge      4
        rotate_L    4,Q1,Q0

        ld          I0.4          ;rung 21
        and_not     I0.3
        and         M0.5
        r_edge      5
        rotate_L    5,Q1,Q0

        ld          I0.4          ;rung 22
        and_not     I0.3
        and         M0.6
        r_edge      6
        rotate_L    6,Q1,Q0

        ld          I0.4          ;rung 23
        and_not     I0.3
        and         M0.7
        r_edge      7
        rotate_L    7,Q1,Q0

        ld          I0.6          ;rung 24
        and_not     I0.7
        and_not     I0.5
        and_not     I0.4
        and_not     I0.3
        and_not     I0.2
        and_not     I0.1
        and_not     I0.0
        Swap        Q1,Q0
;--------------- user program ends here --
```

FIGURE 10.9 (*Continued*)
The user program of UZAM_plc_16i16o_ex18.asm.

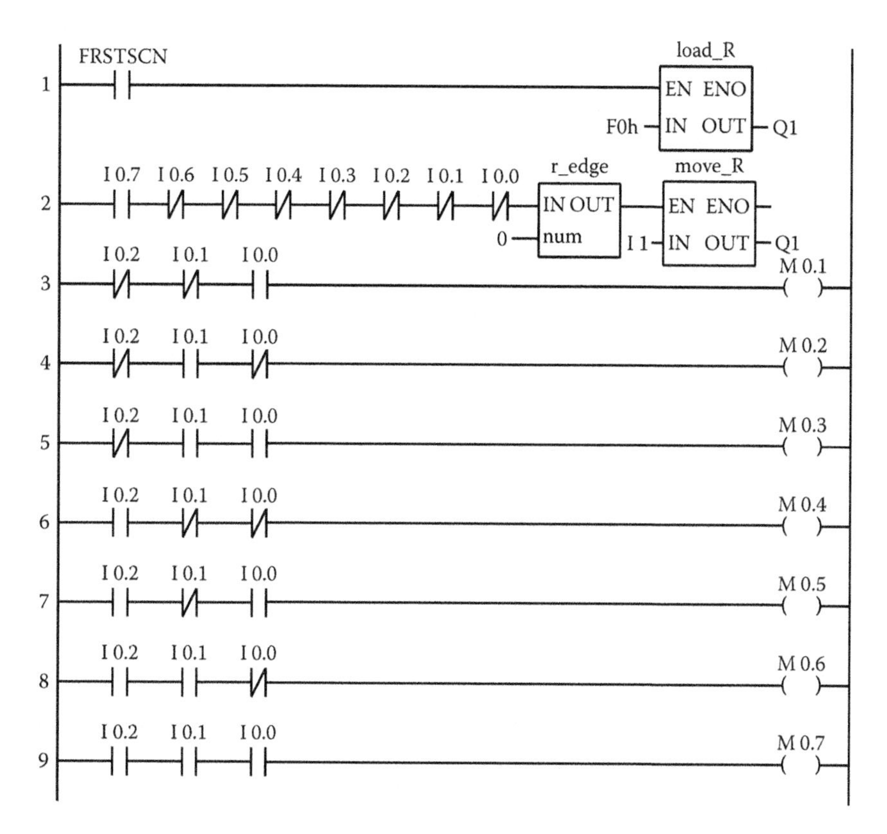

FIGURE 10.10
The ladder diagram of the user program of UZAM_plc_16i16o_ex18.asm. (*Continued*)

marker bits M0.1, M0.2, …, M0.7. Rotate right operations defined in these rungs are applied to the 8-bit input variable Q1. The result of the rotate right operations defined in these rungs will be stored in Q0. The only difference for these seven rotate right operations is the number of rotation. It can be seen that for each rung one rising edge detector is used. This is to make sure that when the related rotate right operation is chosen, it will be carried out only once. In order to choose one of these seven rotate right operations, the input bits I0.4 and I0.3 must be as follows: I0.4 = 0, I0.3 = 1.

In the seven rungs between 17 and 23, we define seven different rotate left operations according to the 3-to-8 decoder outputs represented by the marker bits M0.1, M0.2, …, M0.7. Rotate left operations defined in these rungs are applied to the 8-bit input variable Q1. The result of the rotate left operations defined in these rungs will be stored in Q0. The only difference for these seven rotate left operations is the number of rotation. It can be seen that for each rung one rising edge detector is used. This is to make sure that when the related rotate left operation is chosen, it will be carried out only once. In

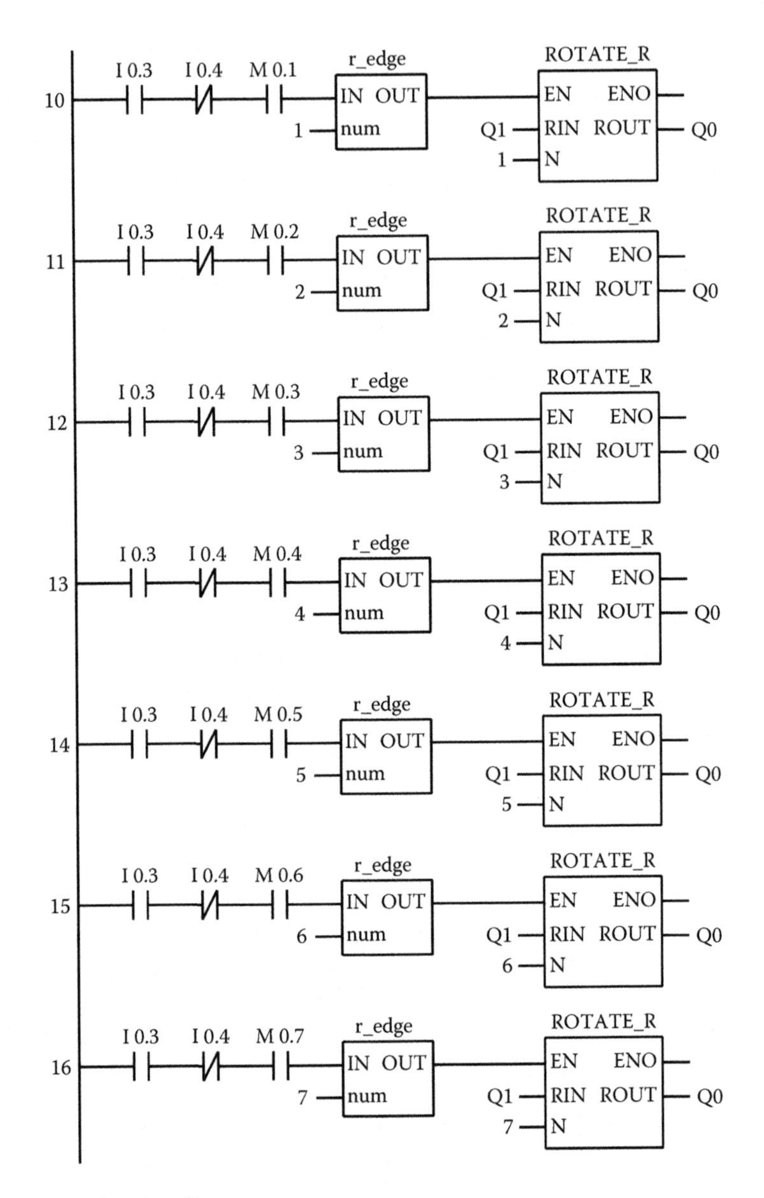

FIGURE 10.10 (***Continued***)
The ladder diagram of the user program of UZAM_plc_16i16o_ex18.asm. (*Continued*)

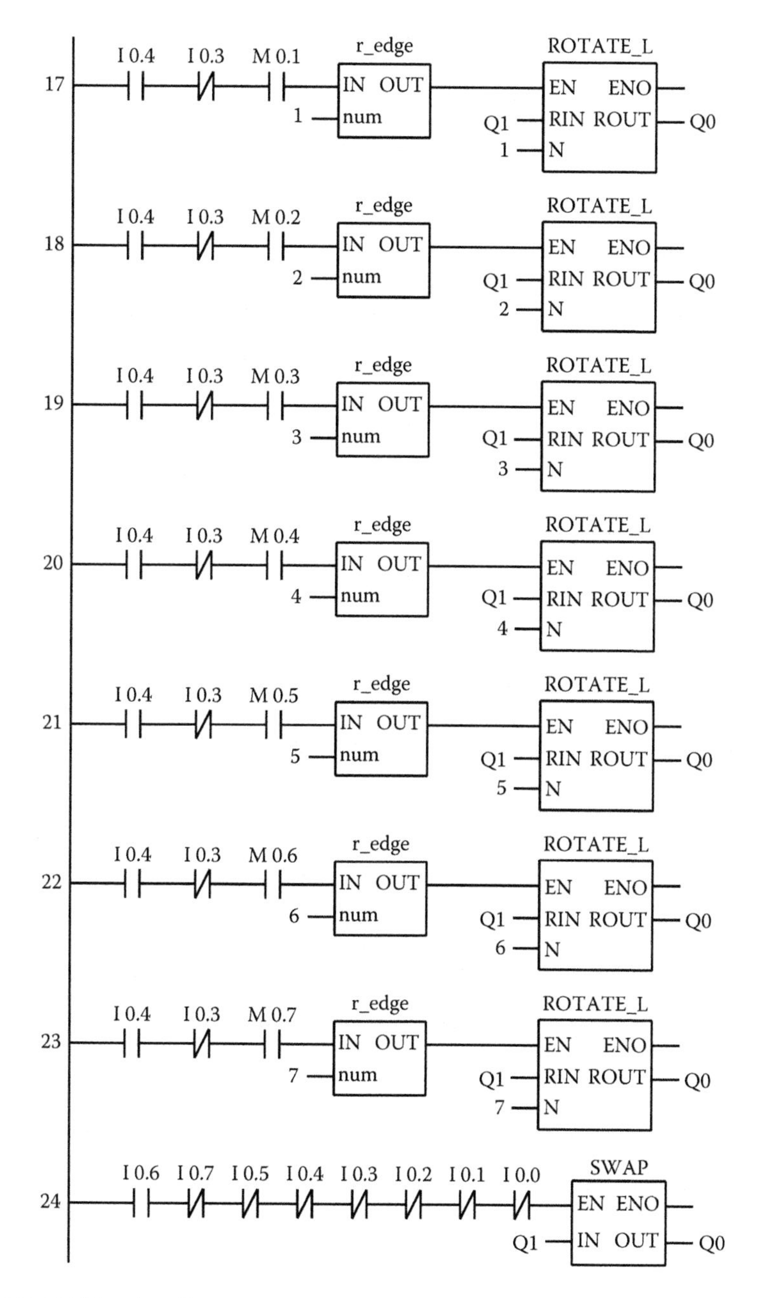

FIGURE 10.10 (*Continued*)
The ladder diagram of the user program of UZAM_plc_16i16o_ex18.asm.

TABLE 10.7

Truth Table of the User Program of UZAM_plc_16i16o_ex18.asm

I0.4	I0.3	I0.2	I0.1	I0.0	Selected Process
0	0	×	×	×	No process is selected
1	1	×	×	×	No process is selected
0	1	0	0	0	No process is selected
0	1	0	0	1	Rotate right Q1 once
0	1	0	1	0	Rotate right Q1 twice
0	1	0	1	1	Rotate right Q1 3 times
0	1	1	0	0	Rotate right Q1 4 times
0	1	1	0	1	Rotate right Q1 5 times
0	1	1	1	0	Rotate right Q1 6 times
0	1	1	1	1	Rotate right Q1 7 times
1	0	0	0	0	No process is selected
1	0	0	0	1	Rotate left Q1 once
1	0	0	1	0	Rotate left Q1 twice
1	0	0	1	1	Rotate left Q1 3 times
1	0	1	0	0	Rotate left Q1 4 times
1	0	1	0	1	Rotate left Q1 5 times
1	0	1	1	0	Rotate left Q1 6 times
1	0	1	1	1	Rotate left Q1 7 times

×: Don't care. Note that the result of the rotate operations will be stored in Q0. In addition, when I0 = 40h, the process Q0 = SWAP Q1 is selected.

order to choose one of these seven rotate left operations, the input bits I0.4 and I0.3 must be set as follows: I0.4 = 1, I0.3 = 0.

In the last rung, the use of the swap function is shown. If the 8-bit input register I0 is set to be 40h, then the "Swap Q1 and store the result in Q0" process is selected.

Table 10.7 shows the truth table of the user program of UZAM_plc_16i16o_ex18.asm.

11

Multiplexer Macros

As a standard combinational component, the multiplexer (MUX), allows the selection of one input signal among n signals, where $n > 1$, and is a power of two. Select lines connected to the multiplexer determine which input signal is selected and passed to the output of the multiplexer. As can be seen from Figure 11.1, in general, an n-to-1 multiplexer has n data input lines, m select lines where $m = \log2\ n$, i.e., $2^m = n$, and one output line. Although not shown in Figure 11.1, in addition to the other inputs, the multiplexer may have an enable line, E, for enabling it. When the multiplexer is disabled with E set to 0 (for active high enable input E), no input signal is selected and passed to the output.

In this chapter, the following multiplexer macros are described for the PIC16F648A-based PLC:

- `mux_2_1` (2 × 1 MUX)
- `mux_2_1_E` (2 × 1 MUX with enable input)
- `mux_4_1` (4 × 1 MUX)
- `mux_4_1_E` (4 × 1 MUX with enable input)
- `mux_8_1` (8 × 1 MUX)
- `mux_8_1_E` (8 × 1 MUX with enable input)

The file definitions.inc, included within the CD-ROM attached to this book, contains all multiplexer macros defined for the PIC16F648A-based PLC. Let us now consider these macros in detail.

11.1 Macro `mux_2_1`

The symbol and the truth table of the macro `mux_2_1` are depicted in Table 11.1. Figure 11.2 shows the macro `mux_2_1` and its flowchart. In this macro, the select input s_0, input signals d_0 and d_1, and the output y are all Boolean variables. When $s_0 = 0$, the input signal d_0 is selected and passed to the output y. When $s_0 = 1$, the input signal d_1 is selected and passed to the output y.

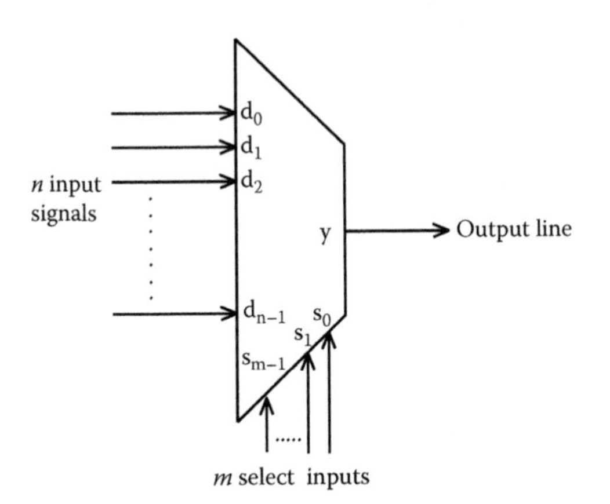

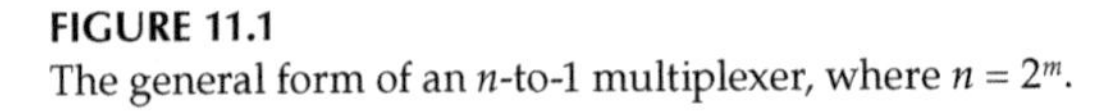

FIGURE 11.1
The general form of an n-to-1 multiplexer, where $n = 2^m$.

11.2 Macro mux_2_1_E

The symbol and the truth table of the macro mux_2_1_E are depicted in Table 11.2. Figure 11.3 shows the macro mux_2_1_E and its flowchart. In this macro, the active high enable input E, the select input s_0, input signals d_0 and d_1, and the output y are all Boolean variables. When this multiplexer is disabled with E set to 0, no input signal is selected and passed to the output. When this multiplexer is enabled with E set to 1, it functions as described for mux_2_1. This means that when E = 1: if $s_0 = 0$, then the input signal d_0 is selected and passed to the output y. When E = 1: if $s_0 = 1$, then the input signal d_1 is selected and passed to the output y.

TABLE 11.1
Symbol and Truth Table of the Macro mux_2_1

Symbol: (d_0, d_1, s_0, y)

s0 =	regs0,bits0
d0 =	regi0,biti0
d1 =	regi1,biti1
y =	rego,bito

Truth Table:

input	output
s0	y
0	d0
1	d1

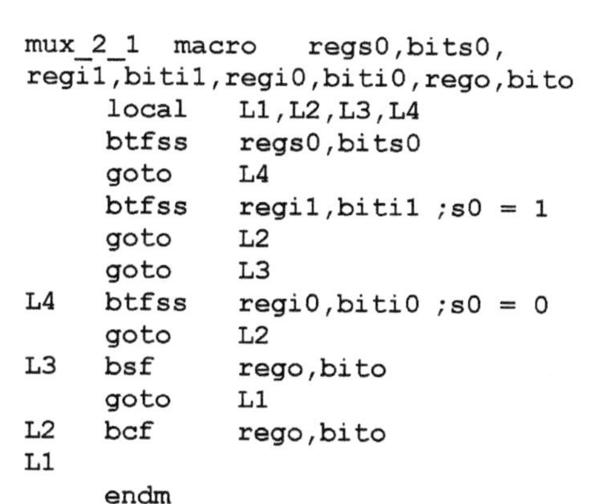

```
mux_2_1  macro   regs0,bits0,
regi1,biti1,regi0,biti0,rego,bito
     local   L1,L2,L3,L4
     btfss   regs0,bits0
     goto    L4
     btfss   regi1,biti1 ;s0 = 1
     goto    L2
     goto    L3
L4   btfss   regi0,biti0 ;s0 = 0
     goto    L2
L3   bsf     rego,bito
     goto    L1
L2   bcf     rego,bito
L1
     endm
```

(a)

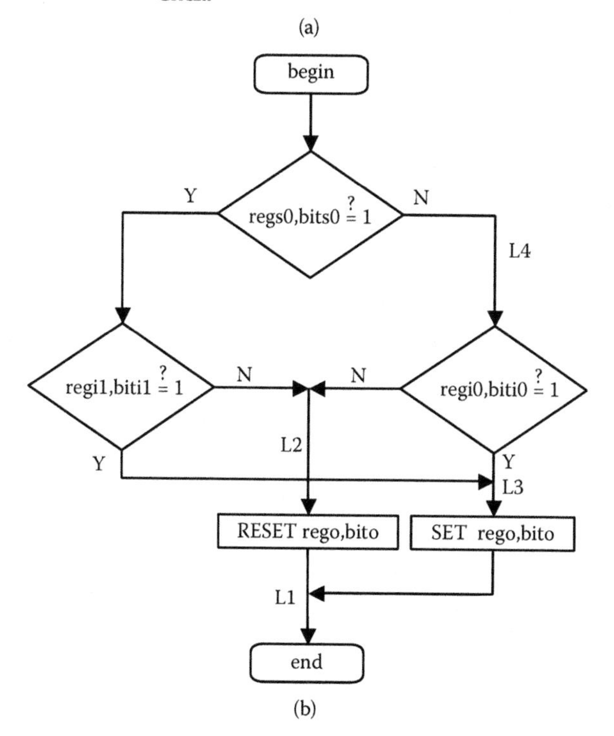

(b)

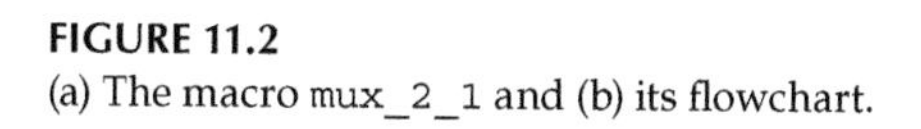

FIGURE 11.2
(a) The macro `mux_2_1` and (b) its flowchart.

11.3 Macro `mux_4_1`

The symbol and the truth table of the macro `mux_4_1` are depicted in Table 11.3. Figure 11.4 shows the macro `mux_4_1` and its flowchart. In this macro, select inputs s_1 and s_0, input signals d_0, d_1, d_2, and d_3, and the

TABLE 11.2
Symbol and Truth Table of the Macro mux_2_1_E

Symbol

E, d_0, d_1, y, s_0

W	E
s0 =	regs0,bits0
d0 =	regi0,biti0
d1 =	regi1,biti1
y =	rego,bito

Truth Table

inputs		output
E	s0	y
0	×	0
1	0	d0
1	1	d1

×: don't care.

output y are all Boolean variables. When $s_1s_0 = 00$ (respectively, 01, 10, 11), the input signal d_0 (respectively, d_1, d_2, d_3) is selected and passed to the output y.

11.4 Macro mux_4_1_E

The symbol and the truth table of the macro mux_4_1_E are depicted in Table 11.4. Figures 11.5 and 11.6 show the macro mux_4_1_E and its flowchart, respectively. In this macro, the active high enable input E, select inputs s_1 and s_0, input signals d_0, d_1, d_2, and d_3, and the output y are all Boolean variables. When this multiplexer is disabled with E set to 0, no input signal is

```
mux_2_1_E  macro    regs0,bits0,
regi1,biti1,regi0,biti0,rego,bito
      local    L1,L2,L3,L4
      movwf    Temp_1
      btfss    Temp_1,0
      goto     L2
      btfss    regs0,bits0
      goto     L4
      btfss    regi1,biti1 ;s0 = 1
      goto     L2
      goto     L3
L4    btfss    regi0,biti0 ;s0 = 0
      goto     L2
L3    bsf      rego,bito
      goto     L1
L2    bcf      rego,bito
L1
      endm
```

(a)

FIGURE 11.3
(a) The macro mux_2_1_E and (b) its flowchart. (*Continued*)

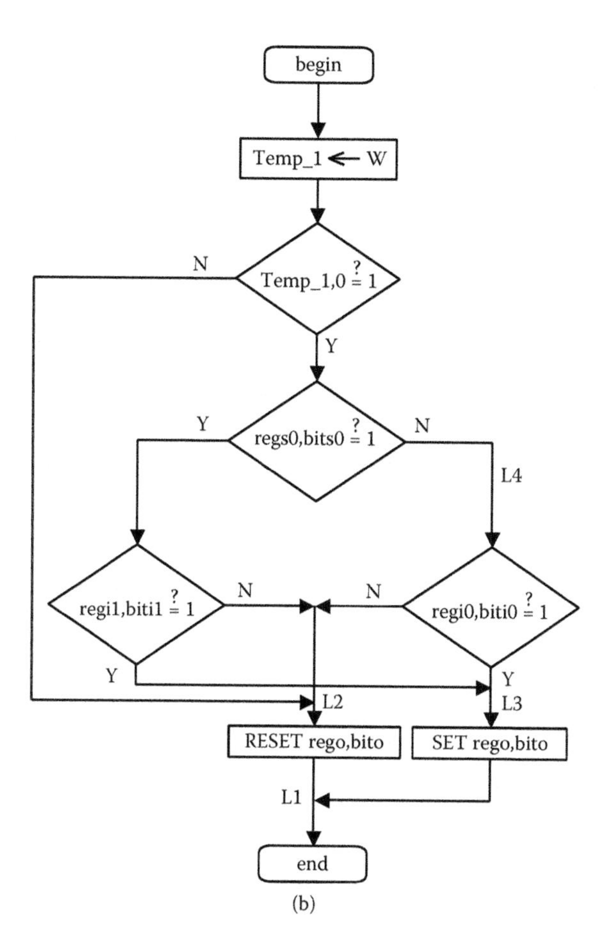

FIGURE 11.3 (*Continued*)
(a) The macro `mux_2_1_E` and (b) its flowchart.

TABLE 11.3

Symbol and Truth Table of the Macro `mux_4_1`

Symbol

s1 =	regs1,bits1
s0 =	regs0,bits0
d3 =	regi3,biti3
d2 =	regi2,biti2
d1 =	regi1,biti1
d0 =	regi0,biti0
y =	rego,bito

Truth Table

inputs		output
s1	s0	y
0	0	d0
0	1	d1
1	0	d2
1	1	d3

```
mux_4_1  macro   regs1,bits1,regs0,bits0,regi3,biti3,
regi2,biti2,regi1,biti1,regi0,biti0,rego,bito
     local   L1,L2,L3,L4,L5,L6
     btfss   regs1,bits1
     goto    L5
     btfss   regs0,bits0
     goto    L6
     btfss   regi3,biti3 ;s1s0 = 11
     goto    L2
     goto    L3
L6   btfss   regi2,biti2 ;s1s0 = 10
     goto    L2
     goto    L3
L5   btfss   regs0,bits0
     goto    L4
     btfss   regi1,biti1 ;s1s0 = 01
     goto    L2
     goto    L3
L4   btfss   regi0,biti0 ;s1s0 = 00
     goto    L2
L3   bsf     rego,bito
     goto    L1
L2   bcf     rego,bito
L1
     endm
```

(a)

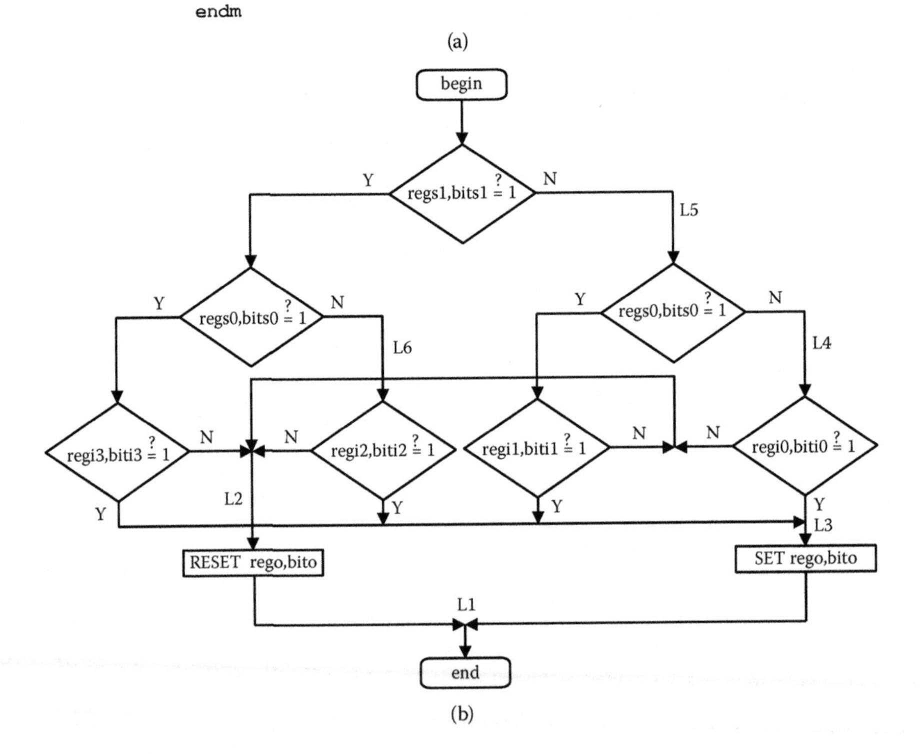

(b)

FIGURE 11.4
(a) The macro `mux_4_1` and (b) its flowchart.

TABLE 11.4

Symbol and Truth Table of the Macro mux_4_1_E

Symbol

W	E
s1 =	regs1,bits1
s0 =	regs0,bits0
d3 =	regi3,biti3
d2 =	regi2,biti2
d1 =	regi1,biti1
d0 =	regi0,biti0
y =	rego,bito

Truth Table

inputs			output
E	s1	s0	y
0	×	×	0
1	0	0	d0
1	0	1	d1
1	1	0	d2
1	1	1	d3

×: don't care.

```
mux_4_1_E  macro   regs1,bits1,regs0,bits0,regi3,biti3,
regi2,biti2,regi1,biti1,regi0,biti0,rego,bito
      local    L1,L2,L3,L4,L5,L6
      movwf    Temp_1
      btfss    Temp_1,0
      goto     L2
      btfss    regs1,bits1
      goto     L5
      btfss    regs0,bits0
      goto     L6
      btfss    regi3,biti3 ;s1s0 = 11
      goto     L2
      goto     L3
L6    btfss    regi2,biti2 ;s1s0 = 10
      goto     L2
      goto     L3
L5    btfss    regs0,bits0
      goto     L4
      btfss    regi1,biti1 ;s1s0 = 01
      goto     L2
      goto     L3
L4    btfss    regi0,biti0 ;s1s0 = 00
      goto     L2
L3    bsf      rego,bito
      goto     L1
L2    bcf      rego,bito
L1
      endm
```

FIGURE 11.5
The macro mux_4_1_E.

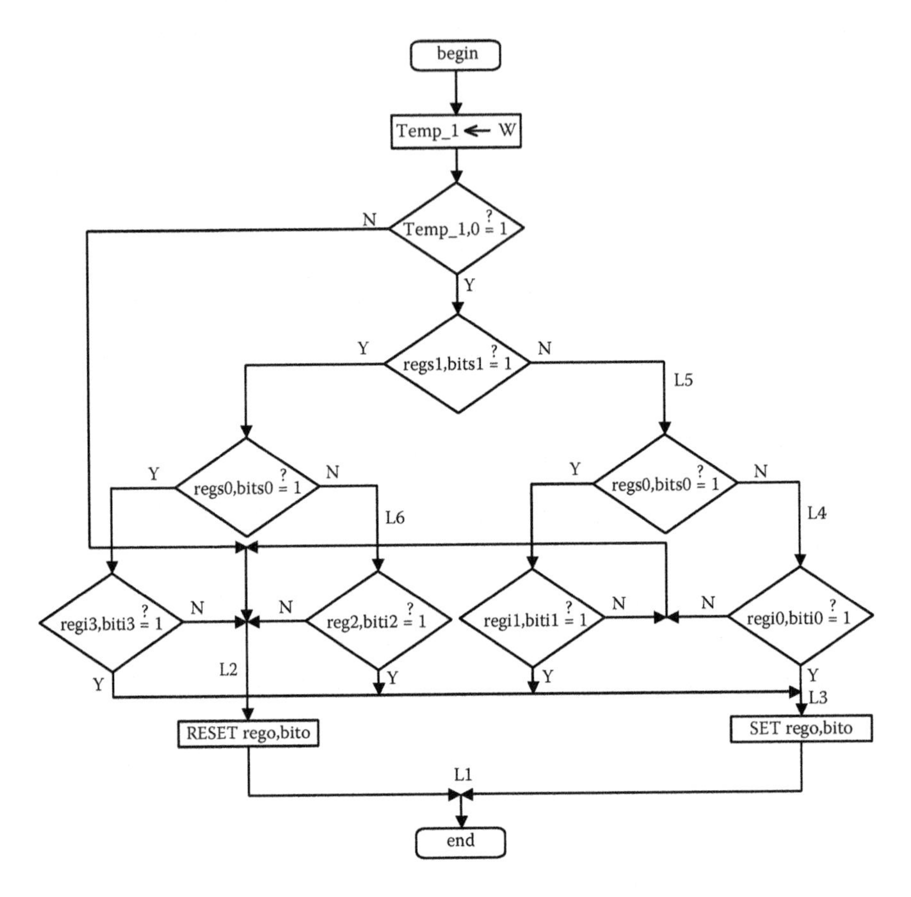

FIGURE 11.6
The flowchart of the macro mux_4_1_E.

selected and passed to the output. When this multiplexer is enabled with E set to 1, it functions as described for mux_4_1. This means that when E = 1: if $s_1s_0 = 00$ (respectively, 01, 10, 11), then the input signal d_0 (respectively, d_1, d_2, d_3) is selected and passed to the output y.

11.5 Macro mux_8_1

The symbol and the truth table of the macro mux_8_1 are depicted in Table 11.5. Figures 11.7 and 11.8 show the macro mux_8_1 and its flowchart, respectively. In this macro, select inputs s_2, s_1, and s_0, input signals d_0, d_1, d_2, d_3, d_4, d_5, d_6, and d_7, and the output y are all Boolean variables. When $s_2s_1s_0$ = 000 (respectively, 001, 010, 011, 100, 101, 110, 111), the input signal d_0 (respectively, d_1, d_2, d_3, d_4, d_5, d_6, d_7) is selected and passed to the output y.

TABLE 11.5

Symbol and Truth Table of the Macro `mux_8_1`

Symbol

s2 =	regs2,bits2
s1 =	regs1,bits1
s0 =	regs0,bits0
d7 =	regi7,biti7
d6 =	regi6,biti6
d5 =	regi5,biti5
d4 =	regi4,biti4
d3 =	regi3,biti3
d2 =	regi2,biti2
d1 =	regi1,biti1
d0 =	regi0,biti0
y =	rego,bito

Truth Table

inputs			output
s2	s1	s0	y
0	0	0	d0
0	0	1	d1
0	1	0	d2
0	1	1	d3
1	0	0	d4
1	0	1	d5
1	1	0	d6
1	1	1	d7

11.6 Macro `mux_8_1_E`

The symbol and the truth table of the macro `mux_8_1_E` are depicted in Table 11.6. Figures 11.9 and 11.10 show the macro `mux_8_1_E` and its flowchart, respectively. In this macro, the active high enable input E, select inputs s_2, s_1, and s_0, input signals d_0, d_1, d_2, d_3, d_4, d_5, d_6, and d_7, and the output y are all Boolean variables. When this multiplexer is disabled with E set to 0, no input signal is selected and passed to the output. When this multiplexer is enabled with E set to 1, it functions as described for `mux_8_1`. This means that when E = 1: if $s_2s_1s_0$ = 000 (respectively, 001, 010, 011, 100, 101, 110, 111), then the input signal d_0 (respectively, d_1, d_2, d_3, d_4, d_5, d_6, d_7) is selected and passed to the output y.

11.7 Examples for Multiplexer Macros

In this section, we will consider three examples, namely, UZAM_plc_16i16o_exX.asm (*X* = 19, 20, 21), to show the usage of multiplexer macros. In order to test one of these examples, please take the related file UZAM_plc_16i16o_exX.asm (*X* = 19, 20, 21) from the CD-ROM attached to this book, and then

```
mux_8_1  macro   regs2,bits2,regs1,bits1,
regs0,bits0,regi7,biti7,regi6,biti6,regi5,
biti5,regi4,biti4,regi3,biti3,regi2,biti2,
regi1,biti1,regi0,biti0,rego,bito
     local   L1,L2,L3,L4,L5,L6,L7,L8,L9,L10
     btfss   regs2,bits2
     goto    L7
     btfss   regs1,bits1
     goto    L9
     btfss   regs0,bits0
     goto    L10
     btfss   regi7,biti7      ;s2s1s0 = 111
     goto    L2
     goto    L3
L10  btfss   regi6,biti6      ;s2s1s0 = 110
     goto    L2
     goto    L3
L9   btfss   regs0,bits0
     goto    L8
     btfss   regi5,biti5      ;s2s1s0 = 101
     goto    L2
     goto    L3
L8   btfss   regi4,biti4      ;s2s1s0 = 100
     goto    L2
     goto    L3
L7   btfss   regs1,bits1
     goto    L5
     btfss   regs0,bits0
     goto    L6
     btfss   regi3,biti3      ;s2s1s0 = 011
     goto    L2
     goto    L3
L6   btfss   regi2,biti2      ;s2s1s0 = 010
     goto    L2
     goto    L3
L5   btfss   regs0,bits0
     goto    L4
     btfss   regi1,biti1      ;s2s1s0 = 001
     goto    L2
     goto    L3
L4   btfss   regi0,biti0      ;s2s1s0 = 000
     goto    L2
L3   bsf     rego,bito
     goto    L1
L2   bcf     rego,bito
L1
     endm
```

FIGURE 11.7
The macro mux_8_1.

TABLE 11.6

Symbol and Truth Table of the Macro `mux_8_1_E`

Symbol

W	E
s2 =	regs2,bits2
s1 =	regs1,bits1
s0 =	regs0,bits0
d7 =	regi7,biti7
d6 =	regi6,biti6
d5 =	regi5,biti5
d4 =	regi4,biti4
d3 =	regi3,biti3
d2 =	regi2,biti2
d1 =	regi1,biti1
d0 =	regi0,biti0
y =	rego,bito

Truth Table

inputs				output
E	s2	s1	s0	y
0	×	×	×	0
1	0	0	0	d0
1	0	0	1	d1
1	0	1	0	d2
1	0	1	1	d3
1	1	0	0	d4
1	1	0	1	d5
1	1	1	0	d6
1	1	1	1	d7

×: don't care.

open the program by MPLAB IDE and compile it. After that, by using the PIC programmer software, take the compiled file UZAM_plc_16i16o_exX.hex (X = 19, 20, 21), and by your PIC programmer hardware, send it to the program memory of the PIC16F648A microcontroller within the PIC16F648A-based PLC. To do this, switch the 4PDT in PROG position and the power switch in OFF position. After loading the file UZAM_plc_16i16o_exX.hex (X = 19, 20, 21), switch the 4PDT in RUN and the power switch in ON position. Please check the program's accuracy by cross-referencing it with the related macros.

Let us now consider these example programs: The first example program, UZAM_plc_16i16o_ex19.asm, is shown in Figure 11.11. It shows the usage of two multiplexer macros `mux_2_1` and `mux_2_1_E`. The schematic diagram of the user program of UZAM_plc_16i16o_ex19.asm, shown in Figure 11.11, is depicted in Figure 11.12.

In the first rung, the multiplexer macro `mux_2_1` (2 × 1 multiplexer) is used. In this multiplexer, input signals are d_0 = I0.1 and d_1 = I0.2, while the output is y = Q0.0 and the select input is s_0 = I0.0.

In the second rung, another multiplexer macro `mux_2_1` is used. In this multiplexer, input signals are d_0 = T1.5 (838.8608 ms) and d_1 = T1.4 (419.4304 ms), while the output is y = Q0.3 and the select input is s_0 = I0.7.

In the third rung, the macro `mux_2_1_E` (2 × 1 multiplexer with active high enable input) is used. In this multiplexer, input signals are d_0 = I1.2 and d_1 = I1.3, while the output is y = Q1.0 and the select input is s_0 = I1.1. In addition, the active high enable input E is defined to be E = I1.0.

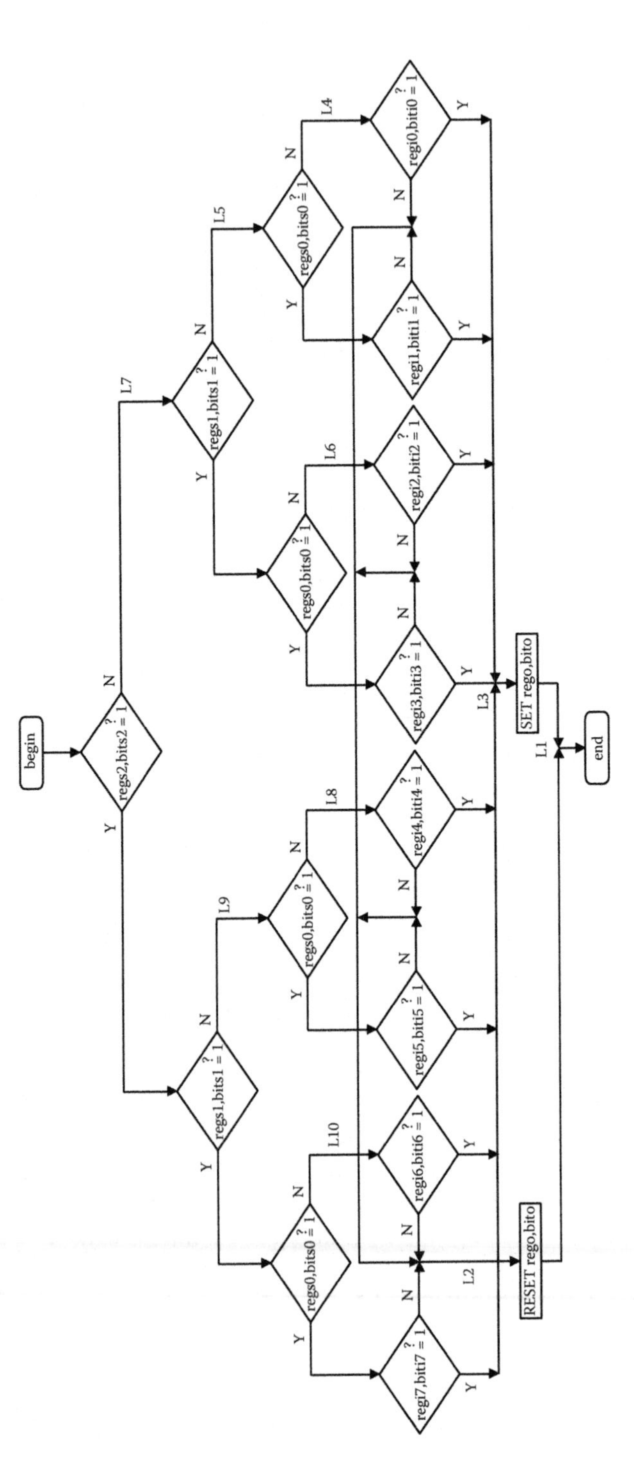

FIGURE 11.8
The flowchart of the macro mux_8_1.

```
mux_8_1_E  macro   regs2,bits2,regs1,bits1,
regs0,bits0,regi7,biti7,regi6,biti6,
regi5,biti5,regi4,biti4,regi3,biti3,
regi2,biti2,regi1,biti1,regi0,biti0,rego,bito
      local    L1,L2,L3,L4,L5,L6,L7,L8,L9,L10
      movwf    Temp_1
      btfss    Temp_1,0
      goto     L2
      btfss    regs2,bits2
      goto     L7
      btfss    regs1,bits1
      goto     L9
      btfss    regs0,bits0
      goto     L10
      btfss    regi7,biti7       ;s2s1s0 = 111
      goto     L2
      goto     L3
L10   btfss    regi6,biti6       ;s2s1s0 = 110
      goto     L2
      goto     L3
L9    btfss    regs0,bits0
      goto     L8
      btfss    regi5,biti5       ;s2s1s0 = 101
      goto     L2
      goto     L3
L8    btfss    regi4,biti4       ;s2s1s0 = 100
      goto     L2
      goto     L3
L7    btfss    regs1,bits1
      goto     L5
      btfss    regs0,bits0
      goto     L6
      btfss    regi3,biti3       ;s2s1s0 = 011
      goto     L2
      goto     L3
L6    btfss    regi2,biti2       ;s2s1s0 = 010
      goto     L2
      goto     L3
L5    btfss    regs0,bits0
      goto     L4
      btfss    regi1,biti1       ;s2s1s0 = 001
      goto     L2
      goto     L3
L4    btfss    regi0,biti0       ;s2s1s0 = 000
      goto     L2
L3    bsf      rego,bito
      goto     L1
L2    bcf      rego,bito
L1
      endm
```

FIGURE 11.9
The macro mux_8_1_E.

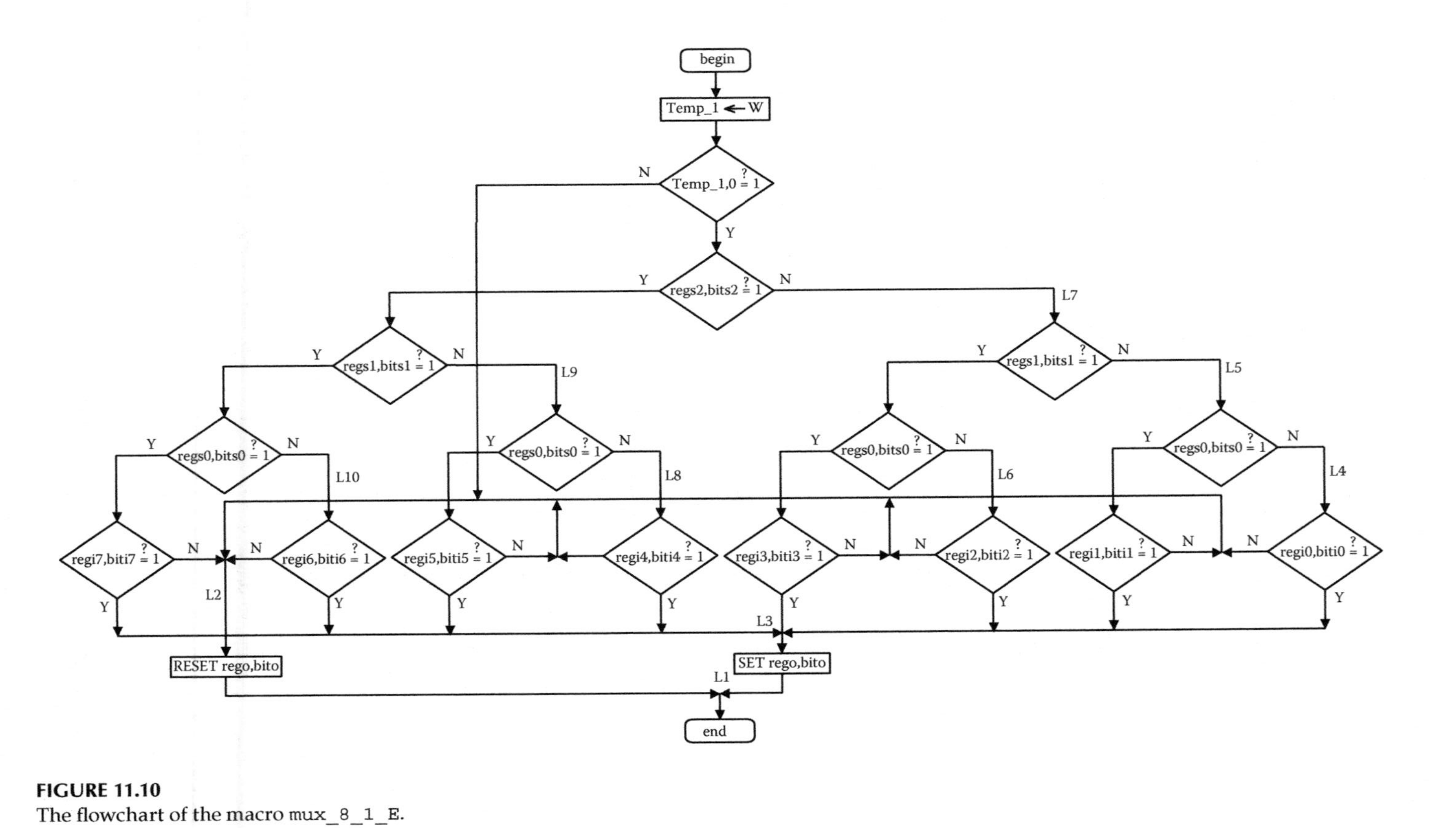

FIGURE 11.10
The flowchart of the macro `mux_8_1_E`.

```
.
;--------------- user program starts here -------
        mux_2_1      I0.0,I0.2,I0.1,Q0.0      ;rung 1

        mux_2_1      I0.7,T1.4,T1.5,Q0.3      ;rung 2


        ld           I1.0                     ;rung 3
        mux_2_1_E    I1.1,I1.3,I1.2,Q1.0

        ld_not       I1.4                     ;rung 4
        mux_2_1_E    I1.5,T1.4,T1.5,Q1.7
;--------------- user program ends here ---------
.
```

FIGURE 11.11
The user program of UZAM_plc_16i16o_ex19.asm.

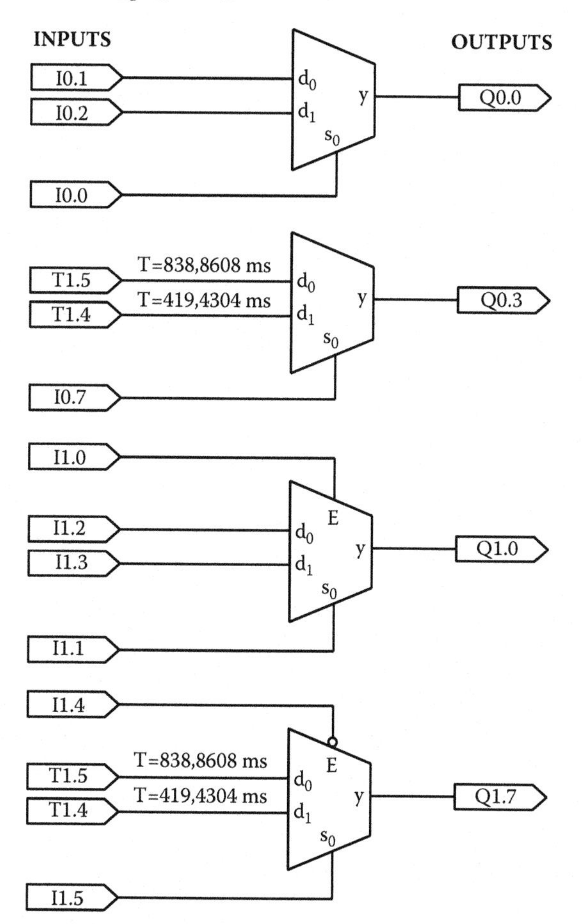

FIGURE 11.12
The schematic diagram of the user program of UZAM_plc_16i16o_ex19.asm.

```
.
;--------------- user program starts here -----------------
        mux_4_1      I0.1,I0.0,I0.5,I0.4,I0.3,I0.2,Q0.0  ;rung 1

        ld           I1.0                                 ;rung 2
        mux_4_1_E    I1.2,I1.1,T1.3,T1.2,T1.1,T1.0,Q1.0
;--------------- user program ends here --------------------
.
```

FIGURE 11.13
The user program of UZAM_plc_16i16o_ex20.asm.

In the fourth and last rung, another multiplexer macro `mux_2_1_E` is used. In this multiplexer, input signals are d_0 = T1.5 (838.8608 ms) and d_1 = T1.4 (419.4304 ms), while the output is y = Q1.7 and the select input is s_0 = I1.5. In addition, the active high enable input E is defined to be E = inverted I1.4. Note that this arrangement forces the enable input E to be active low.

The second example program, UZAM_plc_16i16o_ex20.asm, is shown in Figure 11.13. It shows the usage of two multiplexer macros `mux_4_1` and `mux_4_1_E`. The schematic diagram of the user program of UZAM_plc_16i16o_ex20.asm, shown in Figure 11.13, is depicted in Figure 11.14. In the first rung, the multiplexer macro `mux_4_1` (4 × 1 multiplexer) is used. In this multiplexer, input signals are d_0 = I0.2, d_1 = I0.3, d_2 = I0.4, and d_3 = I0.5, select

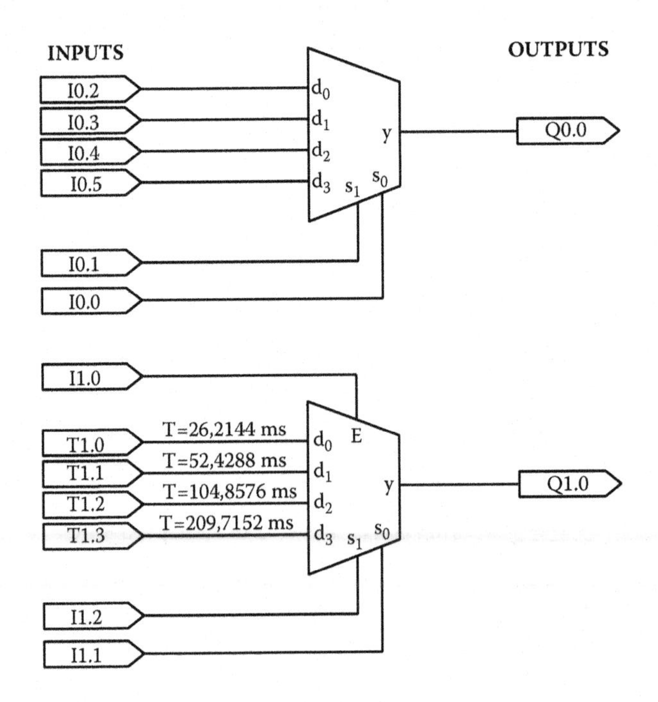

FIGURE 11.14
The schematic diagram of the user program of UZAM_plc_16i16o_ex20.asm.

```
.
;--------------- user program starts here -----------------------
ld         I0.0                                           ;rung 1
mux_8_1_E I0.3,I0.2,I0.1,I1.7,I1.6,I1.5,I1.4,I1.3,I1.2,I1.1,I1.0,Q0.0
;--------------- user program ends here -------------------------
.
```

FIGURE 11.15
The user program of UZAM_plc_16i16o_ex21.asm.

inputs are s_1 = I0.1 and s_0 = I0.0, and the output is y = Q0.0. In the second rung, the multiplexer macro mux_4_1_E (4 × 1 multiplexer with active high enable input) is used. In this multiplexer, input signals are d_0 = T1.0 (26.2144 ms), d_1 = T1.1 (52.4288 ms), d_2 = T1.2 (104.8576 ms), and d_3 = T1.3 (209.7152 ms), select inputs are s_1 = I1.2 and s_0 = I1.1, and the output is y = Q1.0. In addition, the active high enable input E is defined to be E = I1.0.

The third example program, UZAM_plc_16i16o_ex21.asm, is shown in Figure 11.15. It shows the usage of the multiplexer macro mux_8_1_E. The schematic diagram of the user program of UZAM_plc_16i16o_ex21.asm, shown in Figure 11.15, is depicted in Figure 11.16.

In this example, the multiplexer macro mux_8_1_E (8 × 1 multiplexer with active high enable input) is used. In this multiplexer, input signals are d_0 = I1.0, d_1 = I1.1, d_2 = I1.2, d_3 = I1.3, d_4 = I1.4, d_5 = I1.5, d_6 = I1.6, and d_7 = I1.7, select inputs are s_2 = I0.3, s_1 = I0.2, and s_0 = I0.1, and the output is y = Q0.0. In addition, the active high enable input E is defined to be E = I0.0.

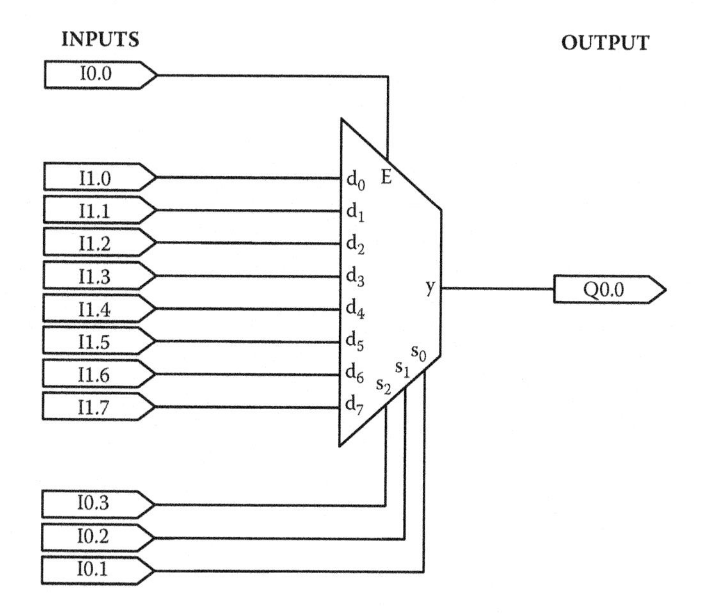

FIGURE 11.16
The schematic diagram of the user program of UZAM_plc_16i16o_ex21.asm.

12

Demultiplexer Macros

A demultiplexer (DMUX) is used when a circuit is to send a signal to one of many devices. This description sounds similar to the description given for a decoder, but a decoder is used to select among many devices, while a demultiplexer is used to send a signal among many devices. However, any decoder having an enable line can function as a demultiplexer. If the enable line of a decoder is used as a data input, then the data can be routed to any one of the outputs, and thus in that case the decoder can be used as a demultiplexer. As the name infers, a demultiplexer performs the opposite function as that of a multiplexer. A single *input signal* can be connected to any one of the output lines provided by the choice of an appropriate select signal. The general form of a 1-to-n demultiplexer can be seen from Figure 12.1. If there are m *select inputs*, then the number of *output lines* to which the data can be routed is $n = 2^m$. Although not shown in Figure 12.1, in addition to the other inputs, the demultiplexer may have an enable line, E, for enabling it. When the demultiplexer is disabled with E set to 0 (for active high enable input E), no output line is selected, and therefore the input signal is not passed to any output line.

In this chapter, the following demultiplexer macros are described for the PIC16F648A-based PLC: `Dmux_1_2` (1 × 2 DMUX), `Dmux_1_2_E` (1 × 2 DMUX with enable input), `Dmux_1_4` (1 × 4 DMUX), `Dmux_1_4_E` (1 × 4 DMUX with enable input), `Dmux_1_8` (1 × 8 DMUX), and `Dmux_1_8_E` (1 × 8 DMUX with enable input).

The file definitions.inc, included within the CD-ROM attached to this book, contains all demultiplexer macros defined for the PIC16F648A-based PLC. Let us now consider these macros in detail.

12.1 Macro `Dmux_1_2`

The symbol and the truth table of the macro `Dmux_1_2` are depicted in Table 12.1. Figure 12.2 shows the macro `Dmux_1_2` and its flowchart. In this macro, the select input s_0, output signals y_0 and y_1, and the input signal i are all Boolean variables. When the select input $s_0 = 0$, the input signal i is passed to the output line y_0. When the select input $s_0 = 1$, the input signal i is passed to the output line y_1.

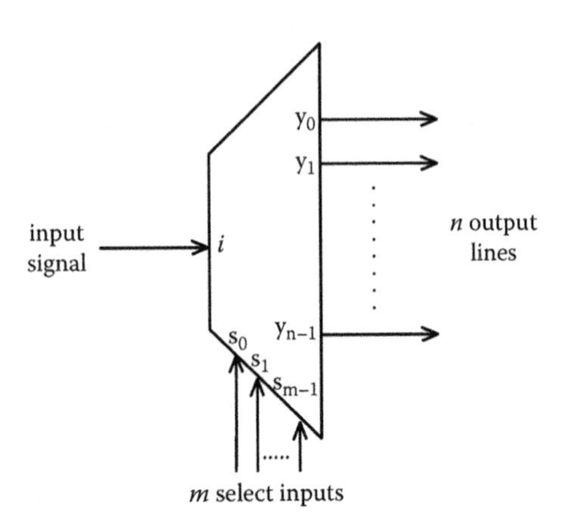

FIGURE 12.1
The general form of a 1-to-n demultiplexer, where $n = 2^m$.

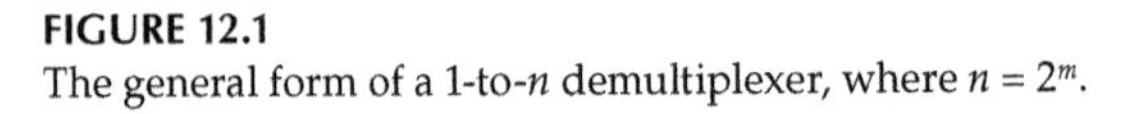

12.2 Macro `Dmux_1_2_E`

The symbol and the truth table of the macro `Dmux_1_2_E` are depicted in Table 12.2. Figure 12.3 shows the macro `Dmux_1_2_E` and its flowchart. In this macro, the active high enable input E, the select input s_0, output signals y_0 and y_1, and the input signal i are all Boolean variables. When this demultiplexer is disabled with E set to 0, no output line is selected and the input signal is not passed to any output. When this demultiplexer is enabled with E set to 1, it functions as described for `Dmux_1_2`. This means that when E = 1: if the select input $s_0 = 0$, then the input signal i is passed to the output line y_0. When E = 1: if the select input $s_0 = 1$, then the input signal i is passed to the output line y_1.

TABLE 12.1
Symbol and Truth Table of the Macro `Dmux_1_2`

Symbol

(Symbol drawing labels: i, y_0, y_1, s_1)

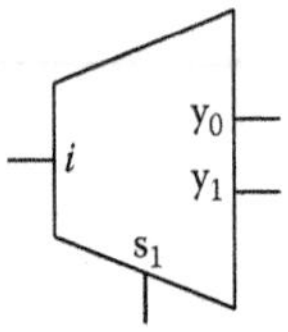

i =	regi,biti
s0 =	regs0,bits0
y0 =	rego0,bito0
y1 =	rego1,bito1

Truth Table

input	outputs	
s0	y0	y1
0	i	0
1	0	i

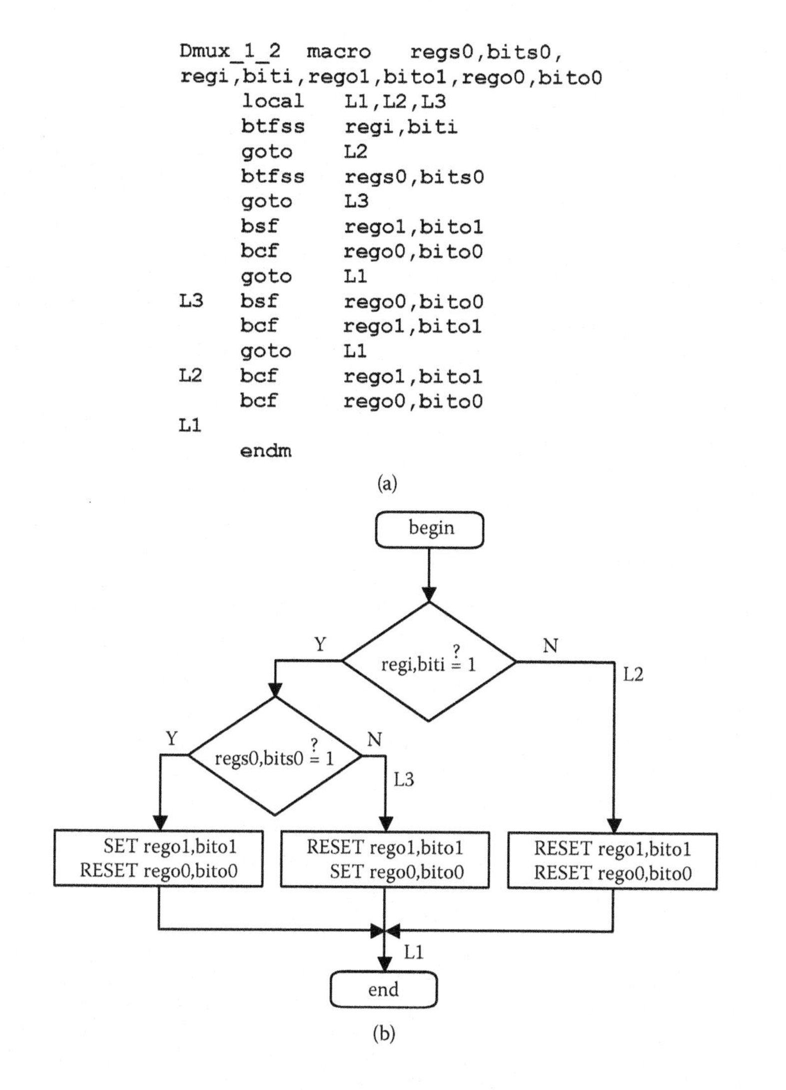

```
Dmux_1_2  macro    regs0,bits0,
regi,biti,rego1,bito1,rego0,bito0
     local    L1,L2,L3
     btfss    regi,biti
     goto     L2
     btfss    regs0,bits0
     goto     L3
     bsf      rego1,bito1
     bcf      rego0,bito0
     goto     L1
L3   bsf      rego0,bito0
     bcf      rego1,bito1
     goto     L1
L2   bcf      rego1,bito1
     bcf      rego0,bito0
L1
     endm
```

(a)

FIGURE 12.2
(a) The macro `Dmux_1_2` and (b) its flowchart.

12.3 Macro `Dmux_1_4`

The symbol and the truth table of the macro `Dmux_1_4` are depicted in Table 12.3. Figure 12.4 shows the macro `Dmux_1_4` and its flowchart. In this macro, select inputs s_1 and s_0, output signals y_0, y_1, y_2, and y_3, and the input signal *i* are all Boolean variables. When the select inputs are $s_1s_0 = 00$ (respectively, 01, 10, 11), the input signal *i* is passed to the output line y_0 (respectively, y_1, y_2, y_3).

TABLE 12.2
Symbol and Truth Table of the Macro `Dmux_1_2_E`

Symbol

E, i, y_0, y_1, s_0

W	E
i =	regi,biti
s0 =	regs0,bits0
y0 =	rego0,bito0
y1 =	rego1,bito1

Truth Table

inputs		outputs	
E	s0	y0	y1
0	×	0	0
1	0	i	0
1	1	0	i

×: don't care.

12.4 Macro `Dmux_1_4_E`

The symbol and the truth table of the macro `Dmux_1_4_E` are depicted in Table 12.4. Figures 12.5 and 12.6 show the macro `Dmux_1_4_E` and its flowchart, respectively. In this macro, the active high enable input E, select inputs s_1 and s_0, output signals y_0, y_1, y_2, and y_3, and the input signal i are all Boolean variables. When this demultiplexer is disabled with E set to 0, no output line is selected and the input signal is not passed to any output. When this demultiplexer is enabled with E set to 1, it functions as described

```
Dmux_1_2_E  macro    regs0,bits0,
regi,biti,rego1,bito1,rego0,bito0
            local    L1,L2,L3
            movwf    Temp_1
            btfss    Temp_1,0
            goto     L2
            btfss    regi,biti
            goto     L2
            btfss    regs0,bits0
            goto     L3
            bsf      rego1,bito1
            bcf      rego0,bito0
            goto     L1
L3          bsf      rego0,bito0
            bcf      rego1,bito1
            goto     L1
L2          bcf      rego1,bito1
            bcf      rego0,bito0
L1
            endm
```

(a)

FIGURE 12.3
(a) The macro `Dmux_1_2_E` and (b) its flowchart. (*Continued*)

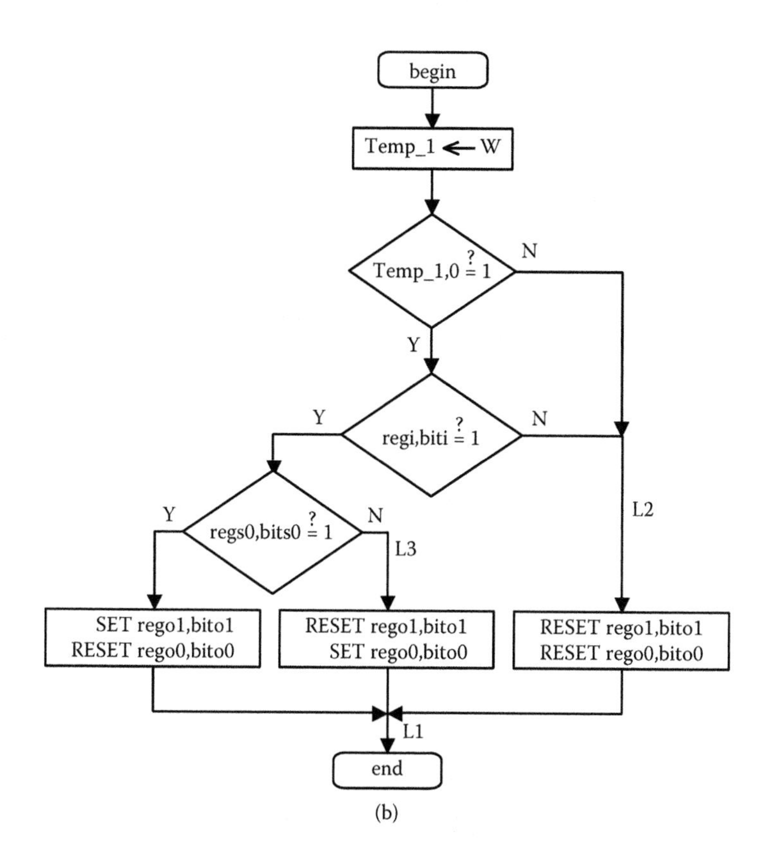

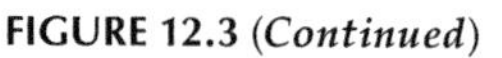
FIGURE 12.3 (*Continued*)
(a) The macro `Dmux_1_2_E` and (b) its flowchart.

TABLE 12.3

Symbol and Truth Table of the Macro `Dmux_1_4`

Symbol

i =	regi,biti
s1 =	regs1,bits1
s0 =	regs0,bits0
y3 =	rego3,bito3
y2 =	rego2,bito2
y1 =	rego1,bito1
y0 =	rego0,bito0

Truth Table

inputs		outputs			
s1	s0	y0	y1	y2	y3
0	0	*i*	0	0	0
0	1	0	*i*	0	0
1	0	0	0	*i*	0
1	1	0	0	0	*i*

```
Dmux_1_4  macro   regs1,bits1,
regs0,bits0,regi,biti,
rego3,bito3,rego2,bito2,
rego1,bito1,rego0,bito0
     local    L1,L2,L3,L4,L5
     btfss    regi,biti
     goto     L2
     btfss    regs1,bits1
     goto     L5
     bcf      rego1,bito1
     bcf      rego0,bito0
     btfss    regs0,bits0
     goto     L4
     bsf      rego3,bito3
     bcf      rego2,bito2
     goto     L1
L5   bcf      rego3,bito3
     bcf      rego2,bito2
     btfss    regs0,bits0
     goto     L3
     bsf      rego1,bito1
     bcf      rego0,bito0
     goto     L1
L4   bcf      rego3,bito3
     bsf      rego2,bito2
     goto     L1
L3   bcf      rego1,bito1
     bsf      rego0,bito0
     goto     L1
L2   bcf      rego3,bito3
     bcf      rego2,bito2
     bcf      rego1,bito1
     bcf      rego0,bito0
L1
     endm
```

(a)

FIGURE 12.4
(a) The macro Dmux_1_4 and (b) its flowchart. (*Continued*)

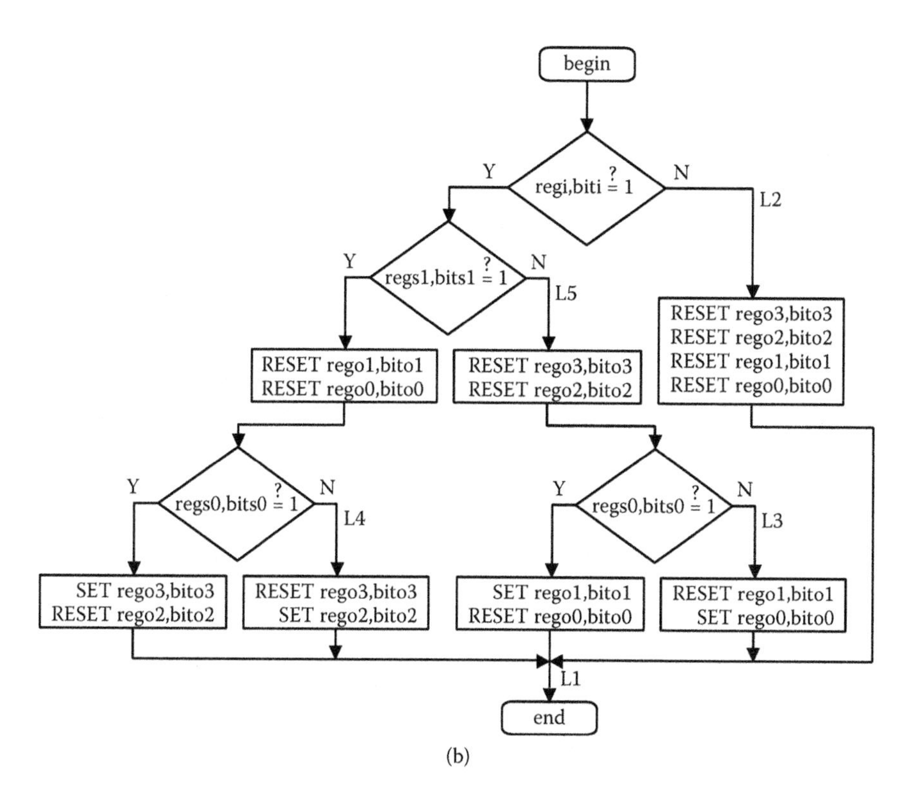

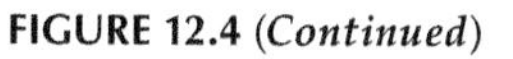
FIGURE 12.4 (*Continued*)
(a) The macro `Dmux_1_4` and (b) its flowchart.

TABLE 12.4

Symbol and Truth Table of the Macro `Dmux_1_4_E`

Symbol

W	E
i =	regi,biti
s1 =	regs1,bits1
s0 =	regs0,bits0
y3 =	rego3,bito3
y2 =	rego2,bito2
y1 =	rego1,bito1
y0 =	rego0,bito0

Truth Table

inputs			outputs			
E	s1	s0	y0	y1	y2	y3
0	×	×	0	0	0	0
1	0	0	i	0	0	0
1	0	1	0	i	0	0
1	1	0	0	0	i	0
1	1	1	0	0	0	i

×: don't care.

```
Dmux_1_4_E  macro   regs1,bits1,
regs0,bits0,regi,biti,
rego3,bito3,rego2,bito2,
rego1,bito1,rego0,bito0
     local   L1,L2,L3,L4,L5
     movwf   Temp_1
     btfss   Temp_1,0
     goto    L2
     btfss   regi,biti
     goto    L2
     btfss   regs1,bits1
     goto    L5
     bcf     rego1,bito1
     bcf     rego0,bito0
     btfss   regs0,bits0
     goto    L4
     bsf     rego3,bito3
     bcf     rego2,bito2
     goto    L1
L5   bcf     rego3,bito3
     bcf     rego2,bito2
     btfss   regs0,bits0
     goto    L3
     bsf     rego1,bito1
     bcf     rego0,bito0
     goto    L1
L4   bcf     rego3,bito3
     bsf     rego2,bito2
     goto    L1
L3   bcf     rego1,bito1
     bsf     rego0,bito0
     goto    L1
L2   bcf     rego3,bito3
     bcf     rego2,bito2
     bcf     rego1,bito1
     bcf     rego0,bito0
L1
     endm
```

FIGURE 12.5
The macro Dmux_1_4_E.

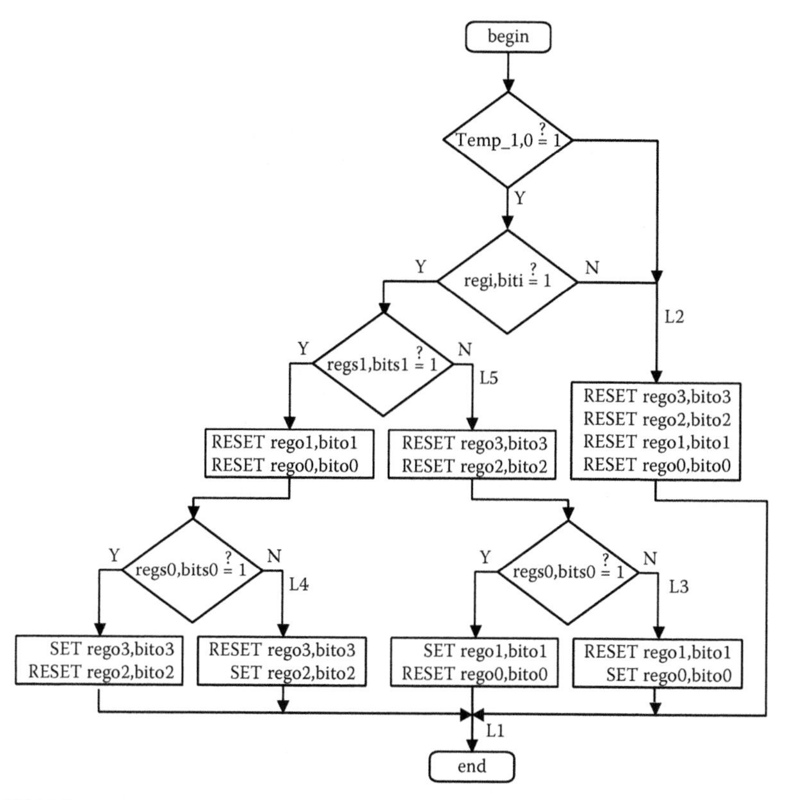

FIGURE 12.6
The flowchart of the macro `Dmux_1_4_E`.

for `Dmux_1_4`. This means that when E = 1: if the select inputs are $s_1s_0 = 00$ (respectively, 01, 10, 11), the input signal i is passed to the output line y_0 (respectively, y_1, y_2, y_3).

12.5 Macro `Dmux_1_8`

The symbol and the truth table of the macro `Dmux_1_8` are depicted in Table 12.5. Figures 12.7 and 12.8 show the macro `Dmux_1_8` and its flowchart, respectively. In this macro, the select inputs s_2, s_1, and s_0, output signals $y_0, y_1, y_2, y_3, y_4, y_5, y_6$, and y_7, and the input signal i are all Boolean variables. When the select inputs are $s_2s_1s_0 = 000$ (respectively, 001, 010, 011, 100, 101, 110, 111), the input signal i is passed to the output line y_0 (respectively, $y_1, y_2, y_3, y_4, y_5, y_6, y_7$).

TABLE 12.5

Symbol and Truth Table of the Macro `Dmux_1_8`

Symbol

i =	regi,biti
s2 =	regs2,bits2
s1 =	regs1,bits1
s0 =	regs0,bits0
y7 =	rego7,bito7
y6 =	rego6,bito6
y5 =	rego5,bito5
y4 =	rego4,bito4
y3 =	rego3,bito3
y2 =	rego2,bito2
y1 =	rego1,bito1
y0 =	rego0,bito0

Truth Table

inputs			outputs							
s2	s1	s0	y0	y1	y2	y3	y4	y5	y6	y7
0	0	0	i	0	0	0	0	0	0	0
0	0	1	0	i	0	0	0	0	0	0
0	1	0	0	0	i	0	0	0	0	0
0	1	1	0	0	0	i	0	0	0	0
1	0	0	0	0	0	0	i	0	0	0
1	0	1	0	0	0	0	0	i	0	0
1	1	0	0	0	0	0	0	0	i	0
1	1	1	0	0	0	0	0	0	0	i

×: don't care.

12.6 Macro `Dmux_1_8_E`

The symbol and the truth table of the macro `Dmux_1_8_E` are depicted in Table 12.6. Figures 12.9 and 12.10 show the macro `Dmux_1_8_E` and its flowchart, respectively. In this macro, the active high enable input E, select inputs s_2, s_1, and s_0, output signals y_0, y_1, y_2, y_3, y_4, y_5, y_6, and y_7, and the input signal i are all Boolean variables. When this demultiplexer is disabled with E set to 0, no output line is selected, and the input signal is not passed to any output. When this demultiplexer is enabled with E set to 1, it functions as described for `Dmux_1_8`. This means that when E = 1: if the select inputs are $s_2s_1s_0 = 000$ (respectively, 001, 010, 011, 100, 101, 110, 111), the input signal i is passed to the output line y_0 (respectively, y_1, y_2, y_3, y_4, y_5, y_6, and y_7).

12.7 Examples for Demultiplexer Macros

In this section, we will consider three examples, namely, UZAM_plc_16i16o_exX.asm (X = 22, 23, 24), to show the usage of demultiplexer macros. In order to test one of these examples, please take the related file UZAM_plc_16i16o_exX.asm (X = 22, 23, 24) from the CD-ROM attached to this book, and then open the program by MPLAB IDE and compile it. After that, by using the PIC programmer software, take the compiled file UZAM_plc_16i16o_exX.hex (X = 22, 23, 24), and by your PIC programmer hardware, send it to the program

```
Dmux_1_8  macro   regs2,bits2,regs1,bits1,regs0,bits0,
regi,biti,rego7,bito7,rego6,bito6,rego5,bito5,rego4,bito4,
rego3,bito3,rego2,bito2,rego1,bito1,rego0,bito0
      local   L1,L2,L3,L4,L5,L6,L7,L8,L9
      btfss   regi,biti
      goto    L2
      btfss   regs2,bits2
      goto    L9
      bcf     rego3,bito3
      bcf     rego2,bito2
      bcf     rego1,bito1
      bcf     rego0,bito0
      btfss   regs1,bits1
      goto    L8
      bcf     rego5,bito5
      bcf     rego4,bito4
      btfss   regs0,bits0
      goto    L7
      bsf     rego7,bito7
      bcf     rego6,bito6
      goto    L1
L9    bcf     rego7,bito7
      bcf     rego6,bito6
      bcf     rego5,bito5
      bcf     rego4,bito4
      btfss   regs1,bits1
      goto    L5
      bcf     rego1,bito1
      bcf     rego0,bito0
      btfss   regs0,bits0
      goto    L4
      bsf     rego3,bito3
      bcf     rego2,bito2
      goto    L1
L8    bcf     rego7,bito7
      bcf     rego6,bito6
      btfss   regs0,bits0
      goto    L6
      bsf     rego5,bito5
      bcf     rego4,bito4
      goto    L1
L7    bcf     rego7,bito7
      bsf     rego6,bito6
      goto    L1
L6    bcf     rego5,bito5
      bsf     rego4,bito4
      goto    L1
L5    bcf     rego3,bito3
      bcf     rego2,bito2
      btfss   regs0,bits0
      goto    L3
      bsf     rego1,bito1
      bcf     rego0,bito0
      goto    L1
L4    bcf     rego3,bito3
      bsf     rego2,bito2
      goto    L1
L3    bcf     rego1,bito1
      bsf     rego0,bito0
      goto    L1
L2    bcf     rego7,bito7
      bcf     rego6,bito6
      bcf     rego5,bito5
      bcf     rego4,bito4
      bcf     rego3,bito3
      bcf     rego2,bito2
      bcf     rego1,bito1
      bcf     rego0,bito0
L1
      endm
```

FIGURE 12.7
The macro Dmux_1_8.

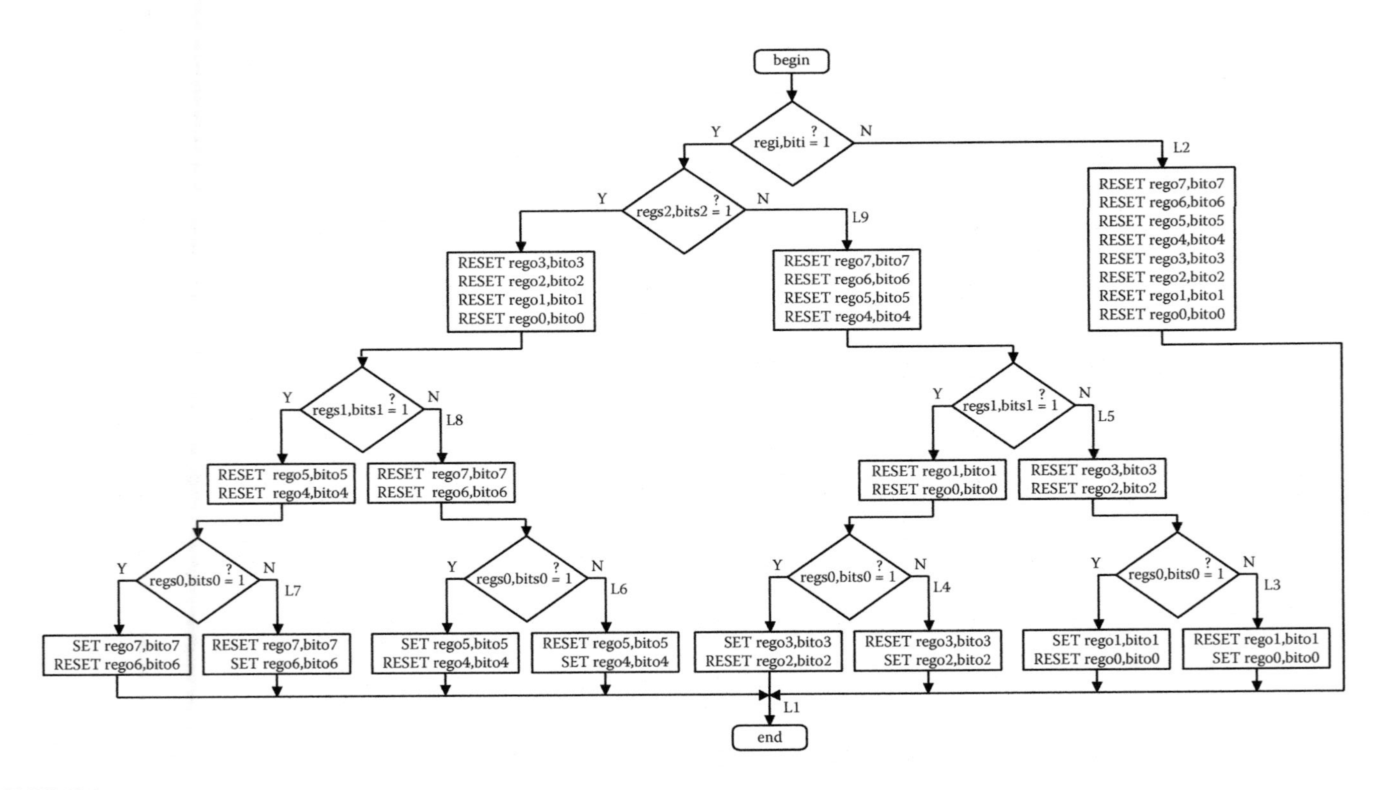

FIGURE 12.8
The flowchart of the macro Dmux_1_8.

TABLE 12.6

Symbol and Truth Table of the Macro `Dmux_1_8_E`

Symbol

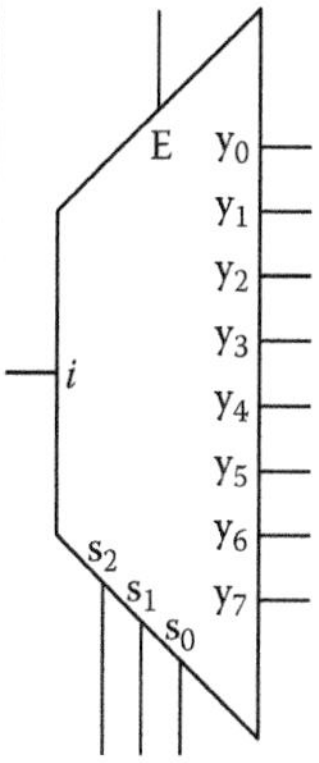

W	E
i =	regi,biti
s2 =	regs2,bits2
s1 =	regs1,bits1
s0 =	regs0,bits0
y7 =	rego7,bito7
y6 =	rego6,bito6
y5 =	rego5,bito5
y4 =	rego4,bito4
y3 =	rego3,bito3
y2 =	rego2,bito2
y1 =	rego1,bito1
y0 =	rego0,bito0

Truth Table

inputs				outputs							
E	s2	s1	s0	y0	y1	y2	y3	y4	y5	y6	y7
0	×	×	×	0	0	0	0	0	0	0	0
1	0	0	0	i	0	0	0	0	0	0	0
1	0	0	1	0	i	0	0	0	0	0	0
1	0	1	0	0	0	i	0	0	0	0	0
1	0	1	1	0	0	0	i	0	0	0	0
1	1	0	0	0	0	0	0	i	0	0	0
1	1	0	1	0	0	0	0	0	i	0	0
1	1	1	0	0	0	0	0	0	0	i	0
1	1	1	1	0	0	0	0	0	0	0	i

×: don't care.

memory of PIC16F648A microcontroller within the PIC16F648A-based PLC. To do this, switch the 4PDT in PROG position and the power switch in OFF position. After loading the file UZAM_plc_16i16o_ex*X*.hex (*X* = 22, 23, 24), switch the 4PDT in RUN and the power switch in ON position. Please check the program's accuracy by cross-referencing it with the related macros.

Let us now consider these example programs: The first example program, UZAM_plc_16i16o_ex22.asm, is shown in Figure 12.11. It shows the usage of two demultiplexer macros `Dmux_1_2` and `Dmux _1_2_E`. The schematic diagram of the user program of UZAM_plc_16i16o_ex22.asm, shown in Figure 12.11, is depicted in Figure 12.12.

In the first rung, the demultiplexer macro `Dmux_1_2` (1 × 2 demultiplexer) is used. In this demultiplexer, the input signal is i = I0.1, and the select input is s_0 = I0.0, while the output lines are y_0 = Q0.0 and y_1 = Q0.1.

In the second rung, another demultiplexer macro `Dmux_1_2` (1 × 2 demultiplexer) is used. In this demultiplexer, the input signal is i = T1.4 (419.4304 ms), and the select input is s_0 = I0.7, while the output lines are y_0 = Q0.6 and y_1 = Q0.7.

In the third rung, the macro `Dmux_1_2_E` (1 × 2 demultiplexer with active high enable input) is used. In this demultiplexer, the input signal is i = I1.2, and the select input is s_0 = I1.1, while the output lines are y_0 = Q1.0 and y_1 = Q1.1. In addition, the active high enable input E is defined to be E = I1.0.

In the fourth and last rung, another macro `Dmux_1_2_E` (1 × 2 demultiplexer with active high enable input) is used. In this demultiplexer, the input signal is i = T1.3 (209.7152 ms), and the select input is s_0 = I1.6, while the

```
Dmux_1_8_E  macro    regs2,bits2,regs1,bits1,regs0,bits0,
regi,biti,rego7,bito7,rego6,bito6,rego5,bito5,
rego4,bito4,rego3,bito3,rego2,bito2,rego1,bito1,rego0,bito0
    local   L1,L2,L3,L4,L5,L6,L7,L8,L9
    movwf   Temp_1
    btfss   Temp_1,0
    goto    L2
    btfss   regi,biti
    goto    L2
    btfss   regs2,bits2
    goto    L9
    bcf     rego3,bito3
    bcf     rego2,bito2
    bcf     rego1,bito1
    bcf     rego0,bito0
    btfss   regs1,bits1
    goto    L8
    bcf     rego5,bito5
    bcf     rego4,bito4
    btfss   regs0,bits0
    goto    L7
    bsf     rego7,bito7
    bcf     rego6,bito6
    goto    L1
L9  bcf     rego7,bito7
    bcf     rego6,bito6
    bcf     rego5,bito5
    bcf     rego4,bito4
    btfss   regs1,bits1
    goto    L5
    bcf     rego1,bito1
    bcf     rego0,bito0
    btfss   regs0,bits0
    goto    L4
    bsf     rego3,bito3
    bcf     rego2,bito2
    goto    L1
L8  bcf     rego7,bito7
    bcf     rego6,bito6
    btfss   regs0,bits0
    goto    L6
    bsf     rego5,bito5
    bcf     rego4,bito4
    goto    L1
L7  bcf     rego7,bito7
    bsf     rego6,bito6
    goto    L1
L6  bcf     rego5,bito5
    bsf     rego4,bito4
    goto    L1
L5  bcf     rego3,bito3
    bcf     rego2,bito2
    btfss   regs0,bits0
    goto    L3
    bsf     rego1,bito1
    bcf     rego0,bito0
    goto    L1
L4  bcf     rego3,bito3
    bsf     rego2,bito2
    goto    L1
L3  bcf     rego1,bito1
    bsf     rego0,bito0
    goto    L1
L2  bcf     rego7,bito7
    bcf     rego6,bito6
    bcf     rego5,bito5
    bcf     rego4,bito4
    bcf     rego3,bito3
    bcf     rego2,bito2
    bcf     rego1,bito1
    bcf     rego0,bito0
L1
    endm
```

FIGURE 12.9
The macro Dmux_1_8_E.

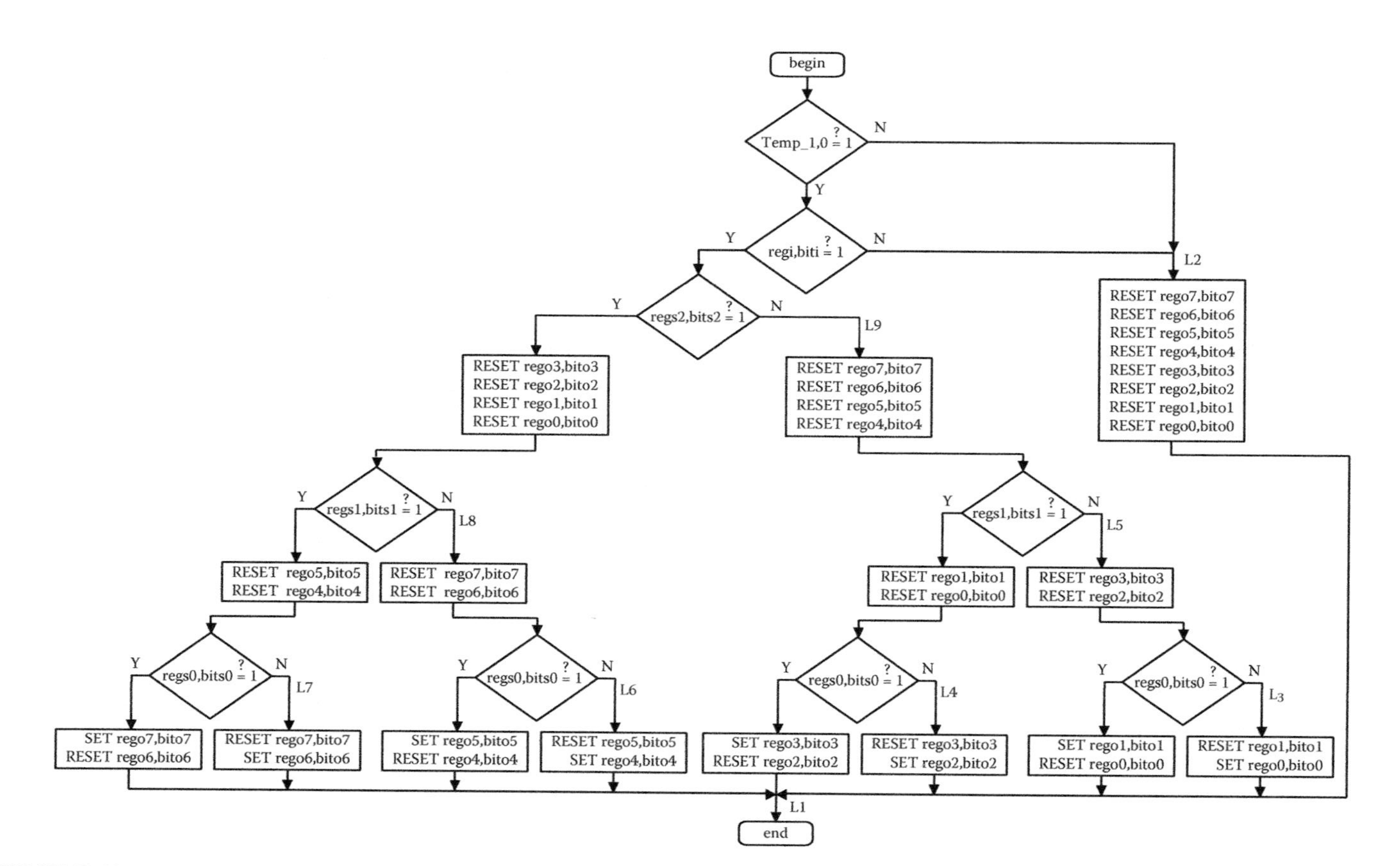

FIGURE 12.10
The flowchart of the macro Dmux_1_8_E.

```
.
;--------------- user program starts here ---
    Dmux_1_2  I0.0,I0.1,Q0.1,Q0.0        ;rung 1

    Dmux_1_2  I0.7,T1.4,Q0.7,Q0.6        ;rung 2

    ld        I1.0                       ;rung 3
    Dmux_1_2_E      I1.1,I1.2,Q1.1,Q1.0

    ld_not    I1.7                       ;rung 4
    Dmux_1_2_E      I1.6,T1.3,Q1.7,Q1.6
;--------------- user program ends here -----
.
```

FIGURE 12.11
The user program of UZAM_plc_16i16o_ex22.asm.

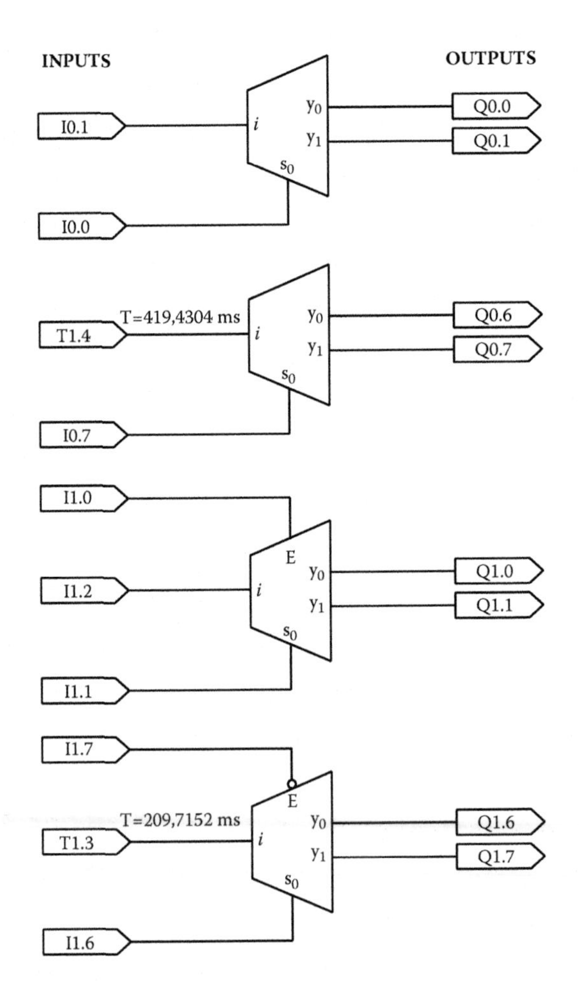

FIGURE 12.12
The schematic diagram of the user program of UZAM_plc_16i16o_ex22.asm.

```
.
;--------------- user program starts here -------------------
        Dmux_1_4   I0.1,I0.0,I0.2,Q0.3,Q0.2,Q0.1,Q0.0        ;rung 1

        Dmux_1_4   I0.7,I0.6,T1.2,Q0.7,Q0.6,Q0.5,Q0.4        ;rung 2


        ld         I1.0                                      ;rung 3
        Dmux_1_4_E      I1.2,I1.1,I1.3,Q1.3,Q1.2,Q1.1,Q1.0

        ld_not     I1.7                                      ;rung 4
        Dmux_1_4_E      I1.6,I1.5,T1.3,Q1.7,Q1.6,Q1.5,Q1.4
;--------------- user program ends here --------------------
.
```

FIGURE 12.13
The user program of UZAM_plc_16i16o_ex23.asm.

output lines are y_0 = Q1.6 and y_1 = Q1.7. In addition, the active high enable input E is defined to be E = inverted I1.7. Note that this arrangement forces the enable input E to be active low.

The second example program, UZAM_plc_16i16o_ex23.asm, is shown in Figure 12.13. It shows the usage of two demultiplexer macros `Dmux_1_4` and `Dmux _1_4_E`. The schematic diagram of the user program of UZAM_plc_16i16o_ex23.asm, shown in Figure 12.13, is depicted in Figure 12.14.

In the first rung, the demultiplexer macro `Dmux_1_4` (1 × 4 demultiplexer) is used. In this demultiplexer, the input signal is i = I0.2, and the select inputs are s_1 = I0.1 and s_0 = I0.0, while the output lines are y_0 = Q0.0, y_1 = Q0.1, y_2 = Q0.2, and y_3 = Q0.3.

In the second rung, another demultiplexer macro `Dmux_1_4` (1 × 4 demultiplexer) is used. In this demultiplexer, the input signal is i = T1.2 (104.8576 ms), and the select inputs are s_1 = I0.7 and s_0 = I0.6, while the output lines are y_0 = Q0.4, y_1 = Q0.5, y_2 = Q0.6, and y_3 = Q0.7.

In the third rung, the macro `Dmux_1_4_E` (1 × 4 demultiplexer with active high enable input) is used. In this demultiplexer, the input signal is i = I1.3, and the select inputs are s_1 = I1.2 and s_0 = I1.1, while the output lines are y_0 = Q1.0, y_1 = Q1.1, y_2 = Q1.2, and y_3 = Q1.3. In addition, the active high enable input E is defined to be E = I1.0.

In the fourth and last rung, another macro `Dmux_1_4_E` (1 × 4 demultiplexer with active high enable input) is used. In this demultiplexer, the input signal is i = T1.3 (209.7152 ms), and the select inputs are s_1 = I1.6 and s_0 = I1.5, while the output lines are y_0 = Q1.4, y_1 = Q1.5, y_2 = Q1.6, and y_3 = Q1.7. In addition, the active high enable input E is defined to be E = inverted I1.7. Note that this arrangement forces the enable input E to be active low.

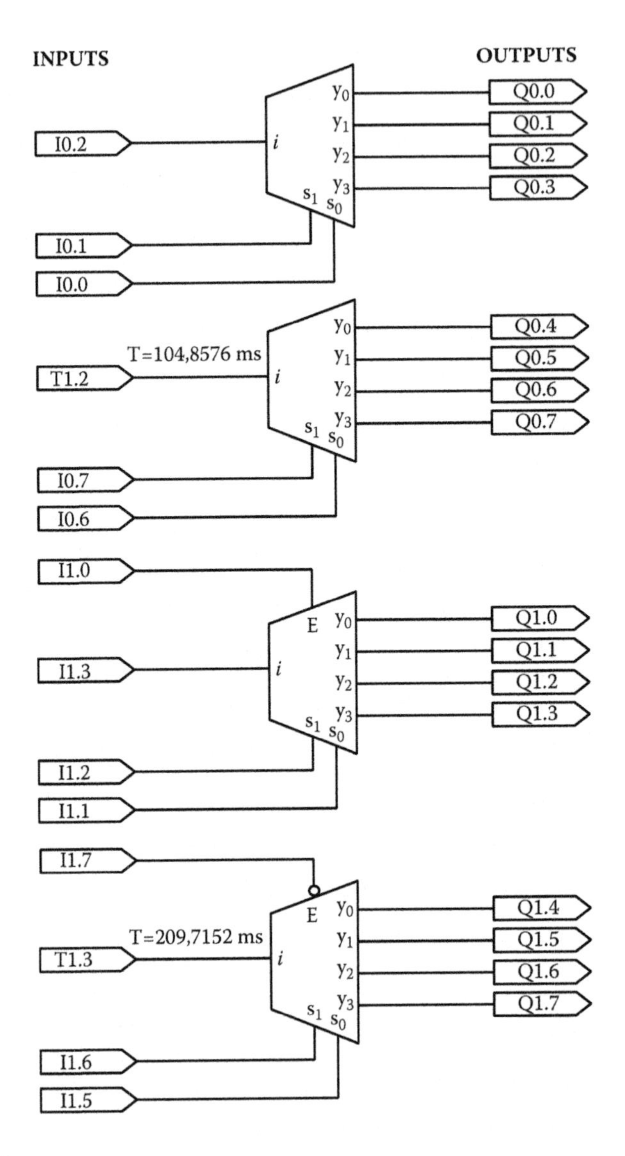

FIGURE 12.14
The schematic diagram of the user program of UZAM_plc_16i16o_ex23.asm.

```
.
;--------------- user program starts here ---------------------------------
Dmux_1_8 I0.2,I0.1,I0.0,I0.3,Q0.7,Q0.6,Q0.5,Q0.4,Q0.3,Q0.2,Q0.1,Q0.0 ;rung 1

ld       I1.0                                                          ;rung 2
Dmux_1_8_E I1.3,I1.2,I1.1,T1.3,Q1.7,Q1.6,Q1.5,Q1.4,Q1.3,Q1.2,Q1.1,Q1.0
;--------------- user program ends here -----------------------------------
.
```

FIGURE 12.15
The user program of UZAM_plc_16i16o_ex24.asm.

The third example program, UZAM_plc_16i16o_ex24.asm, is shown in Figure 12.15. It shows the usage of two demultiplexer macros `Dmux_1_8` and `Dmux _1_8_E`. The schematic diagram of the user program of UZAM_plc_16i16o_ex24.asm, shown in Figure 12.15, is depicted in Figure 12.16.

In the first rung, the demultiplexer macro `Dmux_1_8` (1 × 8 demultiplexer) is used. In this demultiplexer, the input signal is i = I0.3, and the

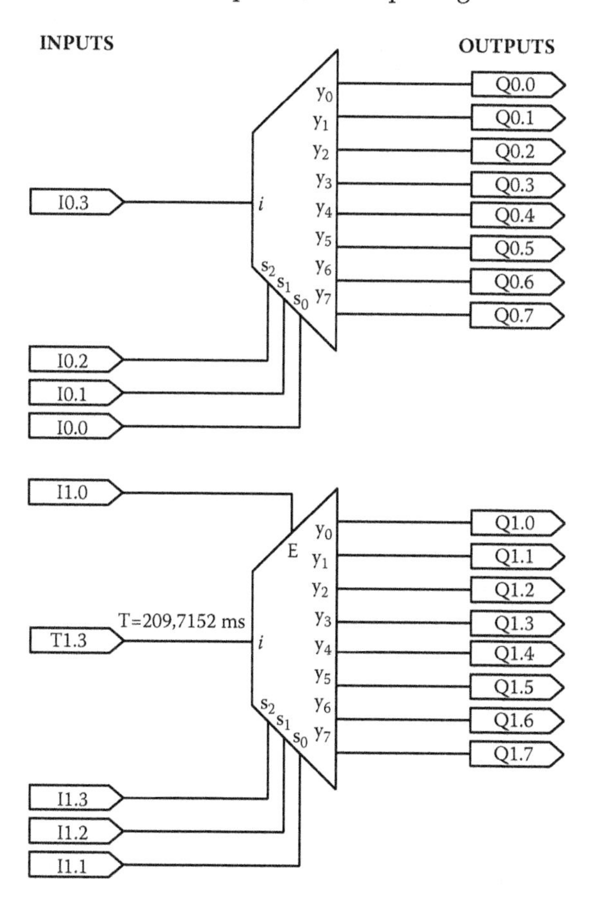

FIGURE 12.16
The schematic diagram of the user program of UZAM_plc_16i16o_ex24.asm.

select inputs are s_2 = I0.2, s_1 = I0.1, and s_0 = I0.0, while the output lines are y_0 = Q0.0, y_1 = Q0.1, y_2 = Q0.2, y_3 = Q0.3, y_4 = Q0.4, y_5 = Q0.5, y_6 = Q0.6, and y_7 = Q0.7.

In the second and last rung, the macro `Dmux_1_8_E` (1 × 8 demultiplexer with active high enable input) is used. In this demultiplexer, the input signal is i = T1.3 (209.7152 ms), and the select inputs are s_2 = I1.3, s_1 = I1.2, and s_0 = I1.1, while the output lines are y_0 = Q1.0, y_1 = Q1.1, y_2 = Q1.2, y_3 = Q1.3, y_4 = Q1.4, y_5 = Q1.5, y_6 = Q1.6, and y_7 = Q1.7. In addition, the active high enable input E is defined to be E = I1.0.

13

Decoder Macros

A decoder is a circuit that changes a code into a set of signals. It is called a decoder because it does the reverse of encoding. A common type of decoder is the line decoder, which takes an m-bit binary input datum and decodes it into 2^m data lines. As a standard combinational component, a decoder asserts one out of n output lines, depending on the value of an m-bit binary input datum. The general form of an m-to-n decoder can be seen in Figure 13.1. In general, an m-to-n decoder has m input lines, $i_{m-1}, \ldots, i_1, i_0$, and n output lines, $d_{n-1}, \ldots, d_1, d_0$, where $n = 2^m$. Although not shown in Figure 13.1, in addition, it may have an enable line, E, for enabling the decoder. When the decoder is disabled with E set to 0 (for active high enable input E), all the output lines are de-asserted. When the decoder is enabled, then the output line whose index is equal to the value of the input binary data is asserted (set to 1 for active high), while the rest of the output lines are de-asserted (set to 0 for active high). A decoder is used in a system having multiple components, and we want only one component to be selected or enabled at any time.

In this chapter, the following decoder macros are described for the PIC16F648A-based PLC:

- `decod_1_2` (1 × 2 decoder)
- `decod_1_2_AL` (1 × 2 decoder with active low outputs)
- `decod_1_2_E` (1 × 2 decoder with enable input)
- `decod_1_2_E_AL` (1 × 2 decoder with enable input and active low outputs)
- `decod_2_4` (2 × 4 decoder)
- `decod_2_4_AL` (2 × 4 decoder with active low outputs)
- `decod_2_4_E` (2 × 4 decoder with enable input)
- `decod_2_4_E_AL` (2 × 4 decoder with enable input and active low outputs)
- `decod_3_8` (3 × 8 decoder)
- `decod_3_8_AL` (3 × 8 decoder with active low outputs)
- `decod_3_8_E` (3 × 8 decoder with enable input)
- `decod_3_8_E_AL` (3 × 8 decoder with enable input and active low outputs)

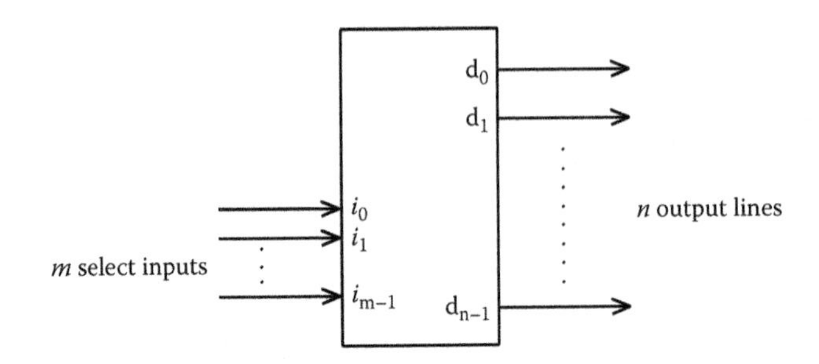

FIGURE 13.1
The general form of an m-to-n decoder, where $n = 2^m$.

The file definitions.inc, included within the CD-ROM attached to this book, contains all decoder macros defined for the PIC16F648A-based PLC. Let us now consider these macros in detail.

13.1 Macro `decod_1_2`

The symbol and the truth table of the macro `decod_1_2` are depicted in Table 13.1. Figure 13.2 shows the macro `decod_1_2` and its flowchart. This macro defines a 1 × 2 decoder with active high outputs. In this macro, the select input A and output signals d_0 and d_1 are all Boolean variables. In this decoder, when the select input is A = 0, the output line d_0 is asserted (set to 1) and the output line d_1 is de-asserted (set to 0). Similarly, when the select input is A = 1, the output line d_1 is asserted (set to 1) and the output line d_0 is de-asserted (set to 0).

TABLE 13.1
Symbol and Truth Table of the Macro `decod_1_2`

Symbol			Truth Table		
1×2 DECODER (inputs: A; outputs: d_0, d_1)	A =	regs0,bits0	input	outputs	
	d0 =	regd0,bitd0	A	d0	d1
	d1 =	regd1,bitd1	0	1	0
			1	0	1

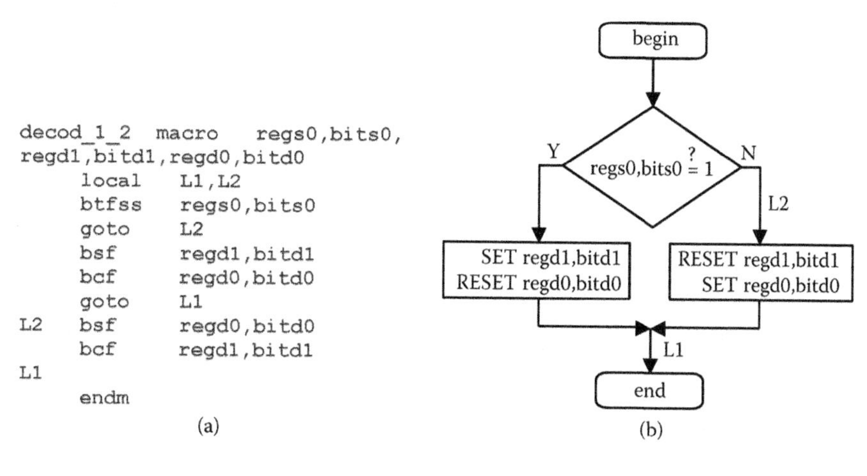

FIGURE 13.2
(a) The macro `decod_1_2` and (b) its flowchart.

13.2 Macro `decod_1_2_AL`

The symbol and the truth table of the macro `decod_1_2_AL` are depicted in Table 13.2. Figure 13.3 shows the macro `decod_1_2_AL` and its flowchart. This macro defines a 1 × 2 decoder with active low outputs. In this macro, the select input A and active low output signals d_0 and d_1 are all Boolean variables. In this decoder, when the select input is A = 0, the output line d_0 is asserted (set to 0) and the output line d_1 is de-asserted (set to 1). Similarly, when the select input is A = 1, the output line d_1 is asserted (set to 0) and the output line d_0 is de-asserted (set to 1).

TABLE 13.2
Symbol and Truth Table of the Macro `decod_1_2_AL`

Symbol:

1×2 DECODER (input A; active low outputs d_0, d_1)

A =	regs0,bits0
d0 =	regd0,bitd0
d1 =	regd1,bitd1

Truth Table:

input	outputs	
A	d0	d1
0	0	1
1	1	0

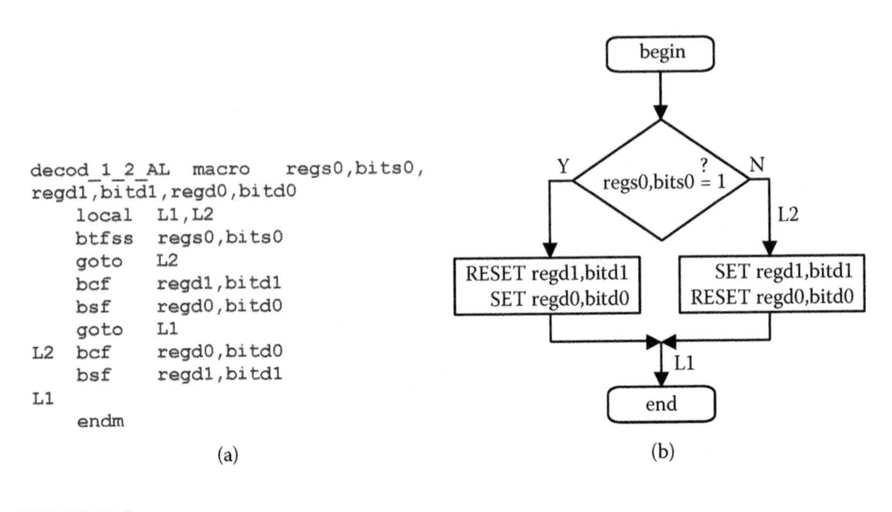

FIGURE 13.3
(a) The macro decod_1_2_AL and (b) its flowchart.

13.3 Macro decod_1_2_E

The symbol and the truth table of the macro decod_1_2_E are depicted in Table 13.3. Figure 13.4 shows the macro decod_1_2_E and its flowchart. This macro defines a 1 × 2 decoder with enable input and active high outputs. In this macro, the active high enable input E, the select input A, and active high output signals d_0 and d_1 are all Boolean variables. In addition to the decod_1_2, this decoder macro has an active high enable line, E, for enabling it. When this decoder is disabled with E set to 0, all output lines are de-asserted (set to 0). When this decoder is enabled with E set to 1, it functions as described for decod_1_2. This means that when E = 1: if the select input is A = 0, then the output line d_0 is asserted (set to 1) and the output line d_1 is de-asserted (set to 0). Similarly, when E = 1: if the select input

TABLE 13.3

Symbol and Truth Table of the Macro decod_1_2_E

Symbol:

1×2 DECODER (inputs A, E; outputs d_0, d_1)

W	E
A =	regs0,bits0
d0 =	regd0,bitd0
d1 =	regd1,bitd1

Truth Table:

inputs		outputs	
E	A	d0	d1
0	×	0	0
1	0	1	0
1	1	0	1

×: don't care.

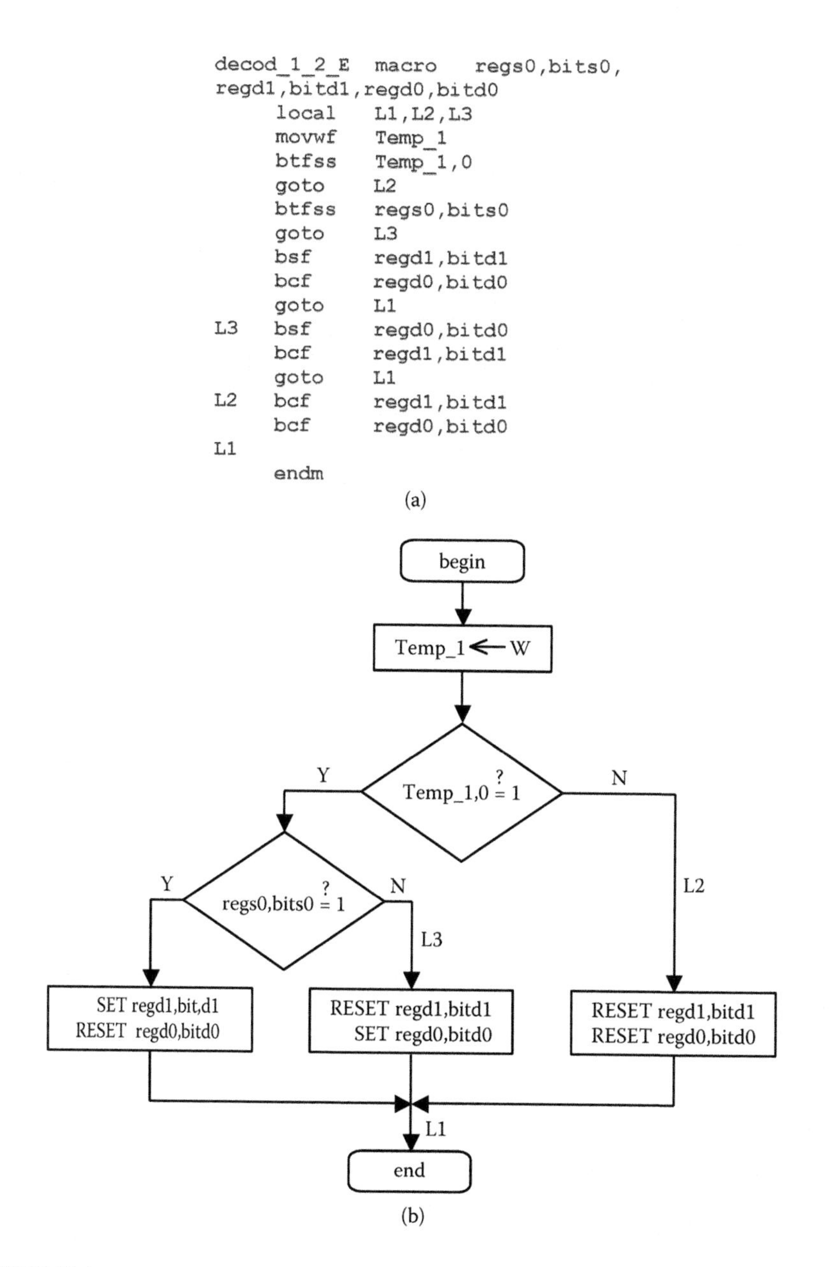

```
decod_1_2_E  macro   regs0,bits0,
regd1,bitd1,regd0,bitd0
     local    L1,L2,L3
     movwf    Temp_1
     btfss    Temp_1,0
     goto     L2
     btfss    regs0,bits0
     goto     L3
     bsf      regd1,bitd1
     bcf      regd0,bitd0
     goto     L1
L3   bsf      regd0,bitd0
     bcf      regd1,bitd1
     goto     L1
L2   bcf      regd1,bitd1
     bcf      regd0,bitd0
L1
     endm
```


FIGURE 13.4
(a) The macro `decod_1_2_E` and (b) its flowchart.

is A = 1, then the output line d_1 is asserted (set to 1) and the output line d_0 is de-asserted (set to 0).

13.4 Macro decod_1_2_E_AL

The symbol and the truth table of the macro decod_1_2_E_AL are depicted in Table 13.4. Figure 13.5 shows the macro decod_1_2_E_AL and its flowchart. This macro defines a 1 × 2 decoder with enable input and active low outputs. In this macro, the active high enable input E, the select input A, and active low output signals d_0 and d_1 are all Boolean variables. In addition to the decod_1_2_AL, this decoder macro has an active high enable line, E, for enabling it. When this decoder is disabled with E set to 0,

TABLE 13.4

Symbol and Truth Table of the Macro decod_1_2_E_AL

Symbol

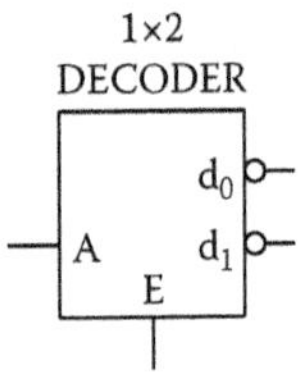

W	E
A =	regs0,bits0
d0 =	regd0,bitd0
d1 =	regd1,bitd1

Truth Table

inputs		outputs	
E	A	d0	d1
0	×	1	1
1	0	0	1
1	1	1	0

×: don't care.

```
decod_1_2_E_AL   macro   regs0,bits0,
regd1,bitd1,regd0,bitd0
          local    L1,L2,L3
          movwf    Temp_1
          btfss    Temp_1,0
          goto     L2
          btfss    regs0,bits0
          goto     L3
          bcf      regd1,bitd1
          bsf      regd0,bitd0
          goto     L1
L3        bcf      regd0,bitd0
          bsf      regd1,bitd1
          goto     L1
L2        bsf      regd1,bitd1
          bsf      regd0,bitd0
L1
          endm
```

(a)

FIGURE 13.5
(a) The macro decod_1_2_E_AL and (b) its flowchart. (*Continued*)

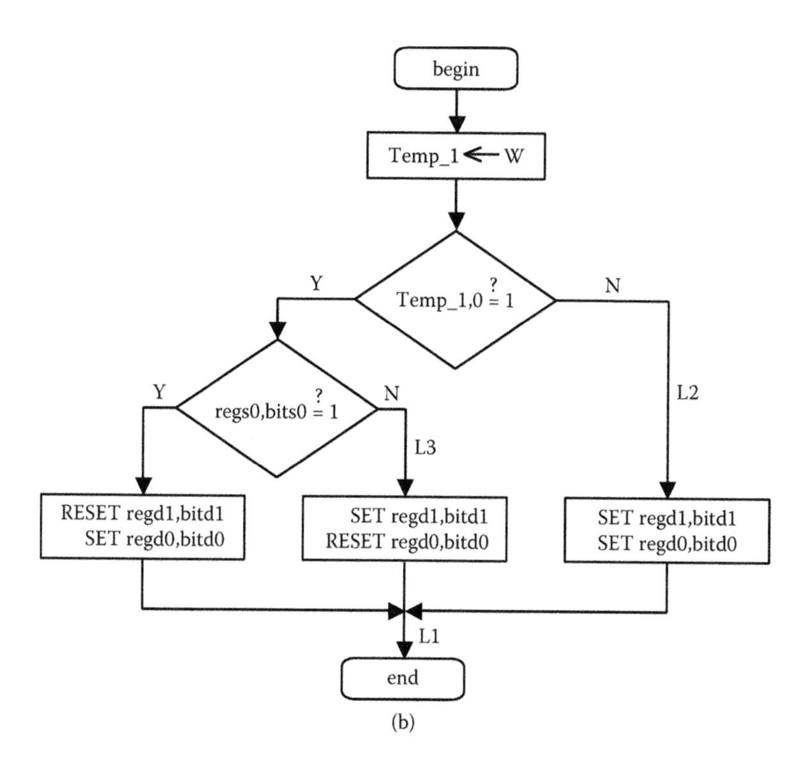

FIGURE 13.5 (*Continued*)
(a) The macro `decod_1_2_E_AL` and (b) its flowchart.

all output lines are de-asserted (set to 1). When this decoder is enabled with E set to 1, it functions as described for `decod_1_2_AL`. This means that when E = 1: if the select input is A = 0, then the output line d_0 is asserted (set to 0) and the output line d_1 is de-asserted (set to 1). Similarly, when E = 1: if the select input is A = 1, then the output line d_1 is asserted (set to 0) and the output line d_0 is de-asserted (set to 1).

13.5 Macro decod_2_4

The symbol and the truth table of the macro `decod_2_4` are depicted in Table 13.5. Figure 13.6 shows the macro `decod_2_4` and its flowchart. This macro defines a 2 × 4 decoder with active high outputs. In this macro, select inputs A and B, and active high output signals d_0, d_1, d_2, and d_3 are all Boolean variables. In this decoder, when the select inputs are AB = 00 (respectively, 01, 10, 11), the output line, d_0 (respectively, d_1, d_2, d_3), is asserted (set to 1) and all other output lines are de-asserted (set to 0).

TABLE 13.5

Symbol and Truth Table of the Macro decod_2_4

Symbol			Truth Table					
2×4 DECODER (inputs A, B; outputs d_0, d_1, d_2, d_3)	A =	regs1,bits1	inputs		outputs			
	B =	regs0,bits0	A	B	d0	d1	d2	d3
	d3 =	regd3,bitd3	0	0	1	0	0	0
	d2 =	regd2,bitd2	0	1	0	1	0	0
	d1 =	regd1,bitd1	1	0	0	0	1	0
	d0 =	regd0,bitd0	1	1	0	0	0	1

13.6 Macro decod_2_4_AL

The symbol and the truth table of the macro decod_2_4_AL are depicted in Table 13.6. Figure 13.7 shows the macro decod_2_4_AL and its flowchart. This macro defines a 2 × 4 decoder with active low outputs. In this macro, select inputs A and B, and active low output signals d_0, d_1, d_2, and d_3 are all Boolean variables. In this decoder, when the select inputs are

```
decod_2_4  macro   regs1,bits1,
regs0,bits0,regd3,bitd3,
regd2,bitd2,regd1,bitd1,regd0,bitd0
          local    L1,L2,L3,L4
          btfss    regs1,bits1
          goto     L4
          bcf      regd1,bitd1
          bcf      regd0,bitd0
          btfss    regs0,bits0
          goto     L3
          bsf      regd3,bitd3
          bcf      regd2,bitd2
          goto     L1
L4        bcf      regd3,bitd3
          bcf      regd2,bitd2
          btfss    regs0,bits0
          goto     L2
          bsf      regd1,bitd1
          bcf      regd0,bitd0
          goto     L1
L3        bcf      regd3,bitd3
          bsf      regd2,bitd2
          goto     L1
L2        bcf      regd1,bitd1
          bsf      regd0,bitd0
L1
          endm
```

(a)

FIGURE 13.6
(a) The macro decod_2_4 and (b) its flowchart. (*Continued*)

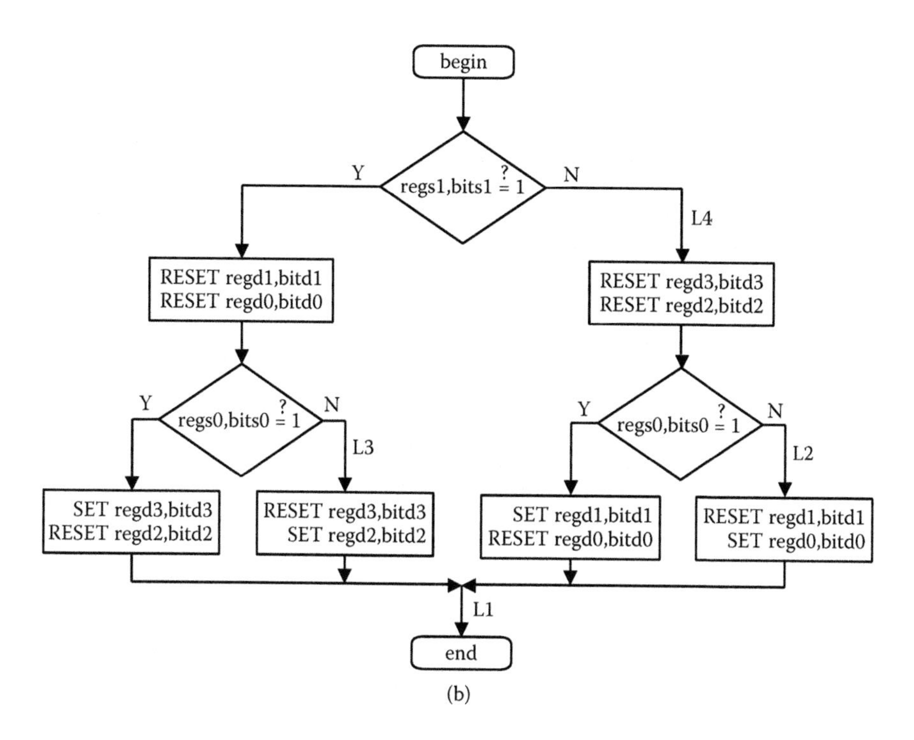

FIGURE 13.6 (*Continued*)
(a) The macro `decod_2_4` and (b) its flowchart.

TABLE 13.6

Symbol and Truth Table of the Macro `decod_2_4_AL`

Symbol

2×4 DECODER (inputs A, B; active-low outputs d_0, d_1, d_2, d_3)

A =	regs1,bits1
B =	regs0,bits0
d3 =	regd3,bitd3
d2 =	regd2,bitd2
d1 =	regd1,bitd1
d0 =	regd0,bitd0

Truth Table

inputs		outputs			
A	B	d0	d1	d2	d3
0	0	0	1	1	1
0	1	1	0	1	1
1	0	1	1	0	1
1	1	1	1	1	0

```
decod_2_4_AL  macro    regs1,bits1,
regs0,bits0,regd3,bitd3,regd2,bitd2,
regd1,bitd1,regd0,bitd0
      local    L1,L2,L3,L4
      btfss    regs1,bits1
      goto     L4
      bsf      regd1,bitd1
      bsf      regd0,bitd0
      btfss    regs0,bits0
      goto     L3
      bcf      regd3,bitd3
      bsf      regd2,bitd2
      goto     L1
L4    bsf      regd3,bitd3
      bsf      regd2,bitd2
      btfss    regs0,bits0
      goto     L2
      bcf      regd1,bitd1
      bsf      regd0,bitd0
      goto     L1
L3    bsf      regd3,bitd3
      bcf      regd2,bitd2
      goto     L1
L2    bsf      regd1,bitd1
      bcf      regd0,bitd0
L1
      endm
```

(a)

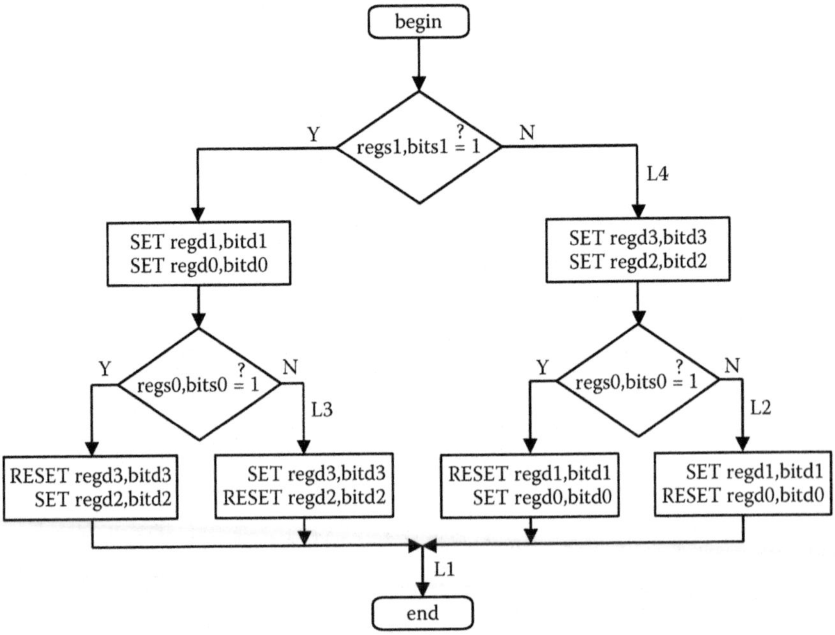

(b)

FIGURE 13.7
(a) The macro `decod_2_4_AL` and (b) its flowchart.

TABLE 13.7
Symbol and Truth Table of the Macro `decod_2_4_E`

W	E
A =	regs1,bits1
B =	regs0,bits0
d3 =	regd3,bitd3
d2 =	regd2,bitd2
d1 =	regd1,bitd1
d0 =	regd0,bitd0

inputs			outputs			
E	A	B	d0	d1	d2	d3
0	×	×	0	0	0	0
1	0	0	1	0	0	0
1	0	1	0	1	0	0
1	1	0	0	0	1	0
1	1	1	0	0	0	1

×: don't care.

AB = 00 (respectively, 01, 10, 11), the output line, d_0 (respectively, d_1, d_2, d_3), is asserted (set to 0) and all other output lines are de-asserted (set to 1).

13.7 Macro decod_2_4_E

The symbol and the truth table of the macro `decod_2_4_E` are depicted in Table 13.7. Figures 13.8 and 13.9 show the macro `decod_2_4_E` and its flowchart, respectively. This macro defines a 2 × 4 decoder with enable input and active high outputs. In this macro, the active high enable input E, select inputs A and B, and active high output signals d_0, d_1, d_2, and d_3 are all Boolean variables. In addition to the `decod_2_4`, this decoder macro has an active high enable line, E, for enabling it. When this decoder is disabled with E set to 0, all active high output lines are de-asserted (set to 0). When this decoder is enabled with E set to 1, it functions as described for `decod_2_4`. This means that when E = 1: if the select inputs are AB = 00 (respectively, 01, 10, 11), then the output line, d_0 (respectively, d_1, d_2, d_3), is asserted (set to 1) and all other output lines are de-asserted (set to 0).

13.8 Macro decod_2_4_E_AL

The symbol and the truth table of the macro `decod_2_4_E_AL` are depicted in Table 13.8. Figures 13.10 and 13.11 show the macro `decod_2_4_E_AL` and its flowchart, respectively. This macro defines a 2 × 4 decoder with enable input and active low outputs. In this macro, the active high enable

```
decod_2_4_E  macro   regs1,bits1,
regs0,bits0,regd3,bitd3,regd2,bitd2,
regd1,bitd1,regd0,bitd0
      local   L1,L2,L3,L4,L5
      movwf   Temp_1
      btfss   Temp_1,0
      goto    L2
      btfss   regs1,bits1
      goto    L5
      bcf     regd1,bitd1
      bcf     regd0,bitd0
      btfss   regs0,bits0
      goto    L4
      bsf     regd3,bitd3
      bcf     regd2,bitd2
      goto    L1
L5    bcf     regd3,bitd3
      bcf     regd2,bitd2
      btfss   regs0,bits0
      goto    L3
      bsf     regd1,bitd1
      bcf     regd0,bitd0
      goto    L1
L4    bcf     regd3,bitd3
      bsf     regd2,bitd2
      goto    L1
L3    bcf     regd1,bitd1
      bsf     regd0,bitd0
      goto    L1
L2    bcf     regd3,bitd3
      bcf     regd2,bitd2
      bcf     regd1,bitd1
      bcf     regd0,bitd0
L1
      endm
```

FIGURE 13.8
The macro decod_2_4_E.

input E, select inputs A and B, and active low output signals d_0, d_1, d_2, and d_3 are all Boolean variables. In addition to the decod_2_4_AL, this decoder macro has an active high enable line, E, for enabling it. When this decoder is disabled with E set to 0, all active low output lines are de-asserted (set to 1). When this decoder is enabled with E set to 1, it functions as described for decod_2_4_AL. This means that when E = 1: if the select inputs are AB = 00 (respectively, 01, 10, 11), then the output line, d_0 (respectively, d_1, d_2, d_3), is asserted (set to 0) and all other output lines are de-asserted (set to 1).

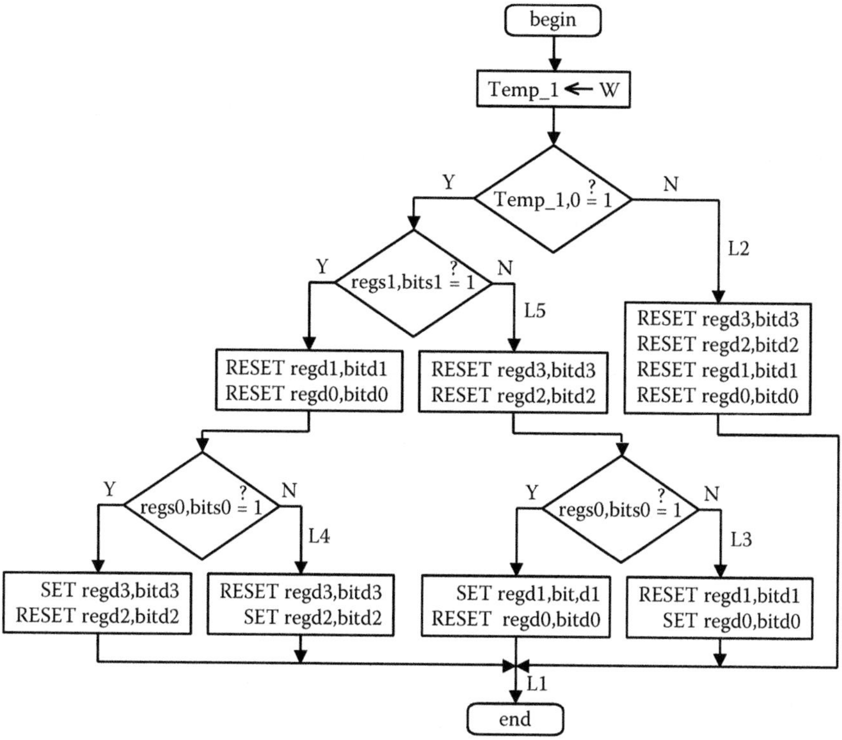

FIGURE 13.9
The flowchart of the macro `decod_2_4_E`.

TABLE 13.8
Symbol and Truth Table of the Macro `decod_2_4_E_AL`

Symbol

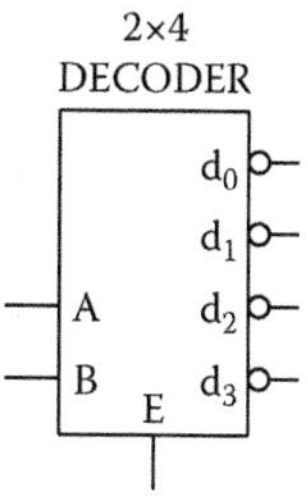

W	E
A =	regs1,bits1
B =	regs0,bits0
d3 =	regd3,bitd3
d2 =	regd2,bitd2
d1 =	regd1,bitd1
d0 =	regd0,bitd0

Truth Table

inputs			outputs			
E	A	B	d0	d1	d2	d3
0	×	×	1	1	1	1
1	0	0	0	1	1	1
1	0	1	1	0	1	1
1	1	0	1	1	0	1
1	1	1	1	1	1	0

×: don't care.

```
decod_2_4_E_AL  macro   regs1,bits1,
regs0,bits0,regd3,bitd3,regd2,bitd2,
regd1,bitd1,regd0,bitd0
        local   L1,L2,L3,L4,L5
        movwf   Temp_1
        btfss   Temp_1,0
        goto    L2
        btfss   regs1,bits1
        goto    L5
        bsf     regd1,bitd1
        bsf     regd0,bitd0
        btfss   regs0,bits0
        goto    L4
        bcf     regd3,bitd3
        bsf     regd2,bitd2
        goto    L1
L5      bsf     regd3,bitd3
        bsf     regd2,bitd2
        btfss   regs0,bits0
        goto    L3
        bcf     regd1,bitd1
        bsf     regd0,bitd0
        goto    L1
L4      bsf     regd3,bitd3
        bcf     regd2,bitd2
        goto    L1
L3      bsf     regd1,bitd1
        bcf     regd0,bitd0
        goto    L1
L2      bsf     regd3,bitd3
        bsf     regd2,bitd2
        bsf     regd1,bitd1
        bsf     regd0,bitd0
L1
        endm
```

FIGURE 13.10
The macro decod_2_4_E_AL.

13.9 Macro decod_3_8

The symbol and the truth table of the macro decod_3_8 are depicted in Table 13.9. Figures 13.12 and 13.13 show the macro decod_3_8 and its flowchart, respectively. This macro defines a 3 × 8 decoder with active high outputs. In this macro, select inputs A, B, and C, and active high output signals d_0, d_1, d_2, d_3, d_4, d_5, d_6, and d_7 are all Boolean variables. In this decoder, when the select inputs are ABC = 000 (respectively, 001, 010, 011, 100, 101, 110, 111), the output line, d_0 (respectively, d_1, d_2, d_3, d_4, d_5, d_6, d_7), is asserted (set to 1) and all other output lines are de-asserted (set to 0).

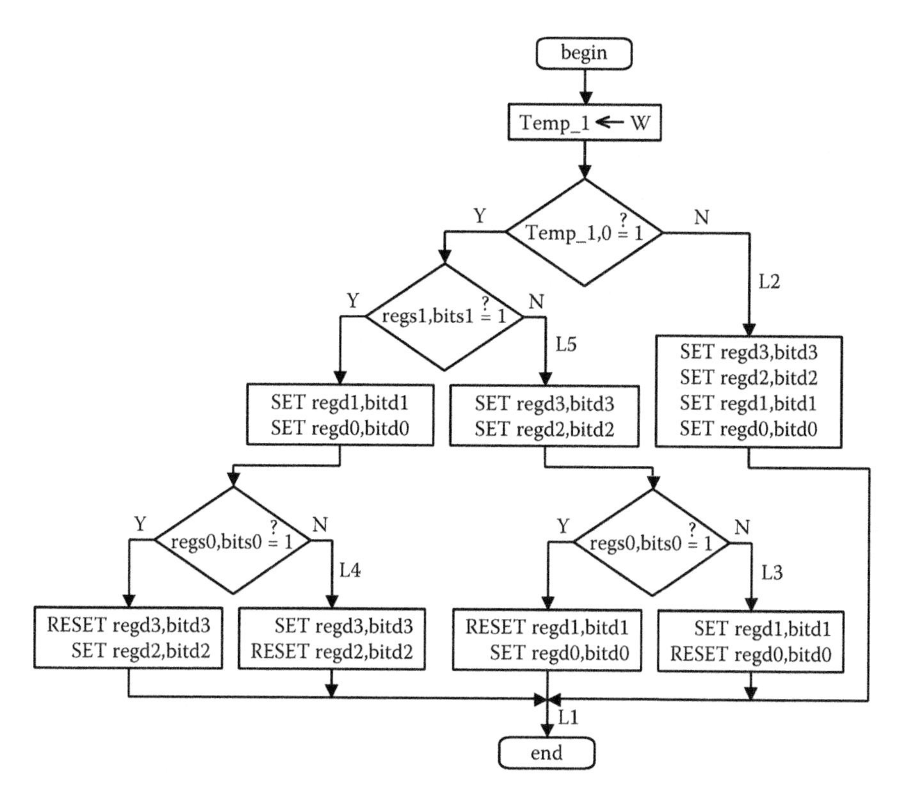

FIGURE 13.11
The flowchart of the macro `decod_2_4_E_AL`.

TABLE 13.9
Symbol and Truth Table of the Macro `decod_3_8`

Symbol

3×8 DECODER

Inputs: A, B, C — Outputs: d_0, d_1, d_2, d_3, d_4, d_5, d_6, d_7

A =	regs2,bits2
B =	regs1,bits1
C =	regs0,bits0
d7 =	rego7,bito7
d6 =	rego6,bito6
d5 =	rego5,bito5
d4 =	rego4,bito4
d3 =	rego3,bito3
d2 =	rego2,bito2
d1 =	rego1,bito1
d0 =	rego0,bito0

Truth Table

inputs			outputs							
A	B	C	d0	d1	d2	d3	d4	d5	d6	d7
0	0	0	1	0	0	0	0	0	0	0
0	0	1	0	1	0	0	0	0	0	0
0	1	0	0	0	1	0	0	0	0	0
0	1	1	0	0	0	1	0	0	0	0
1	0	0	0	0	0	0	1	0	0	0
1	0	1	0	0	0	0	0	1	0	0
1	1	0	0	0	0	0	0	0	1	0
1	1	1	0	0	0	0	0	0	0	1

```
decod_3_8  macro   regs2,bits2,regs1,bits1,
regs0,bits0,regd7,bitd7,regd6,bitd6,
regd5,bitd5,regd4,bitd4,regd3,bitd3,
regd2,bitd2,regd1,bitd1,regd0,bitd0
     local   L1,L2,L3,L4,L5,L6,L7,L8
     btfss   regs2,bits2
     goto    L8
     bcf     regd3,bitd3
     bcf     regd2,bitd2
     bcf     regd1,bitd1
     bcf     regd0,bitd0
     btfss   regs1,bits1
     goto    L7
     bcf     regd5,bitd5
     bcf     regd4,bitd4
     btfss   regs0,bits0
     goto    L6
     bsf     regd7,bitd7
     bcf     regd6,bitd6
     goto    L1
L8   bcf     regd7,bitd7
     bcf     regd6,bitd6
     bcf     regd5,bitd5
     bcf     regd4,bitd4
     btfss   regs1,bits1
     goto    L4
     bcf     regd1,bitd1
     bcf     regd0,bitd0
     btfss   regs0,bits0
     goto    L3
     bsf     regd3,bitd3
     bcf     regd2,bitd2
     goto    L1
L7   bcf     regd7,bitd7
     bcf     regd6,bitd6
     btfss   regs0,bits0
     goto    L5
     bsf     regd5,bitd5
     bcf     regd4,bitd4
     goto    L1
L6   bcf     regd7,bitd7
     bsf     regd6,bitd6
     goto    L1
L5   bcf     regd5,bitd5
     bsf     regd4,bitd4
     goto    L1
L4   bcf     regd3,bitd3
     bcf     regd2,bitd2
     btfss   regs0,bits0
     goto    L2
     bsf     regd1,bitd1
     bcf     regd0,bitd0
     goto    L1
L3   bcf     regd3,bitd3
     bsf     regd2,bitd2
     goto    L1
L2   bcf     regd1,bitd1
     bsf     regd0,bitd0
L1
     endm
```

FIGURE 13.12
The macro decod_3_8.

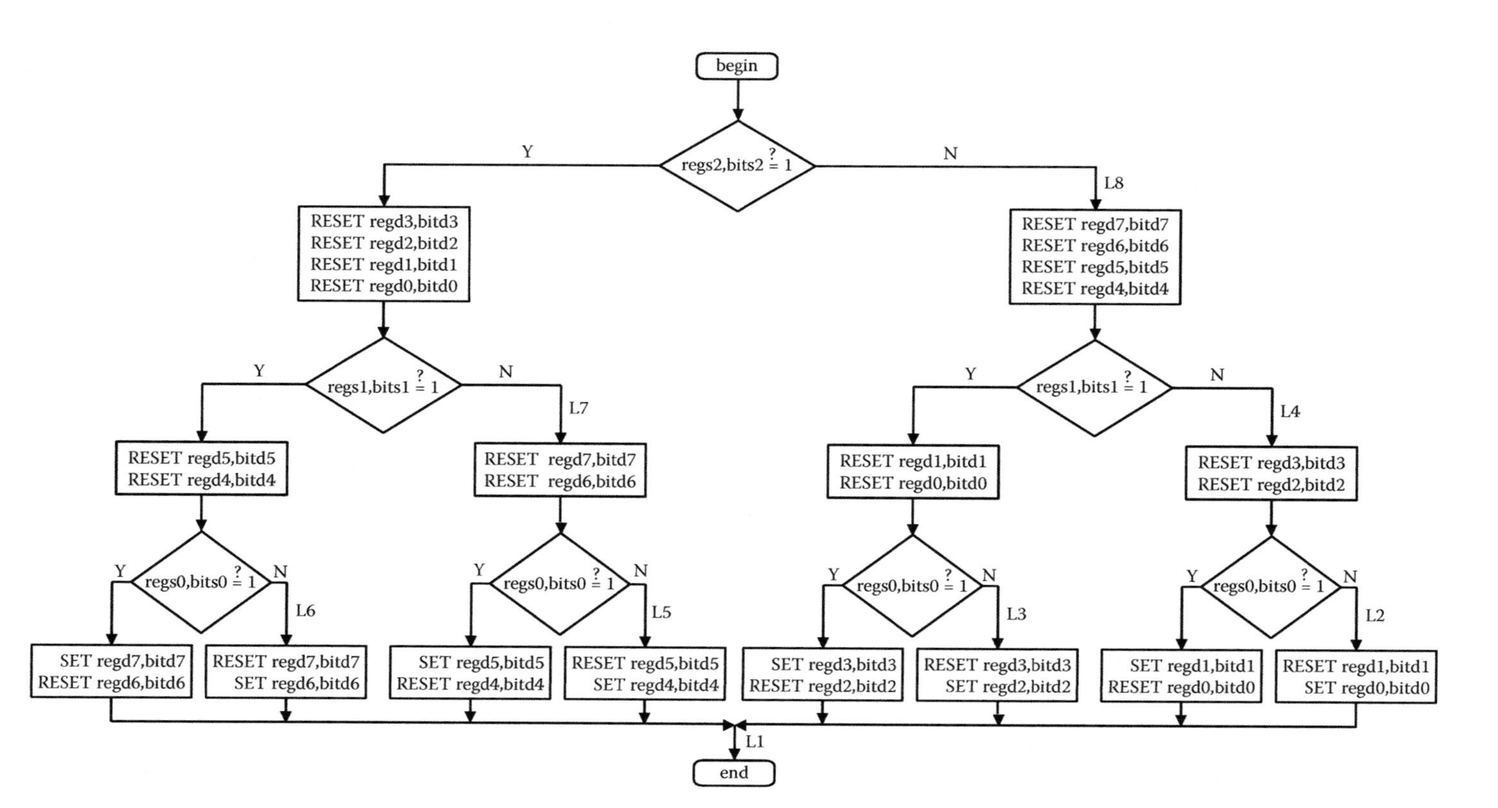

FIGURE 13.13
The flowchart of the macro decod_3_8.

TABLE 13.10
Symbol and Truth Table of the Macro decod_3_8_AL

Symbol:

3×8 DECODER (inputs A, B, C; active low outputs d_0, d_1, d_2, d_3, d_4, d_5, d_6, d_7)

A =	regs2,bits2
B =	regs1,bits1
C =	regs0,bits0
d7 =	rego7,bito7
d6 =	rego6,bito6
d5 =	rego5,bito5
d4 =	rego4,bito4
d3 =	rego3,bito3
d2 =	rego2,bito2
d1 =	rego1,bito1
d0 =	rego0,bito0

Truth Table:

inputs			outputs							
A	B	C	d0	d1	d2	d3	d4	d5	d6	d7
0	0	0	0	1	1	1	1	1	1	1
0	0	1	1	0	1	1	1	1	1	1
0	1	0	1	1	0	1	1	1	1	1
0	1	1	1	1	1	0	1	1	1	1
1	0	0	1	1	1	1	0	1	1	1
1	0	1	1	1	1	1	1	0	1	1
1	1	0	1	1	1	1	1	1	0	1
1	1	1	1	1	1	1	1	1	1	0

13.10 Macro decod_3_8_AL

The symbol and the truth table of the macro decod_3_8_AL are depicted in Table 13.10. Figures 13.14 and 13.15 show the macro decod_3_8_AL and its flowchart, respectively. This macro defines a 3 × 8 decoder with active low outputs. In this macro, select inputs A, B, and C, and active low output signals d_0, d_1, d_2, d_3, d_4, d_5, d_6, and d_7 are all Boolean variables. In this decoder, when the select inputs are ABC = 000 (respectively, 001, 010, 011, 100, 101, 110, 111), the output line, d_0 (respectively, d_1, d_2, d_3, d_4, d_5, d_6, d_7), is asserted (set to 0) and all other output lines are de-asserted (set to 1).

13.11 Macro decod_3_8_E

The symbol and the truth table of the macro decod_3_8_E are depicted in Table 13.11. Figures 13.16 and 13.17 show the macro decod_3_8_E and its flowchart, respectively. This macro defines a 3 × 8 decoder with enable input and active high outputs. In this macro, the active high enable input E, select inputs A, B, and C, and active high output signals d_0, d_1, d_2, d_3, d_4, d_5, d_6, and d_7 are all Boolean variables. In addition to the decod_3_8, this decoder macro has an active high enable line, E, for enabling it. When this decoder is disabled with E set to 0, all active high output lines are

```
decod_3_8_AL macro   regs2,bits2,regs1,
bits1,regs0,bits0,regd7,bitd7,regd6,bitd6,
regd5,bitd5,regd4,bitd4,regd3,bitd3,regd2,
bitd2,regd1,bitd1,regd0,bitd0
     local   L1,L2,L3,L4,L5,L6,L7,L8
     btfss   regs2,bits2
     goto    L8
     bsf     regd3,bitd3
     bsf     regd2,bitd2
     bsf     regd1,bitd1
     bsf     regd0,bitd0
     btfss   regs1,bits1
     goto    L7
     bsf     regd5,bitd5
     bsf     regd4,bitd4
     btfss   regs0,bits0
     goto    L6
     bcf     regd7,bitd7
     bsf     regd6,bitd6
     goto    L1
L8   bsf     regd7,bitd7
     bsf     regd6,bitd6
     bsf     regd5,bitd5
     bsf     regd4,bitd4
     btfss   regs1,bits1
     goto    L4
     bsf     regd1,bitd1
     bsf     regd0,bitd0
     btfss   regs0,bits0
     goto    L3
     bcf     regd3,bitd3
     bsf     regd2,bitd2
     goto    L1
L7   bsf     regd7,bitd7
     bsf     regd6,bitd6
     btfss   regs0,bits0
     goto    L5
     bcf     regd5,bitd5
     bsf     regd4,bitd4
     goto    L1
L6   bsf     regd7,bitd7
     bcf     regd6,bitd6
     goto    L1
L5   bsf     regd5,bitd5
     bcf     regd4,bitd4
     goto    L1
L4   bsf     regd3,bitd3
     bsf     regd2,bitd2
     btfss   regs0,bits0
     goto    L2
     bcf     regd1,bitd1
     bsf     regd0,bitd0
     goto    L1
L3   bsf     regd3,bitd3
     bcf     regd2,bitd2
     goto    L1
L2   bsf     regd1,bitd1
     bcf     regd0,bitd0
L1
     endm
```

FIGURE 13.14
The macro decod_3_8_AL.

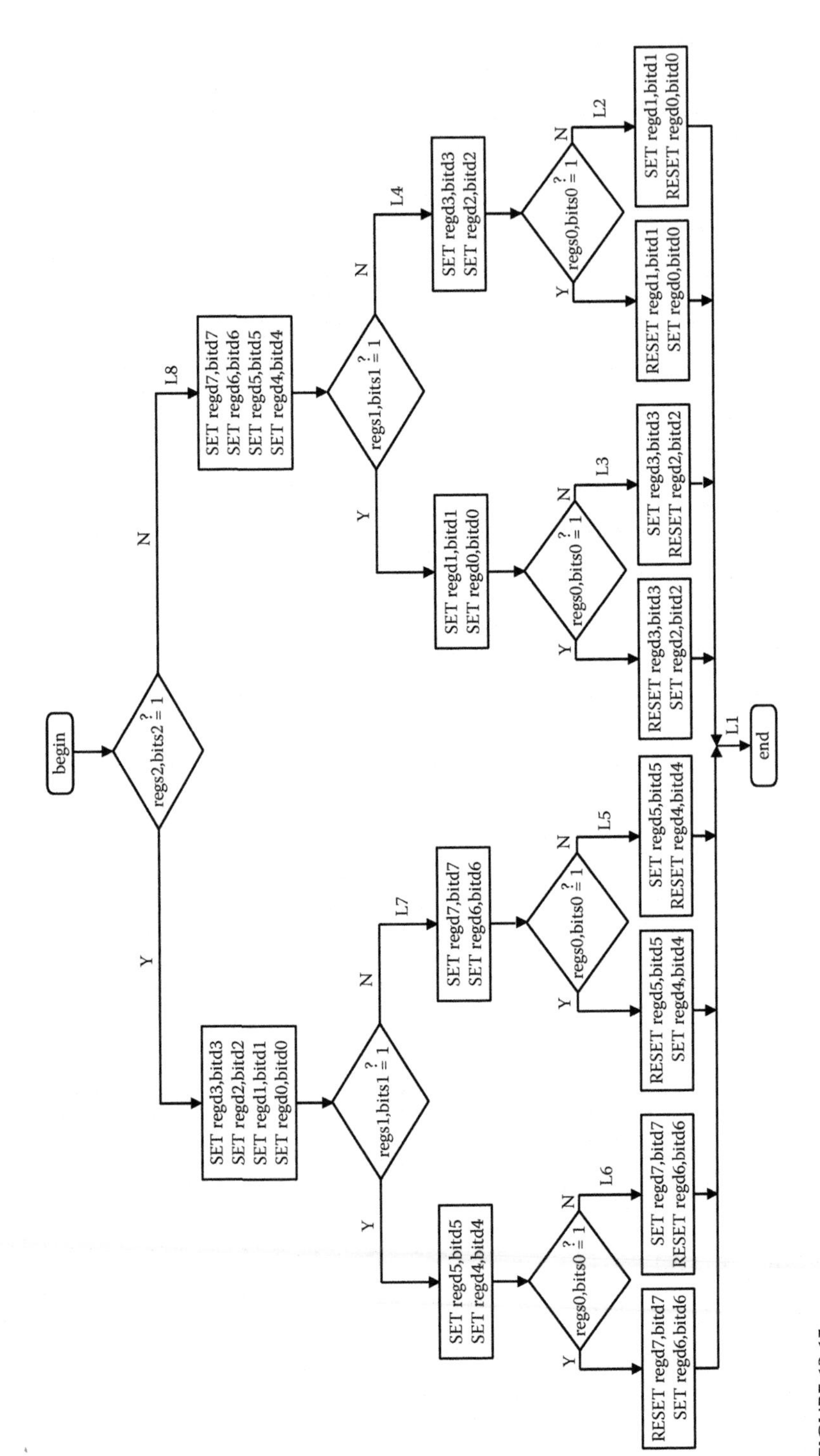

FIGURE 13.15
The flowchart of the macro decod_3_8_AL.

TABLE 13.11

Symbol and Truth Table of the Macro `decod_3_8_E`

Symbol

3×8 DECODER

Inputs: A, B, C, E; outputs: d_0, d_1, d_2, d_3, d_4, d_5, d_6, d_7

W	E
A =	regs2,bits2
B =	regs1,bits1
C =	regs0,bits0
d7 =	regd7,bitd7
d6 =	regd6,bitd6
d5 =	regd5,bitd5
d4 =	regd4,bitd4
d3 =	regd3,bitd3
d2 =	regd2,bitd2
d1 =	regd1,bitd1
d0 =	regd0,bitd0

Truth Table

inputs				outputs							
E	A	B	C	d0	d1	d2	d3	d4	d5	d6	d7
0	×	×	×	0	0	0	0	0	0	0	0
1	0	0	0	1	0	0	0	0	0	0	0
1	0	0	1	0	1	0	0	0	0	0	0
1	0	1	0	0	0	1	0	0	0	0	0
1	0	1	1	0	0	0	1	0	0	0	0
1	1	0	0	0	0	0	0	1	0	0	0
1	1	0	1	0	0	0	0	0	1	0	0
1	1	1	0	0	0	0	0	0	0	1	0
1	1	1	1	0	0	0	0	0	0	0	1

×: don't care.

de-asserted (set to 0). When this decoder is enabled with E set to 1, it functions as described for `decod_3_8`. This means that when E = 1: if the select inputs are ABC = 000 (respectively, 001, 010, 011, 100, 101, 110, 111), then the output line, d_0 (respectively, d_1, d_2, d_3, d_4, d_5, d_6, d_7), is asserted (set to 1) and all other output lines are de-asserted (set to 0).

13.12 Macro `decod_3_8_E_AL`

The symbol and the truth table of the macro `decod_3_8_E_AL` are depicted in Table 13.12. Figures 13.18 and 13.19 show the macro `decod_3_8_E_AL` and its flowchart, respectively. This macro defines a 3 × 8 decoder with enable input and active low outputs. In this macro, the active high enable input E, select inputs A, B, and C, and active low output signals d_0, d_1, d_2, d_3, d_4, d_5, d_6, and d_7 are all Boolean variables. In addition to the `decod_3_8_AL`, this decoder macro has an active high enable line, E, for enabling it. When this decoder is disabled with E set to 0, all active high output lines are de-asserted (set to 1). When this decoder is enabled with E set to 1, it functions as described for `decod_3_8_AL`. This means that when E = 1: if the select inputs are ABC = 000 (respectively, 001, 010, 011, 100, 101, 110, 111), then the output line, d_0 (respectively, d_1, d_2, d_3, d_4, d_5, d_6, d_7), is asserted (set to 0) and all other output lines are de-asserted (set to 1).

```
decod_3_8_E  macro   regs2,bits2,regs1,bits1,
regs0,bits0,regd7,bitd7,regd6,bitd6,
regd5,bitd5,regd4,bitd4,regd3,bitd3,
regd2,bitd2,regd1,bitd1,regd0,bitd0
     local   L1,L2,L3,L4,L5,L6,L7,L8,L9
     movwf   Temp_1
     btfss   Temp_1,0
     goto    L2
     btfss   regs2,bits2
     goto    L9
     bcf     regd3,bitd3
     bcf     regd2,bitd2
     bcf     regd1,bitd1
     bcf     regd0,bitd0
     btfss   regs1,bits1
     goto    L8
     bcf     regd5,bitd5
     bcf     regd4,bitd4
     btfss   regs0,bits0
     goto    L7
     bsf     regd7,bitd7
     bcf     regd6,bitd6
     goto    L1
L9   bcf     regd7,bitd7
     bcf     regd6,bitd6
     bcf     regd5,bitd5
     bcf     regd4,bitd4
     btfss   regs1,bits1
     goto    L5
     bcf     regd1,bitd1
     bcf     regd0,bitd0
     btfss   regs0,bits0
     goto    L4
     bsf     regd3,bitd3
     bcf     regd2,bitd2
     goto    L1
L8   bcf     regd7,bitd7
     bcf     regd6,bitd6
     btfss   regs0,bits0
     goto    L6
     bsf     regd5,bitd5
     bcf     regd4,bitd4
     goto    L1
L7   bcf     regd7,bitd7
     bsf     regd6,bitd6
     goto    L1
L6   bcf     regd5,bitd5
     bsf     regd4,bitd4
     goto    L1
L5   bcf     regd3,bitd3
     bcf     regd2,bitd2
     btfss   regs0,bits0
     goto    L3
     bsf     regd1,bitd1
     bcf     regd0,bitd0
     goto    L1
L4   bcf     regd3,bitd3
     bsf     regd2,bitd2
     goto    L1
L3   bcf     regd1,bitd1
     bsf     regd0,bitd0
     goto    L1
L2   bcf     regd7,bitd7
     bcf     regd6,bitd6
     bcf     regd5,bitd5
     bcf     regd4,bitd4
     bcf     regd3,bitd3
     bcf     regd2,bitd2
     bcf     regd1,bitd1
     bcf     regd0,bitd0
L1
     endm
```

FIGURE 13.16
The macro decod_3_8_E.

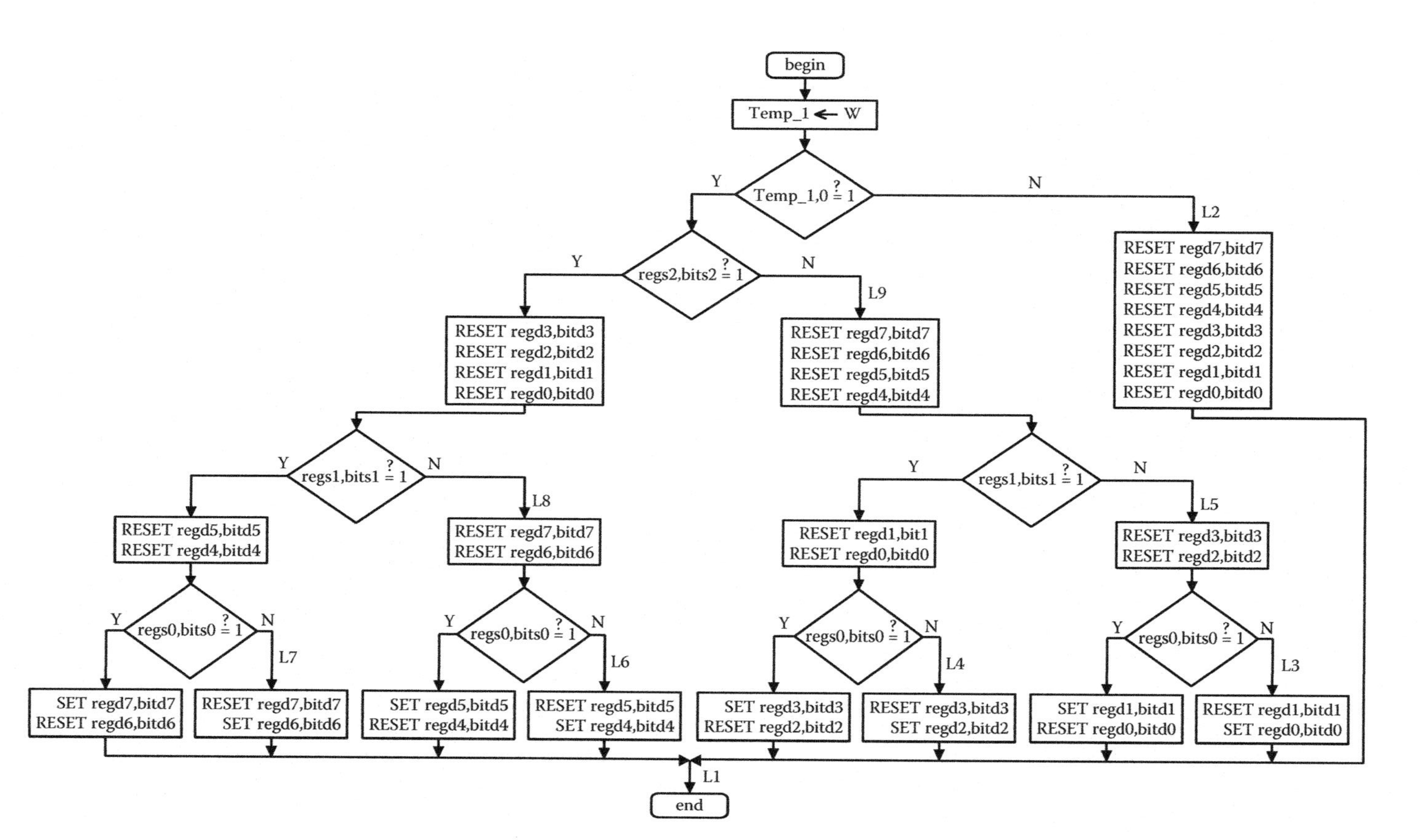

FIGURE 13.17
The flowchart of the macro `decod_3_8_E`.

TABLE 13.12

Symbol and Truth Table of the Macro `decod_3_8_E_AL`

Symbol

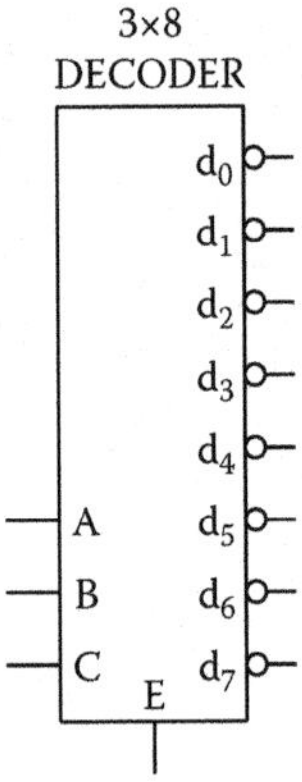

W	E
A =	regs2,bits2
B =	regs1,bits1
C =	regs0,bits0
d7 =	regd7,bitd7
d6 =	regd6,bitd6
d5 =	regd5,bitd5
d4 =	regd4,bitd4
d3 =	regd3,bitd3
d2 =	regd2,bitd2
d1 =	regd1,bitd1
d0 =	regd0,bitd0

Truth Table

inputs				outputs							
E	A	B	C	d0	d1	d2	d3	d4	d5	d6	d7
0	×	×	×	1	1	1	1	1	1	1	1
1	0	0	0	0	1	1	1	1	1	1	1
1	0	0	1	1	0	1	1	1	1	1	1
1	0	1	0	1	1	0	1	1	1	1	1
1	0	1	1	1	1	1	0	1	1	1	1
1	1	0	0	1	1	1	1	0	1	1	1
1	1	0	1	1	1	1	1	1	0	1	1
1	1	1	0	1	1	1	1	1	1	0	1
1	1	1	1	1	1	1	1	1	1	1	0

×: don't care.

13.13 Examples for Decoder Macros

In this section, we will consider four examples, namely, UZAM_plc_16i16o_exX.asm (*X* = 25, 26, 27, 28), to show the usage of decoder macros. In order to test one of these examples, please take the related file UZAM_plc_16i16o_exX.asm (*X* = 25, 26, 27, 28) from the CD-ROM attached to this book, and then open the program by MPLAB IDE and compile it. After that, by using the PIC programmer software, take the compiled file UZAM_plc_16i16o_exX.hex (X = 25, 26, 27, 28), and by your PIC programmer hardware, send it to the program memory of PIC16F648A microcontroller within the PIC16F648A-based PLC. To do this, switch the 4PDT in PROG position and the power switch in OFF position. After loading the file UZAM_plc_16i16o_exX.hex (*X* = 25, 26, 27, 28), switch the 4PDT in RUN and the power switch in ON position. Please check the program's accuracy by cross-referencing it with the related macros.

Let us now consider these example programs: The first example program, UZAM_plc_16i16o_ex25.asm, is shown in Figure 13.20. It shows the usage of four decoder macros, `decod_1_2`, `decod_1_2_AL`, `decod_1_2_E`, and `decod_1_2_E_AL`. The schematic diagram of the user program of UZAM_plc_16i16o_ex25.asm, shown in Figure 13.20, is depicted in Figure 13.21.

In the first rung, the decoder macro `decod_1_2` (1 × 2 decoder) is used. In this decoder, the select input is A = I0.0, while the output lines are d_0 = Q0.0 and d_1 = Q0.1.

```
decod_3_8_E_AL  macro   regs2,bits2,regs1,
bits1,regs0,bits0,regd7,bitd7,regd6,bitd6,
regd5,bitd5,regd4,bitd4,regd3,bitd3,regd2,
bitd2,regd1,bitd1,regd0,bitd0
     local   L1,L2,L3,L4,L5,L6,L7,L8,L9
     movwf   Temp_1
     btfss   Temp_1,0
     goto    L2
     btfss   regs2,bits2
     goto    L9
     bsf     regd3,bitd3
     bsf     regd2,bitd2
     bsf     regd1,bitd1
     bsf     regd0,bitd0
     btfss   regs1,bits1
     goto    L8
     bsf     regd5,bitd5
     bsf     regd4,bitd4
     btfss   regs0,bits0
     goto    L7
     bcf     regd7,bitd7
     bsf     regd6,bitd6
     goto    L1
L9   bsf     regd7,bitd7
     bsf     regd6,bitd6
     bsf     regd5,bitd5
     bsf     regd4,bitd4
     btfss   regs1,bits1
     goto    L5
     bsf     regd1,bitd1
     bsf     regd0,bitd0
     btfss   regs0,bits0
     goto    L4
     bcf     regd3,bitd3
     bsf     regd2,bitd2
     goto    L1
L8   bsf     regd7,bitd7
     bsf     regd6,bitd6
     btfss   regs0,bits0
     goto    L6
     bcf     regd5,bitd5
     bsf     regd4,bitd4
     goto    L1
L7   bsf     regd7,bitd7
     bcf     regd6,bitd6
     goto    L1
L6   bsf     regd5,bitd5
     bcf     regd4,bitd4
     goto    L1
L5   bsf     regd3,bitd3
     bsf     regd2,bitd2
     btfss   regs0,bits0
     goto    L3
     bcf     regd1,bitd1
     bsf     regd0,bitd0
     goto    L1
L4   bsf     regd3,bitd3
     bcf     regd2,bitd2
     goto    L1
L3   bsf     regd1,bitd1
     bcf     regd0,bitd0
     goto    L1
L2   bsf     regd7,bitd7
     bsf     regd6,bitd6
     bsf     regd5,bitd5
     bsf     regd4,bitd4
     bsf     regd3,bitd3
     bsf     regd2,bitd2
     bsf     regd1,bitd1
     bsf     regd0,bitd0
L1
     endm
```

FIGURE 13.18
The macro decod_3_8_E_AL.

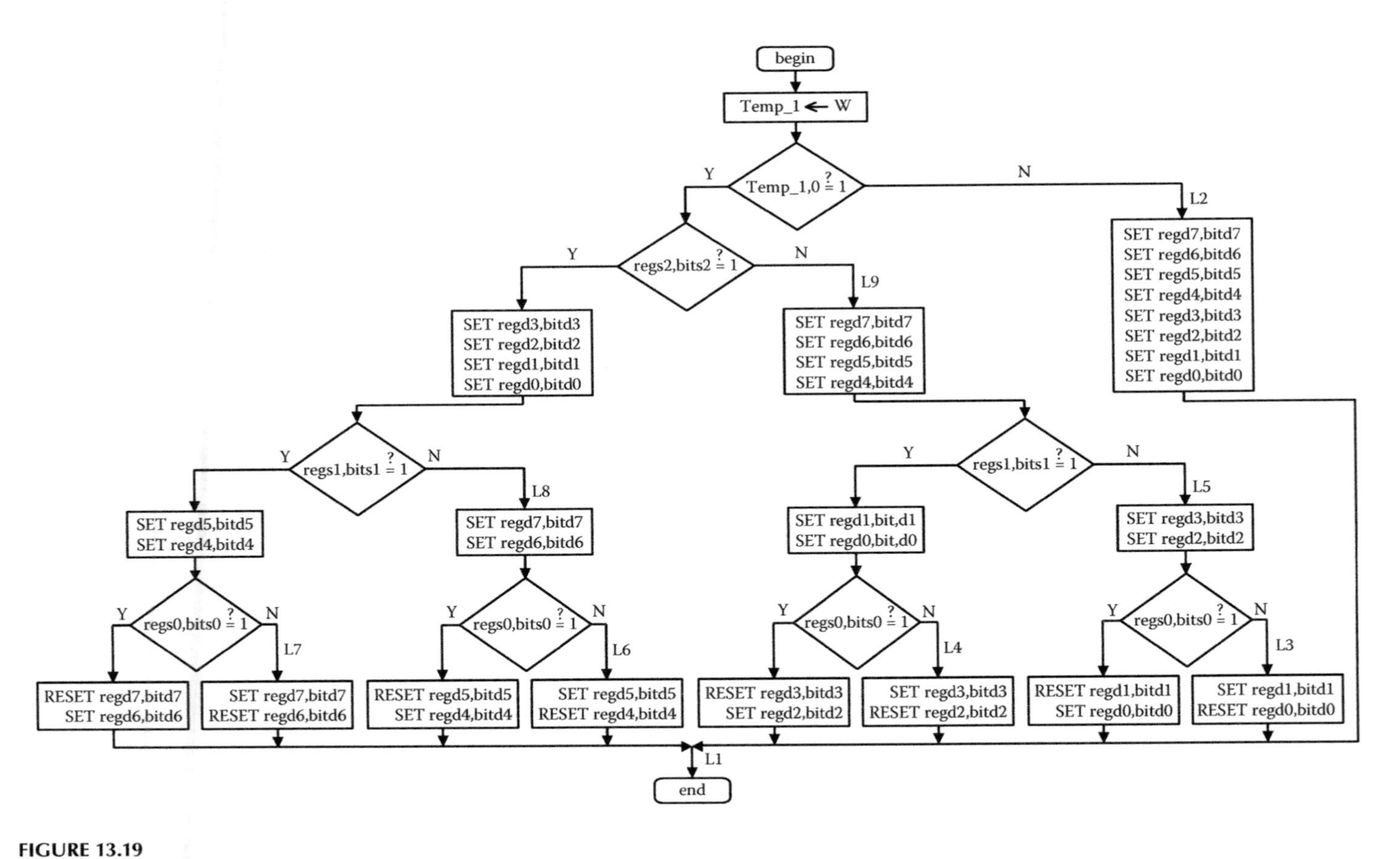

FIGURE 13.19
The flowchart of the macro decod_3_8_E_AL.

```
.
;--------------- user program starts here -----
 decod_1_2      I0.0,Q0.1,Q0.0       ;rung 1

 decod_1_2_AL   I0.7,Q0.7,Q0.6       ;rung 2


 ld             I1.0                 ;rung 3
 decod_1_2_E    I1.1,Q1.1,Q1.0

 ld_not         I1.7                 ;rung 4
 decod_1_2_E_AL I1.6,Q1.7,Q1.6
;--------------- user program ends here -------
.
```

FIGURE 13.20
The user program of UZAM_plc_16i16o_ex25.asm.

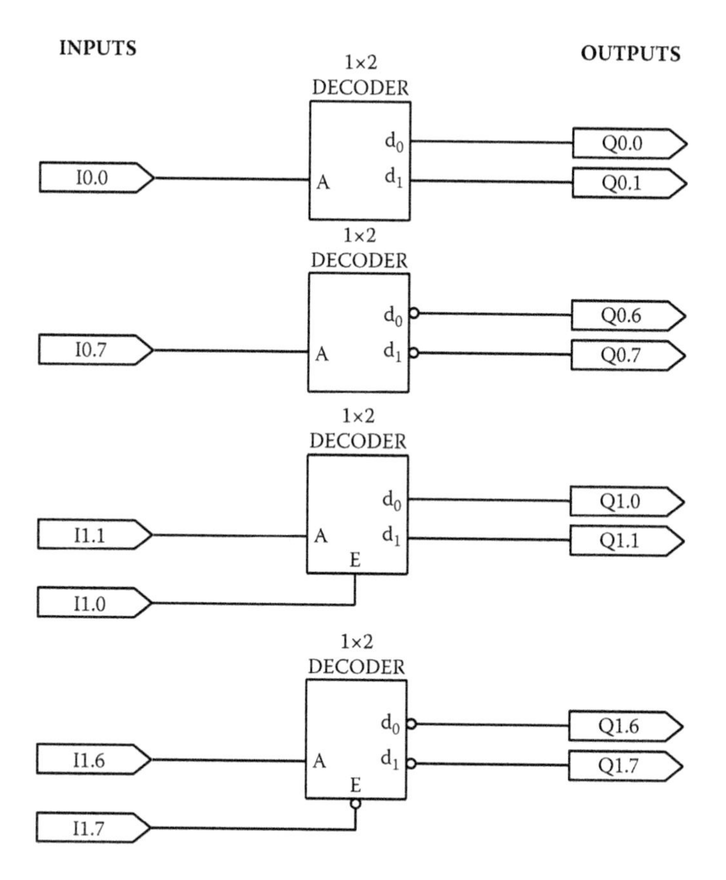

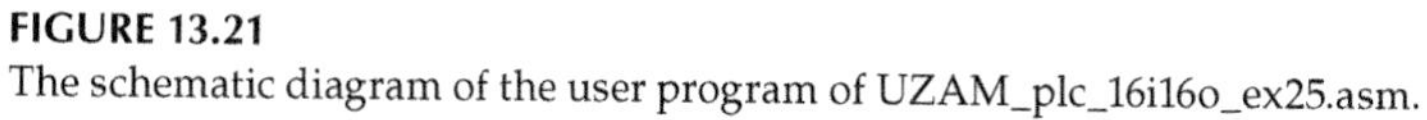

FIGURE 13.21
The schematic diagram of the user program of UZAM_plc_16i16o_ex25.asm.

In the second rung, the decoder macro decod_1_2_AL (1 × 2 decoder with active low outputs) is used. In this decoder, the select input is A = I0.7, while the output lines are d_0 = Q0.6 and d_1 = Q0.7.

In the third rung, the macro decod_1_2_E (1 × 2 decoder with active high enable input) is used. In this decoder, the select input is A = I1.1, while the output lines are d_0 = Q1.0 and d_1 = Q1.1. In addition, the active high enable input E is defined to be E = I1.0.

In the fourth and last rung, the macro decod_1_2_E_AL (1 × 2 decoder with active high enable input and active low outputs) is used. In this decoder, the select input is A = I1.6, while the output lines are d_0 = Q1.6 and d_1 = Q1.7. In addition, the active high enable input E is defined to be E = inverted I1.7. Note that this arrangement forces the enable input E to be active low.

The second example program, UZAM_plc_16i16o_ex26.asm, is shown in Figure 13.22. It shows the usage of four decoder macros, decod_2_4, decod_2_4_AL, decod_2_4_E, and decod_2_4_E_AL. The schematic diagram of the user program of UZAM_plc_16i16o_ex26.asm, shown in Figure 13.22, is depicted in Figure 13.23.

In the first rung, the decoder macro decod_2_4 (2 × 4 decoder) is used. In this decoder, select inputs are A = I0.1 and B = I0.0, while the output lines are d_0 = Q0.0, d_1 = Q0.1, d_2 = Q0.2, and d_3 = Q0.3.

In the second rung, the decoder macro decod_2_4_AL (2 × 4 decoder with active low outputs) is used. In this decoder, select inputs are A = I0.7 and B = I0.6, while the output lines are d_0 = Q0.4, d_1 = Q0.5, d_2 = Q0.6, and d_3 = Q0.7.

In the third rung, the macro decod_2_4_E (2 × 4 decoder with active high enable input) is used. In this decoder, select inputs are A = I1.2 and B = I1.1, while the output lines are d_0 = Q1.0, d_1 = Q1.1, d_2 = Q1.2, and d_3 = Q1.3. In addition, the active high enable input E is defined to be E = I1.0.

In the fourth and last rung, the macro decod_2_4_E_AL (2 × 4 decoder with active high enable input and active low outputs) is used. In this decoder, select inputs are A = I1.6 and B = I1.5, while the output lines are d_0 = Q1.4, d_1 = Q1.5, d_2 = Q1.6, and d_3 = Q1.7. In addition, the active high enable input E is defined to be E = inverted I1.7. Note that this arrangement forces the enable input E to be active low.

```
.
;--------------- user program starts here -------------------
    decod_2_4       I0.1,I0.0,Q0.3,Q0.2,Q0.1,Q0.0       ;rung 1

    decod_2_4_AL    I0.7,I0.6,Q0.7,Q0.6,Q0.5,Q0.4       ;rung 2

    ld              I1.0                                ;rung 3
    decod_2_4_E     I1.2,I1.1,Q1.3,Q1.2,Q1.1,Q1.0

    ld_not          I1.7                                ;rung 4
    decod_2_4_E_AL  I1.6,I1.5,Q1.7,Q1.6,Q1.5,Q1.4
;--------------- user program ends here ---------------------
.
```

FIGURE 13.22
The user program of UZAM_plc_16i16o_ex26.asm.

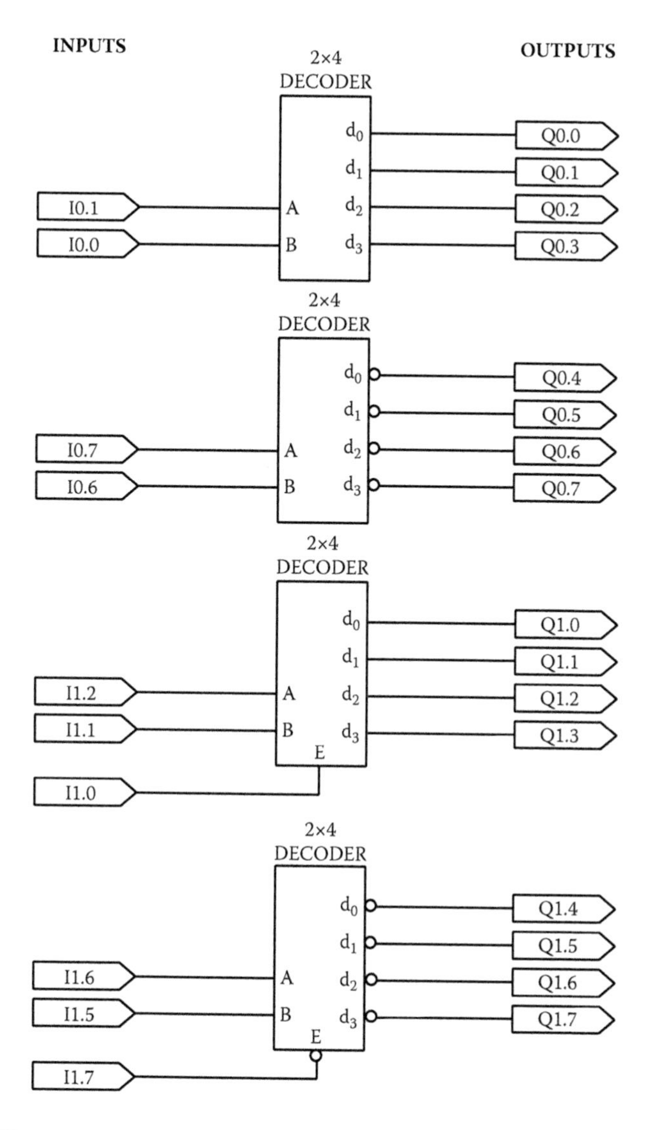

FIGURE 13.23
The schematic diagram of the user program of UZAM_plc_16i16o_ex26.asm.

The third example program, UZAM_plc_16i16o_ex27.asm, is shown in Figure 13.24. It shows the usage of two decoder macros `decod_3_8` and `decod_3_8_AL`. The schematic diagram of the user program of UZAM_plc_16i16o_ex27.asm, shown in Figure 13.24, is depicted in Figure 13.25.

In the first rung, the decoder macro `decod_3_8` (3 × 8 decoder) is used. In this decoder, select inputs are A = I0.2, B = I0.1, and C = I0.0, while the output lines are d_0 = Q0.0, d_1 = Q0.1, d_2 = Q0.2, d_3 = Q0.3, d_4 = Q0.4, d_5 = Q0.5, d_6 = Q0.6, and d_7 = Q0.7.

```
.
;--------------- user program starts here ------------------------
    decod_3_8       I0.2,I0.1,I0.0,Q0.7,Q0.6,Q0.5,Q0.4,Q0.3,Q0.2,Q0.1,Q0.0   ;rung 1
    decod_3_8_AL    I1.2,I1.1,I1.0,Q1.7,Q1.6,Q1.5,Q1.4,Q1.3,Q1.2,Q1.1,Q1.0   ;rung 2
;--------------- user program ends here --------------------------
.
```

FIGURE 13.24
The user program of UZAM_plc_16i16o_ex27.asm.

In the second and last rung, the decoder macro `decod_3_8_AL` (3 × 8 decoder with active low outputs) is used. In this decoder, select inputs are A = I1.2, B = I1.1, and C = I1.0, while the output lines are d_0 = Q1.0, d_1 = Q1.1, d_2 = Q1.2, d_3 = Q1.3, d_4 = Q1.4, d_5 = Q1.5, d_6 = Q1.6, and d_7 = Q1.7.

The fourth example program, UZAM_plc_16i16o_ex28.asm, is shown in Figure 13.26. It shows the usage of two decoder macros, `decod_3_8_E` and `decod_3_8_E_AL`. The schematic diagram of the user program of UZAM_plc_16i16o_ex28.asm, shown in Figure 13.26, is depicted in Figure 13.27.

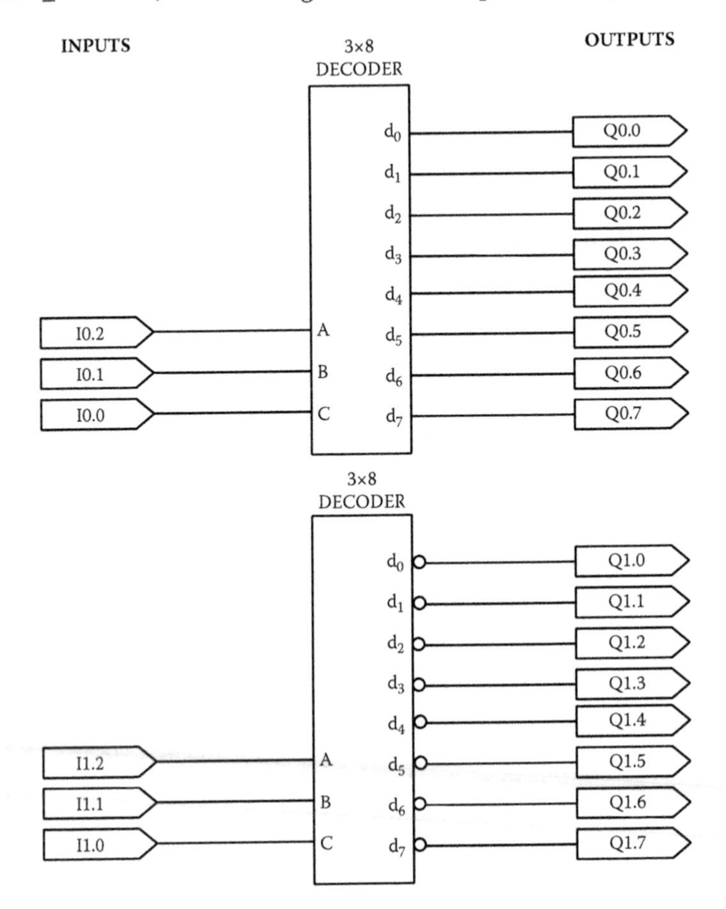

FIGURE 13.25
The schematic diagram of the user program of UZAM_plc_16i16o_ex27.asm.

```
.
;--------------- user program starts here -----------------------
  ld              I0.0                                              ;rung 1
  decod_3_8_E     I0.3,I0.2,I0.1,Q0.7,Q0.6,Q0.5,Q0.4,Q0.3,Q0.2,Q0.1,Q0.0

  ld_not          I1.0                                              ;rung 2
  decod_3_8_E_AL  I1.3,I1.2,I1.1,Q1.7,Q1.6,Q1.5,Q1.4,Q1.3,Q1.2,Q1.1,Q1.0
;--------------- user program ends here --------------------------
.
```

FIGURE 13.26
The user program of UZAM_plc_16i16o_ex28.asm.

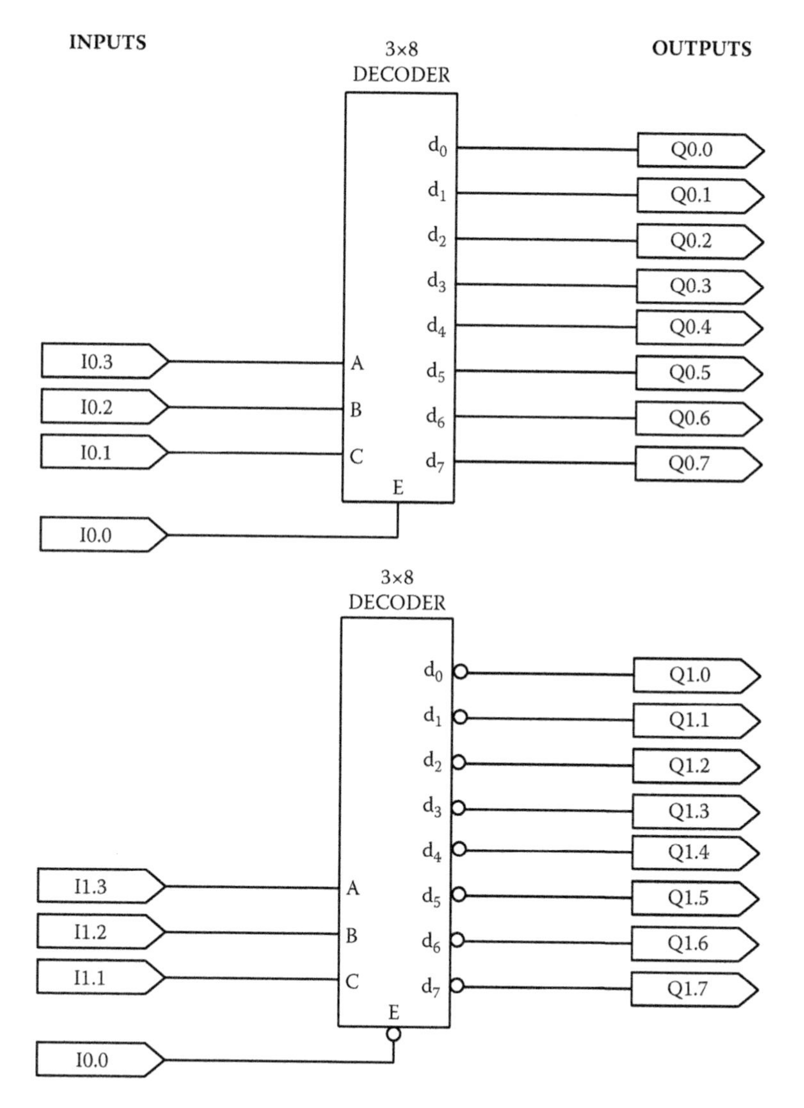

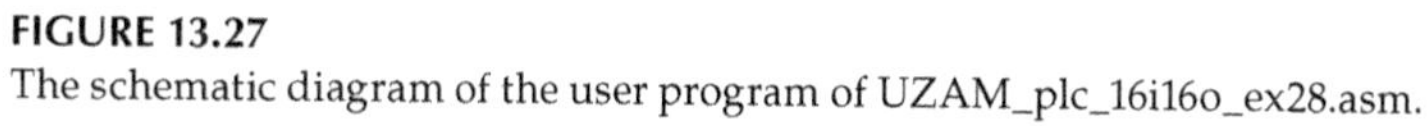

FIGURE 13.27
The schematic diagram of the user program of UZAM_plc_16i16o_ex28.asm.

In the first rung, the decoder macro decod_3_8_E (3 × 8 decoder with active high enable input) is used. In this decoder, select inputs are A = I0.3, B = I0.2, and C = I0.1, while the output lines are d_0 = Q0.0, d_1 = Q0.1, d_2 = Q0.2, d_3 = Q0.3, d_4 = Q0.4, d_5 = Q0.5, d_6 = Q0.6, and d_7 = Q0.7. In addition, the active high enable input E is defined to be E = I0.0.

In the second and last rung, the decoder macro decod_3_8_E _AL (3 × 8 decoder with active high enable input and active low outputs) is used. In this decoder, select inputs are A = I1.3, B = I1.2, C = I1.1, while the output lines are d_0 = Q1.0, d_1 = Q1.1, d_2 = Q1.2, d_3 = Q1.3, d_4 = Q1.4, d_5 = Q1.5, d_6 = Q1.6, and d_7 = Q1.7. In addition, the active high enable input E is defined to be E = inverted I1.0. Note that this arrangement forces the enable input E to be active low.

14

Priority Encoder Macros

An encoder is a circuit that changes a set of signals into a code. As a standard combinational component, an encoder is almost like the inverse of a decoder, where it encodes a 2^n-bit input datum into an n-bit code. As shown by the general form of an m-to-n encoder in Figure 14.1, the encoder has $m = 2^n$ input lines and n output lines. For active high inputs, the operation of the encoder is such that exactly one of the input lines should have a 1, while the remaining input lines should have 0s. The output is the binary value of the index of the input line that has the 1. It is assumed that only one input line can be a 1. Encoders are used to reduce the number of bits needed to represent some given data either in data storage or in data transmission. Encoders are also used in a system with 2^n input devices, each of which may need to request for service. One input line is connected to one input device. The input device requesting for service will assert the input line that is connected to it. The corresponding n-bit output value will indicate to the system which of the 2^n devices is requesting for service. However, this only works correctly if it is guaranteed that only one of the 2^n devices will request for service at any one time. If two or more devices request for service at the same time, then the output will be incorrect. To resolve this problem, a priority is assigned to each of the input lines so that when multiple requests are made, the encoder outputs the index value of the input line with the highest priority. This modified encoder is known as a priority encoder. In this chapter, we are concerned with the priority encoders. Although not shown in Figure 14.1, the priority encoder may have an enable line, E, for enabling it. When the priority encoder is disabled with E set to 0 (for active high enable input E), all the output lines will have 0s (for active high outputs). When the priority encoder is enabled, then the output lines issue the binary data representation of the highest-priority input signal asserted (set to 1 for active high).

In this chapter, the following priority encoder macros are described for the PIC16F648A-based PLC:

- `encod_4_2_p` (4 × 2 priority encoder)
- `encod_4_2_p_E` (4 × 2 priority encoder with enable input)
- `encod_8_3_p` (8 × 3 priority encoder)
- `encod_8_3_p_E` (8 × 3 priority encoder with enable input)

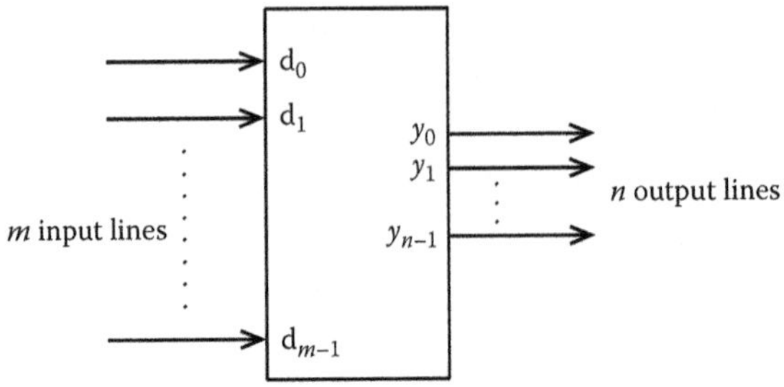

FIGURE 14.1
The general form of an m-to-n encoder, where $m = 2^n$.

`encod_dec_bcd_p` (decimal to binary coded decimal (BCD) priority encoder)

`encod_dec_bcd_p_E` (decimal to BCD priority encoder with enable input)

The file definitions.inc, included within the CD-ROM attached to this book, contains all priority encoder macros defined for the PIC16F648A-based PLC. Let us now consider these macros in detail.

14.1 Macro `encod_4_2_p`

The symbol and the truth table of the macro `encod_4_2_p` are depicted in Table 14.1. Figure 14.2 shows the macro `encod_4_2_p` and its flowchart. This macro defines a 4 × 2 priority encoder. In this macro, active high input signals 3, 2, 1, and 0, and active high output signals A_1 (most significant

TABLE 14.1
Symbol and Truth Table of the Macro `encod_4_2_p`

Symbol

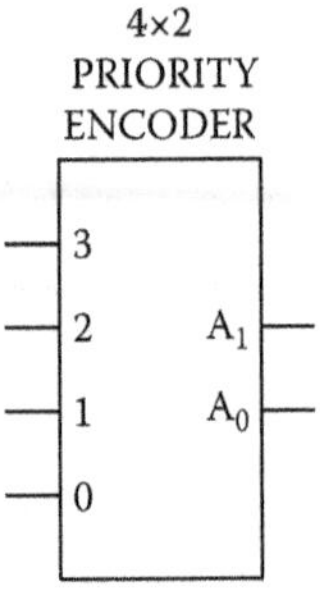

3 =	reg3,bit3
2 =	reg2,bit2
1 =	reg1,bit1
0 =	reg0,bit0
A1 =	regA1,bitA1
A0 =	regA1,bitA0

Truth Table

inputs				**outputs**	
0	**1**	**2**	**3**	**A1**	**A0**
×	×	×	1	1	1
×	×	1	0	1	0
×	1	0	0	0	1
1	0	0	0	0	0

×: don't care

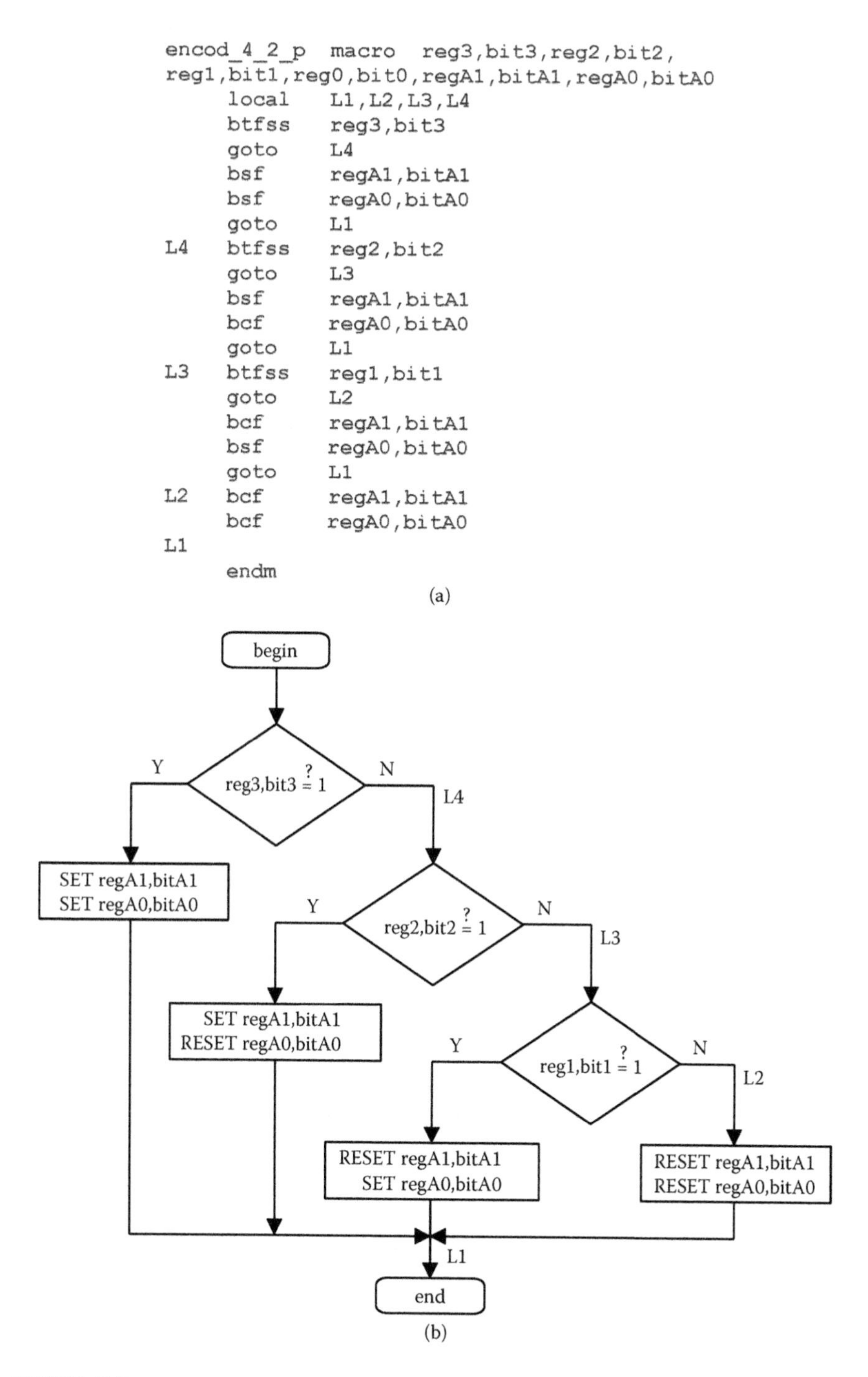

FIGURE 14.2
(a) The macro `encod_4_2_p` and (b) its flowchart.

bit (MSB)) and A_0 (least significant bit (LSB)) are all Boolean variables. The input line 3 has the highest priority, while the input line 0 has the lowest priority. How the macro encod_4_2_p works is shown in the truth table. It can be seen that the output binary code is generated based on the highest-priority input signal present in the four input lines. If the input signals present in the input lines 0, 1, 2, 3 are as follows, ×××1 (respectively, ××10, ×100, 1000), then the output lines generate the following binary code: $A_1A_0 = 11$ (respectively, 10, 01, 00).

14.2 Macro encod_4_2_p_E

The symbol and the truth table of the macro encod_4_2_p_E are depicted in Table 14.2. Figure 14.3 shows the macro encod_4_2_p_E and its flowchart. This macro defines a 4 × 2 priority encoder with enable input. In this macro, the active high enable input E, active high input signals 3, 2, 1, and 0, and active high output signals A_1 (MSB) and A_0 (LSB) are all Boolean variables. The input line 3 has the highest priority, while the input line 0 has the lowest priority. In addition to the encod_4_2_p, this encoder macro has an active high enable line, E, for enabling it. When this encoder is disabled with E set to 0, all output lines are set to 0. When this encoder is enabled with E set to 1, it functions as described for encod_4_2_p. This means that when E = 1: if the input signals present in the input lines 0, 1, 2, 3 are as follows, ×××1 (respectively, ××10, ×100, 1000), then the output lines generate the following binary code: $A_1A_0 = 11$ (respectively, 10, 01, 00).

TABLE 14.2

Symbol and Truth Table of the Macro encod_4_2_p_E

Symbol

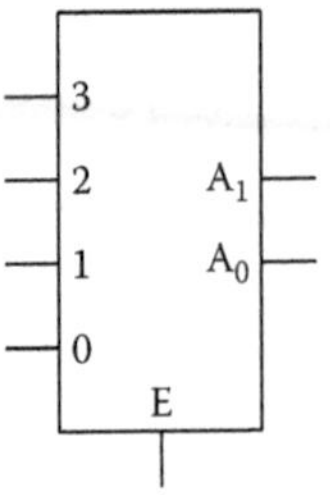

W	E
3 =	reg3,bit3
2 =	reg2,bit2
1 =	reg1,bit1
0 =	reg0,bit0
A1 =	regA1,bitA1
A0 =	regA1,bitA0

Truth Table

inputs					outputs	
E	**0**	**1**	**2**	**3**	**A1**	**A0**
0	×	×	×	×	0	0
1	×	×	×	1	1	1
1	×	×	1	0	1	0
1	×	1	0	0	0	1
1	1	0	0	0	0	0

×: don't care

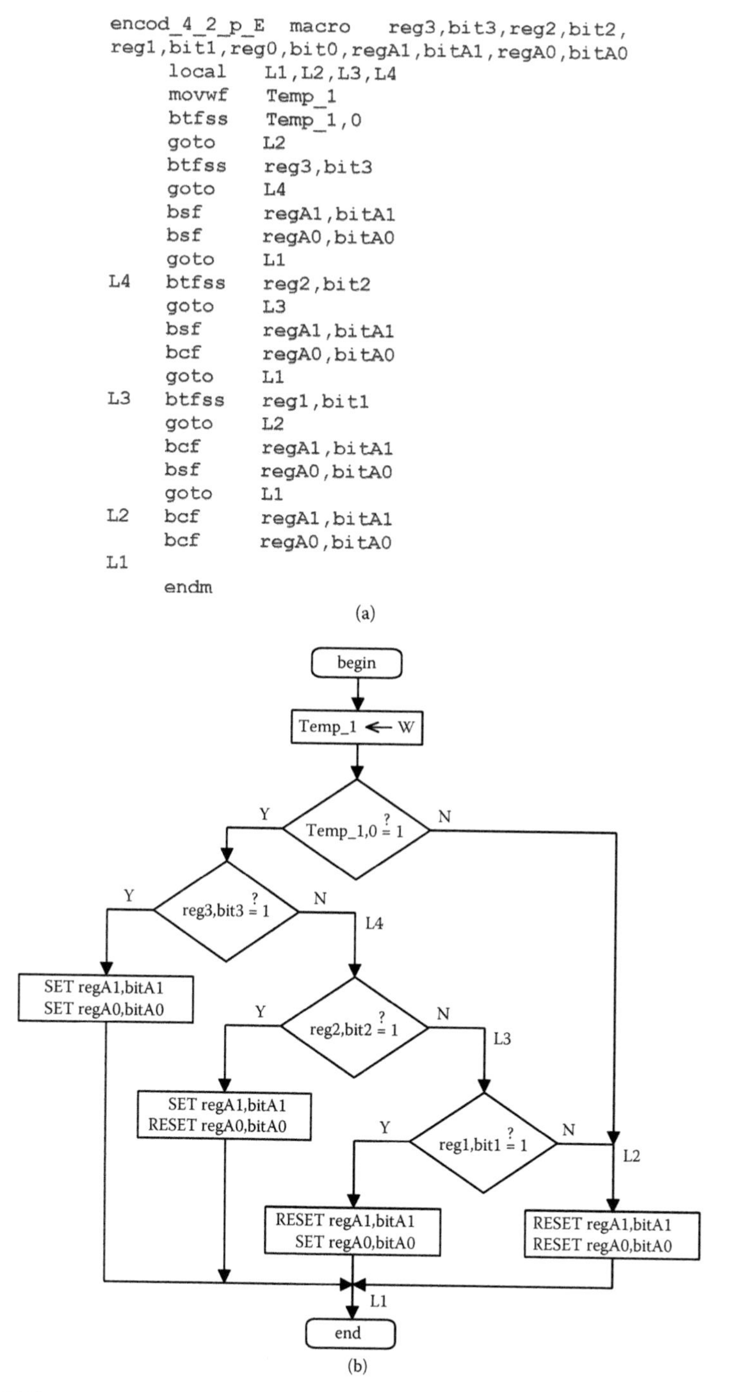

FIGURE 14.3
(a) The macro `encod_4_2_p_E` and (b) its flowchart.

14.3 Macro encod_8_3_p

The symbol and the truth table of the macro encod_8_3_p are depicted in Table 14.3. Figures 14.4 and 14.5 show the macro encod_8_3_p and its flowchart, respectively. This macro defines an 8 × 3 priority encoder. In this macro, active high input signals 7, 6, 5, 4, 3, 2, 1, and 0, and active high output signals A_2 (MSB), A_1, and A_0 (LSB) are all Boolean variables. The input line 7 has the highest priority, while the input line 0 has the lowest priority. How the macro encod_8_3_p works is shown in the truth table. It can be seen that the output binary code is generated based on the highest-priority

TABLE 14.3

Symbol and Truth Table of the Macro encod_8_3_p

Symbol

8×3
PRIORITY
ENCODER

7
6
5
4 A_2
3 A_1
2 A_0
1
0

7 =	reg7,bit7
6 =	reg6,bit6
5 =	reg5,bit5
4 =	reg4,bit4
3 =	reg3,bit3
2 =	reg2,bit2
1 =	reg1,bit1
0 =	reg0,bit0
A2 =	regA2,bitA2
A1 =	regA1,bitA1
A0 =	regA1,bitA0

Truth Table

inputs								outputs		
0	1	2	3	4	5	6	7	A2	A1	A0
×	×	×	×	×	×	×	1	1	1	1
×	×	×	×	×	×	1	0	1	1	0
×	×	×	×	×	1	0	0	1	0	1
×	×	×	×	1	0	0	0	1	0	0
×	×	×	1	0	0	0	0	0	1	1
×	×	1	0	0	0	0	0	0	1	0
×	1	0	0	0	0	0	0	0	0	1
1	0	0	0	0	0	0	0	0	0	0

×: don't care.

```
encod_8_3_p  macro   reg7,bit7,reg6,bit6,
reg5,bit5,reg4,bit4,reg3,bit3,reg2,bit2,
reg1,bit1,reg0,bit0,regA2,bitA2,regA1,
bitA1,regA0,bitA0
        local   L1,L2,L3,L4,L5,L6,L7,L8
        btfss   reg7,bit7
        goto    L8
        bsf     regA2,bitA2
        bsf     regA1,bitA1
        bsf     regA0,bitA0
        goto    L1
L8      btfss   reg6,bit6
        goto    L7
        bsf     regA2,bitA2
        bsf     regA1,bitA1
        bcf     regA0,bitA0
        goto    L1
L7      btfss   reg5,bit5
        goto    L6
        bsf     regA2,bitA2
        bcf     regA1,bitA1
        bsf     regA0,bitA0
        goto    L1
L6      btfss   reg4,bit4
        goto    L5
        bsf     regA2,bitA2
        bcf     regA1,bitA1
        bcf     regA0,bitA0
        goto    L1
L5      btfss   reg3,bit3
        goto    L4
        bcf     regA2,bitA2
        bsf     regA1,bitA1
        bsf     regA0,bitA0
        goto    L1
L4      btfss   reg2,bit2
        goto    L3
        bcf     regA2,bitA2
        bsf     regA1,bitA1
        bcf     regA0,bitA0
        goto    L1
L3      btfss   reg1,bit1
        goto    L2
        bcf     regA2,bitA2
        bcf     regA1,bitA1
        bsf     regA0,bitA0
        goto    L1
L2      bcf     regA2,bitA2
        bcf     regA1,bitA1
        bcf     regA0,bitA0
L1
        endm
```

FIGURE 14.4
The macro encod_8_3_p.

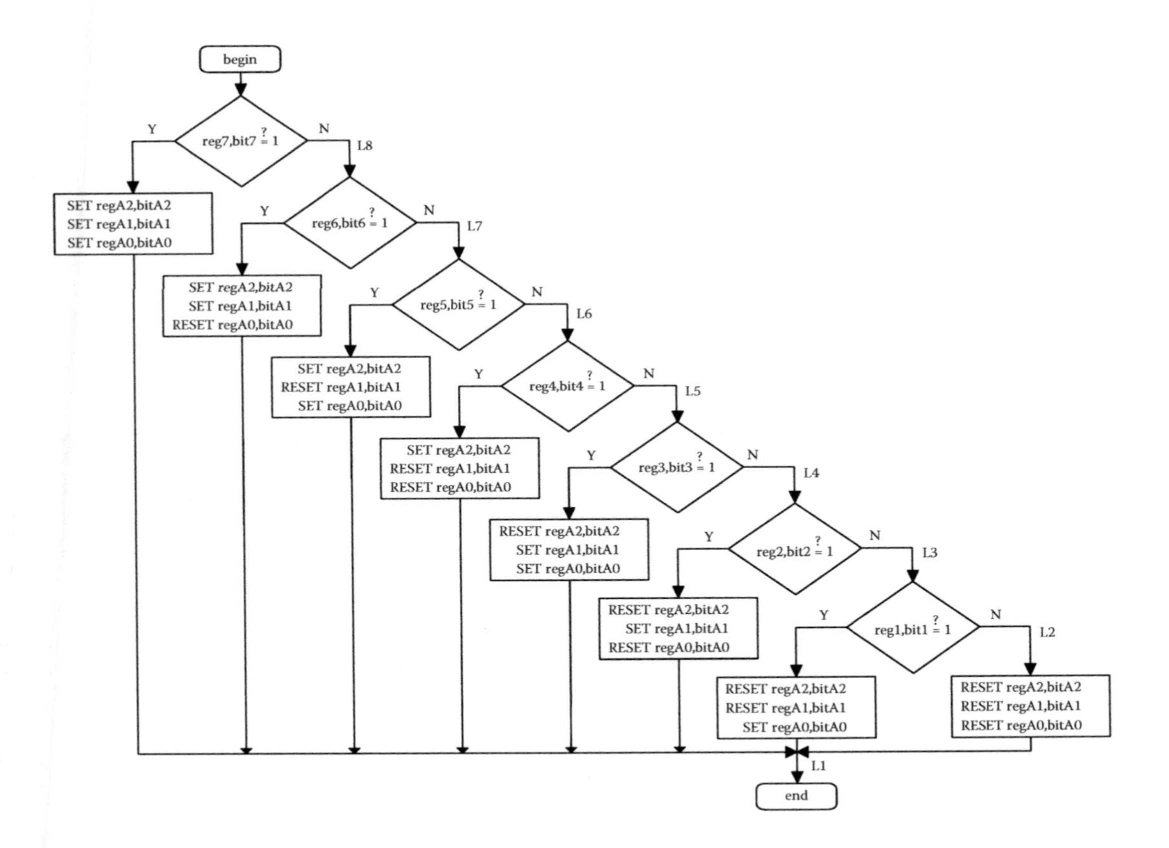

FIGURE 14.5
The flowchart of the macro encod_8_3_p.

input signal present in the eight input lines. If the input signals present in the input lines 0, 1, 2, 3, 4, 5, 6, 7 are as follows, ×××××××1 (respectively, ××××××10, ×××××100, ××××1000, ×××10000, ××100000, ×1000000, 10000000), then the output lines generate the following binary code: $A_2A_1A_0 = 111$ (respectively, 110, 101, 100, 011, 010, 001, 000).

14.4 Macro `encod_8_3_p_E`

The symbol and the truth table of the macro `encod_8_3_p_E` are depicted in Table 14.4. Figures 14.6 and 14.7 show the macro `encod_8_3_p_E` and its flowchart, respectively. This macro defines an 8 × 3 priority encoder with enable input. In this macro, the active high enable input E, active high input signals 7, 6, 5, 4, 3, 2, 1, and 0, and active high output signals A_2 (MSB), A_1, and A_0 (LSB) are all Boolean variables. The input line 7 has the highest priority, while the input line 0 has the lowest priority. In addition to the `encod_8_3_p`, this encoder macro has an active high enable line, E, for enabling it. When this encoder is disabled with E set to 0, all output lines are set to 0. When this encoder is enabled with E set to 1, it functions as described for `encod_8_3_p`. This means that when E = 1: if the input signals present in the input lines 0,1,2,3,4,5,6,7 are as follows, ×××××××1 (respectively, ××××××10, ×××××00, ××××1000, ×××10000, ××100000, ×1000000, 10000000), then the output lines generate the following binary code: $A_2A_1A_0 = 111$ (respectively, 110, 101, 100, 011, 010, 001, 000).

14.5 Macro `encod_dec_bcd_p`

The symbol and the truth table of the macro `encod_dec_bcd_p` are depicted in Table 14.5. Figures 14.8 and 14.9 show the macro `encod_dec_bcd_p` and its flowchart, respectively. This macro defines a decimal to BCD priority encoder. In this macro, active high input signals 9, 8, 7, 6, 5, 4, 3, 2, 1, and 0, and active high output signals A_3 (MSB), A_2, A_1, and A_0 (LSB) are all Boolean variables. The input line 9 has the highest priority, while the input line 0 has the lowest priority. How the macro `encod_dec_bcd_p` works is shown in the truth table. It can be seen that the output binary code is generated based on the highest-priority input signal present in the 10 input lines. If the input signals present in the input lines 0, 1, 2, 3, 4, 5, 6, 7, 8, 9 are as follows, ×××××××××1 (respectively, ××××××××10, ×××××××100, ××××××1000, ×××××10000, ××××100000, ×××1000000, ××10000000, ×100000000, 1000000000), then the output lines generate the following binary code: $A_3A_2A_1A_0 = 1001$ (respectively, 1000, 0111, 0110, 0101, 0100, 0011, 0010, 0001, 0000).

TABLE 14.4

Symbol and Truth Table of the Macro `encod_8_3_p_E`

Symbol

8×3
PRIORITY
ENCODER

7
6
5
4
3
2
1
0
A_2
A_1
A_0
E

W	E
7 =	reg7,bit7
6 =	reg6,bit6
5 =	reg5,bit5
4 =	reg4,bit4
3 =	reg3,bit3
2 =	reg2,bit2
1 =	reg1,bit1
0 =	reg0,bit0
A2 =	regA2,bitA2
A1 =	regA1,bitA1
A0 =	regA1,bitA0

Truth Table

inputs									outputs		
E	0	1	2	3	4	5	6	7	A2	A1	A0
0	×	×	×	×	×	×	×	×	0	0	0
1	×	×	×	×	×	×	×	1	1	1	1
1	×	×	×	×	×	×	1	0	1	1	0
1	×	×	×	×	×	1	0	0	1	0	1
1	×	×	×	×	1	0	0	0	1	0	0
1	×	×	×	1	0	0	0	0	0	1	1
1	×	×	1	0	0	0	0	0	0	1	0
1	×	1	0	0	0	0	0	0	0	0	1
1	1	0	0	0	0	0	0	0	0	0	0

×: don't care.

14.6 Macro `encod_dec_bcd_p_E`

The symbol and the truth table of the macro `encod_dec_bcd_p_E` are depicted in Table 14.6. Figures 14.10 and 14.11 show the macro `encod_dec_bcd_p_E` and its flowchart, respectively. This macro defines a decimal to BCD priority encoder with enable input. In this macro, the active high enable input E, active high input signals 9, 8, 7, 6, 5, 4, 3, 2, 1, and 0, and active high output signals A_3 (MSB), A_2, A_1, and A_0 (LSB) are all Boolean variables. The

```
encod_8_3_p_E  macro   reg7,bit7,reg6,bit6,
reg5,bit5,reg4,bit4,reg3,bit3,reg2,bit2,
reg1,bit1,reg0,bit0,regA2,bitA2,
regA1,bitA1,regA0,bitA0
      local    L1,L2,L3,L4,L5,L6,L7,L8
      movwf    Temp_1
      btfss    Temp_1,0
      goto     L2
      btfss    reg7,bit7
      goto     L8
      bsf      regA2,bitA2
      bsf      regA1,bitA1
      bsf      regA0,bitA0
      goto     L1
L8    btfss    reg6,bit6
      goto     L7
      bsf      regA2,bitA2
      bsf      regA1,bitA1
      bcf      regA0,bitA0
      goto     L1
L7    btfss    reg5,bit5
      goto     L6
      bsf      regA2,bitA2
      bcf      regA1,bitA1
      bsf      regA0,bitA0
      goto     L1
L6    btfss    reg4,bit4
      goto     L5
      bsf      regA2,bitA2
      bcf      regA1,bitA1
      bcf      regA0,bitA0
      goto     L1
L5    btfss    reg3,bit3
      goto     L4
      bcf      regA2,bitA2
      bsf      regA1,bitA1
      bsf      regA0,bitA0
      goto     L1
L4    btfss    reg2,bit2
      goto     L3
      bcf      regA2,bitA2
      bsf      regA1,bitA1
      bcf      regA0,bitA0
      goto     L1
L3    btfss    reg1,bit1
      goto     L2
      bcf      regA2,bitA2
      bcf      regA1,bitA1
      bsf      regA0,bitA0
      goto     L1
L2    bcf      regA2,bitA2
      bcf      regA1,bitA1
      bcf      regA0,bitA0
L1
      endm
```

FIGURE 14.6
The macro encod_8_3_p_E.

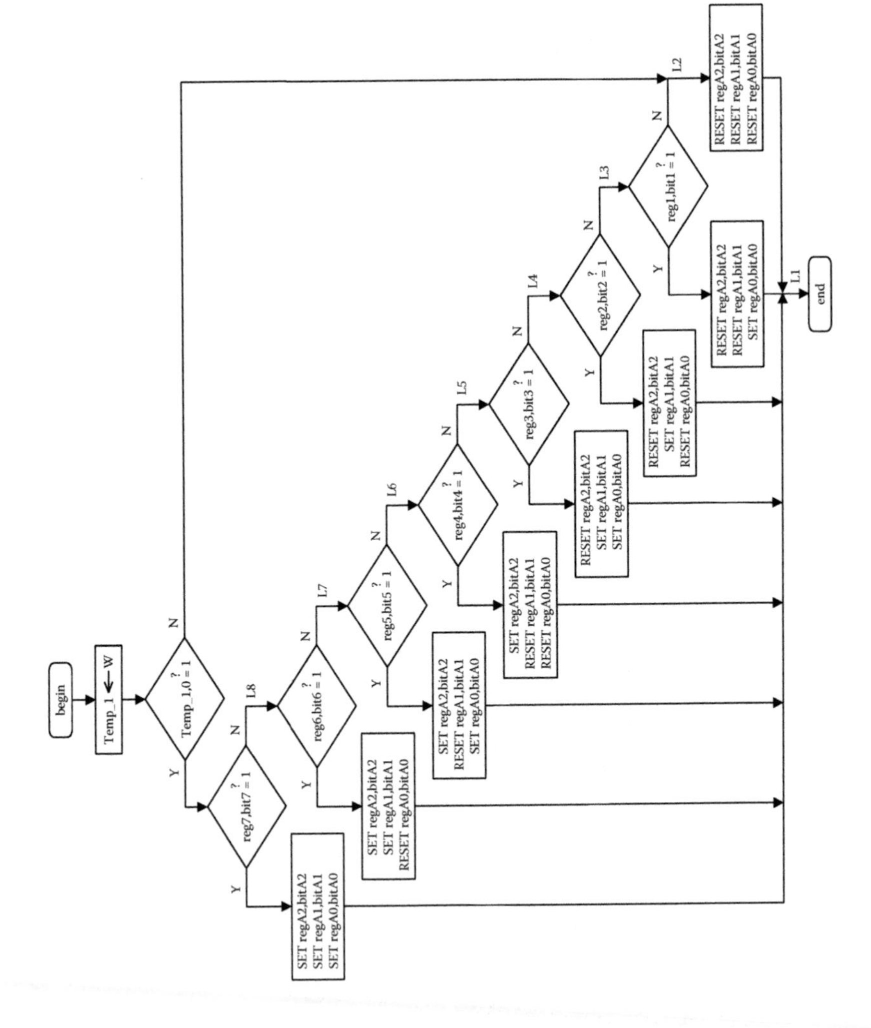

FIGURE 14.7
The flowchart of the macro encod_8_3_p_E.

TABLE 14.5

Symbol and Truth Table of the Macro `encod_dec_bcd_p`

Symbol

DECIMAL TO BCD
PRIORITY ENCODER

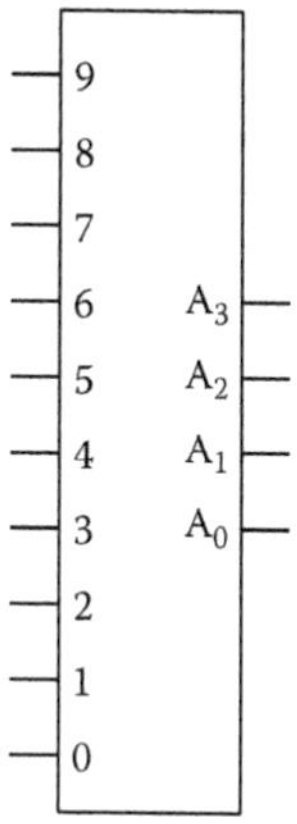

9 =	reg9,bit9
8 =	reg8,bit8
7 =	reg7,bit7
6 =	reg6,bit6
5 =	reg5,bit5
4 =	reg4,bit4
3 =	reg3,bit3
2 =	reg2,bit2
1 =	reg1,bit1
0 =	reg0,bit0
A3 =	regA3,bitA3
A2 =	regA2,bitA2
A1 =	regA1,bitA1
A0 =	regA1,bitA0

Truth Table

inputs										**outputs**			
0	**1**	**2**	**3**	**4**	**5**	**6**	**7**	**8**	**9**	**A3**	**A2**	**A1**	**A0**
×	×	×	×	×	×	×	×	×	1	1	0	0	1
×	×	×	×	×	×	×	×	1	0	1	0	0	0
×	×	×	×	×	×	×	1	0	0	0	1	1	1
×	×	×	×	×	×	1	0	0	0	0	1	1	0
×	×	×	×	×	1	0	0	0	0	0	1	0	1
×	×	×	×	1	0	0	0	0	0	0	1	0	0
×	×	×	1	0	0	0	0	0	0	0	0	1	1
×	×	1	0	0	0	0	0	0	0	0	0	1	0
×	1	0	0	0	0	0	0	0	0	0	0	0	1
1	0	0	0	0	0	0	0	0	0	0	0	0	0

×: don't care.

input line 9 has the highest priority, while the input line 0 has the lowest priority. In addition to the `encod_dec_bcd_p`, this encoder macro has an active high enable line, E, for enabling it. When this encoder is disabled with E set to 0, all output lines are set to 0. When this encoder is enabled with E set to 1, it functions as described for `encod_dec_bcd_p`. This means that when E = 1: if the input signals present in the input lines 0, 1, 2, 3, 4, 5, 6, 7, 8, 9 are as

```
encod_dec_bcd_p  macro   reg9,bit9,reg8,bit8,
reg7,bit7,reg6,bit6,reg5,bit5,reg4,bit4,
reg3,bit3,reg2,bit2,reg1,bit1,reg0,bit0,
regA3,bitA3,regA2,bitA2,regA1,bitA1,regA0,bitA0
        local   L1,L2,L3,L4,L5,L6,L7,L8,L9,L10
        btfss   reg9,bit9
        goto    L10
        bsf     regA3,bitA3
        bcf     regA2,bitA2
        bcf     regA1,bitA1
        bsf     regA0,bitA0
        goto    L1
L10     btfss   reg8,bit8
        goto    L9
        bsf     regA3,bitA3
        bcf     regA2,bitA2
        bcf     regA1,bitA1
        bcf     regA0,bitA0
        goto    L1
L9      btfss   reg7,bit7
        goto    L8
        bcf     regA3,bitA3
        bsf     regA2,bitA2
        bsf     regA1,bitA1
        bsf     regA0,bitA0
        goto    L1
L8      btfss   reg6,bit6
        goto    L7
        bcf     regA3,bitA3
        bsf     regA2,bitA2
        bsf     regA1,bitA1
        bcf     regA0,bitA0
        goto    L1
L7      btfss   reg5,bit5
        goto    L6
        bcf     regA3,bitA3
        bsf     regA2,bitA2
        bcf     regA1,bitA1
        bsf     regA0,bitA0
        goto    L1
L6      btfss   reg4,bit4
        goto    L5
        bcf     regA3,bitA3
        bsf     regA2,bitA2
        bcf     regA1,bitA1
        bcf     regA0,bitA0
        goto    L1
L5      btfss   reg3,bit3
        goto    L4
        bcf     regA3,bitA3
        bcf     regA2,bitA2
        bsf     regA1,bitA1
        bsf     regA0,bitA0
        goto    L1
L4      btfss   reg2,bit2
        goto    L3
        bcf     regA3,bitA3
        bcf     regA2,bitA2
        bsf     regA1,bitA1
        bcf     regA0,bitA0
        goto    L1
L3      btfss   reg1,bit1
        goto    L2
        bcf     regA3,bitA3
        bcf     regA2,bitA2
        bcf     regA1,bitA1
        bsf     regA0,bitA0
        goto    L1
L2      bcf     regA3,bitA3
        bcf     regA2,bitA2
        bcf     regA1,bitA1
        bcf     regA0,bitA0
L1
        endm
```

FIGURE 14.8
The macro encod_dec_bcd_p.

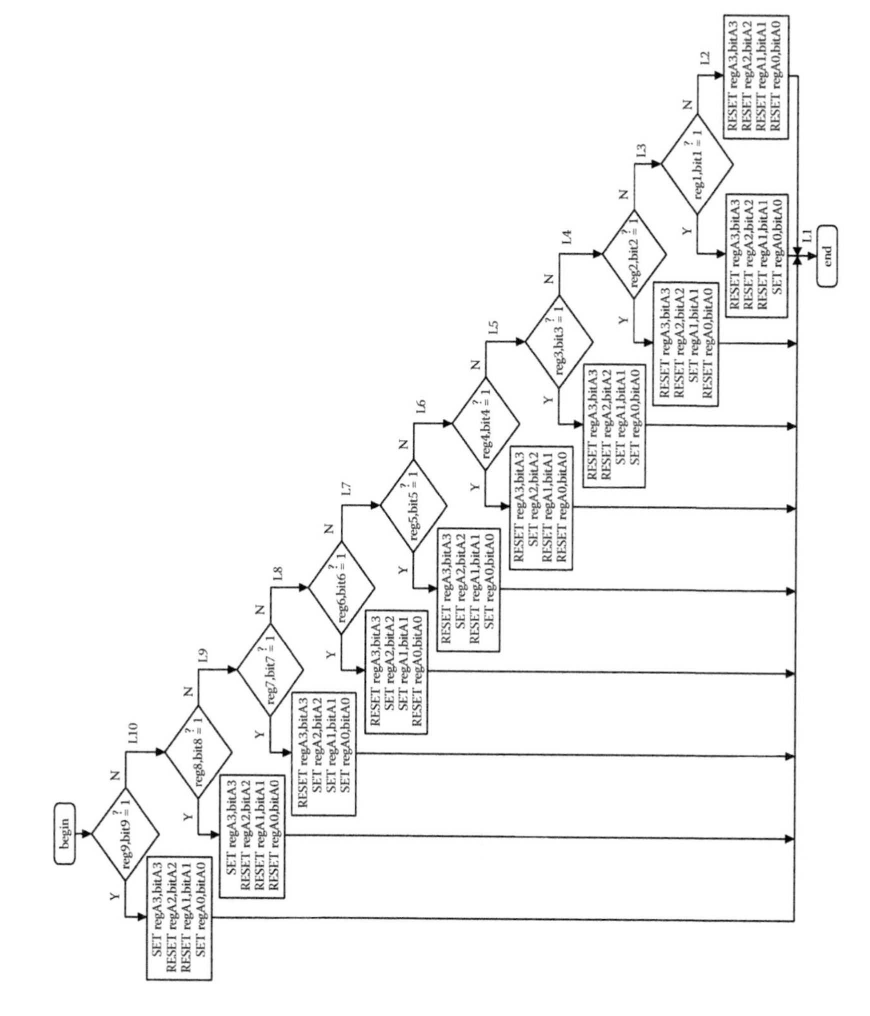

FIGURE 14.9
The flowchart of the macro encod_dec_bcd_p.

TABLE 14.6

Symbol and Truth Table of the Macro `encod_dec_bcd_p_E`

Symbol

DECIMAL TO BCD
PRIORITY ENCODER

9, 8, 7, 6, 5, 4, 3, 2, 1, 0, E, A_3, A_2, A_1, A_0

W	E
9 =	reg9,bit9
8 =	reg8,bit8
7 =	reg7,bit7
6 =	reg6,bit6
5 =	reg5,bit5
4 =	reg4,bit4
3 =	reg3,bit3
2 =	reg2,bit2
1 =	reg1,bit1
0 =	reg0,bit0
A3 =	regA3,bitA3
A2 =	regA2,bitA2
A1 =	regA1,bitA1
A0 =	regA1,bitA0

Truth Table

inputs											outputs			
E	**0**	**1**	**2**	**3**	**4**	**5**	**6**	**7**	**8**	**9**	**A3**	**A2**	**A1**	**A0**
0	×	×	×	×	×	×	×	×	×	×	0	0	0	0
1	×	×	×	×	×	×	×	×	×	1	1	0	0	1
1	×	×	×	×	×	×	×	×	1	0	1	0	0	0
1	×	×	×	×	×	×	×	1	0	0	0	1	1	1
1	×	×	×	×	×	×	1	0	0	0	0	1	1	0
1	×	×	×	×	×	1	0	0	0	0	0	1	0	1
1	×	×	×	×	1	0	0	0	0	0	0	1	0	0
1	×	×	×	1	0	0	0	0	0	0	0	0	1	1
1	×	×	1	0	0	0	0	0	0	0	0	0	1	0
1	×	1	0	0	0	0	0	0	0	0	0	0	0	1
1	1	0	0	0	0	0	0	0	0	0	0	0	0	0

×: don't care.

```
encod_dec_bcd_p_E  macro   reg9,bit9,reg8,bit8,
reg7,bit7,reg6,bit6,reg5,bit5,reg4,bit4,reg3,bit3,
reg2,bit2,reg1,bit1,reg0,bit0,regA3,bitA3,
regA2,bitA2,regA1,bitA1,regA0,bitA0
     local   L1,L2,L3,L4,L5,L6,L7,L8,L9,L10
     movwf   Temp_1
     btfss   Temp_1,0
     goto    L2
     btfss   reg9,bit9
     goto    L10
     bsf     regA3,bitA3
     bcf     regA2,bitA2
     bcf     regA1,bitA1
     bsf     regA0,bitA0
     goto    L1
L10  btfss   reg8,bit8
     goto    L9
     bsf     regA3,bitA3
     bcf     regA2,bitA2
     bcf     regA1,bitA1
     bcf     regA0,bitA0
     goto    L1
L9   btfss   reg7,bit7
     goto    L8
     bcf     regA3,bitA3
     bsf     regA2,bitA2
     bsf     regA1,bitA1
     bsf     regA0,bitA0
     goto    L1
L8   btfss   reg6,bit6
     goto    L7
     bcf     regA3,bitA3
     bsf     regA2,bitA2
     bsf     regA1,bitA1
     bcf     regA0,bitA0
     goto    L1
L7   btfss   reg5,bit5
     goto    L6
     bcf     regA3,bitA3
     bsf     regA2,bitA2
     bcf     regA1,bitA1
     bsf     regA0,bitA0
     goto    L1
L6   btfss   reg4,bit4
     goto    L5
     bcf     regA3,bitA3
     bsf     regA2,bitA2
     bcf     regA1,bitA1
     bcf     regA0,bitA0
     goto    L1
L5   btfss   reg3,bit3
     goto    L4
     bcf     regA3,bitA3
     bcf     regA2,bitA2
     bsf     regA1,bitA1
     bsf     regA0,bitA0
     goto    L1
L4   btfss   reg2,bit2
     goto    L3
     bcf     regA3,bitA3
     bcf     regA2,bitA2
     bsf     regA1,bitA1
     bcf     regA0,bitA0
     goto    L1
L3   btfss   reg1,bit1
     goto    L2
     bcf     regA3,bitA3
     bcf     regA2,bitA2
     bcf     regA1,bitA1
     bsf     regA0,bitA0
     goto    L1
L2   bcf     regA3,bitA3
     bcf     regA2,bitA2
     bcf     regA1,bitA1
     bcf     regA0,bitA0
L1
     endm
```

FIGURE 14.10
The macro encod_dec_bcd_p_E.

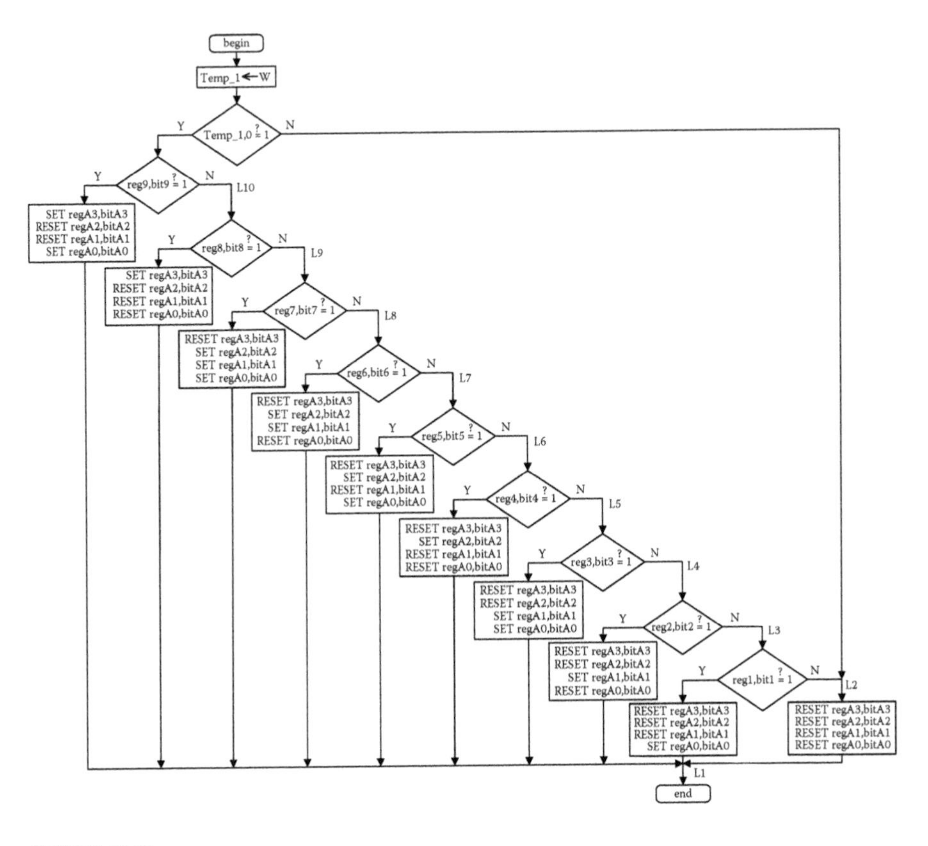

FIGURE 14.11
The flowchart of the macro `encod_dec_bcd_p_E`.

follows, ×××××××××1 (respectively, ××××××××10, ×××××××100, ××××××1000, ×××××10000, ××××100000, ×××1000000, ××10000000, ×100000000, 1000000000), then the output lines generate the following binary code: $A_3A_2A_1A_0$ = 1001 (respectively, 1000, 0111, 0110, 0101, 0100, 0011, 0010, 0001, 0000).

14.7 Examples for Priority Encoder Macros

In this section, we will consider five examples, namely, UZAM_plc_16i16o_ex*X*.asm (*X* = 29, 30, 31, 32, 33), to show the usage of priority encoder macros. In order to test one of these examples, please take the related file UZAM_plc_16i16o_ex*X*.asm (*X* = 29, 30, 31, 32, 33) from the CD-ROM attached to this book, and then open the program by MPLAB IDE and compile it. After that, by using the PIC programmer software, take the compiled file UZAM_plc_16i16o_ex*X*.hex (*X* = 29, 30, 31, 32, 33), and by your PIC programmer

```
.
;--------------- user program starts here -----------------------
        encod_4_2_p         I0.3,I0.2,I0.1,I0.0,Q0.1,Q0.0       ;rung 1

        ld                  I1.7                                ;rung 2
        encod_4_2_p_E       I1.3,I1.2,I1.1,I1.0,Q1.1,Q1.0
;--------------- user program ends here -------------------------
.
```

FIGURE 14.12
The user program of UZAM_plc_16i16o_ex29.asm.

hardware, send it to the program memory of PIC16F648A microcontroller within the PIC16F648A-based PLC. To do this, switch the 4PDT in PROG position and the power switch in OFF position. After loading the file UZAM_plc_16i16o_exX.hex (X = 29, 30, 31, 32, 33), switch the 4PDT in RUN and the power switch in ON position. Please check the program's accuracy by cross-referencing it with the related macros.

Let us now consider these example programs: The first example program, UZAM_plc_16i16o_ex29.asm, is shown in Figure 14.12. It shows the usage of two priority encoder macros, `encod_4_2_p` and `encod_4_2_p_E`. The schematic diagram of the user program of UZAM_plc_16i16o_ex29.asm, shown in Figure 14.12, is depicted in Figure 14.13.

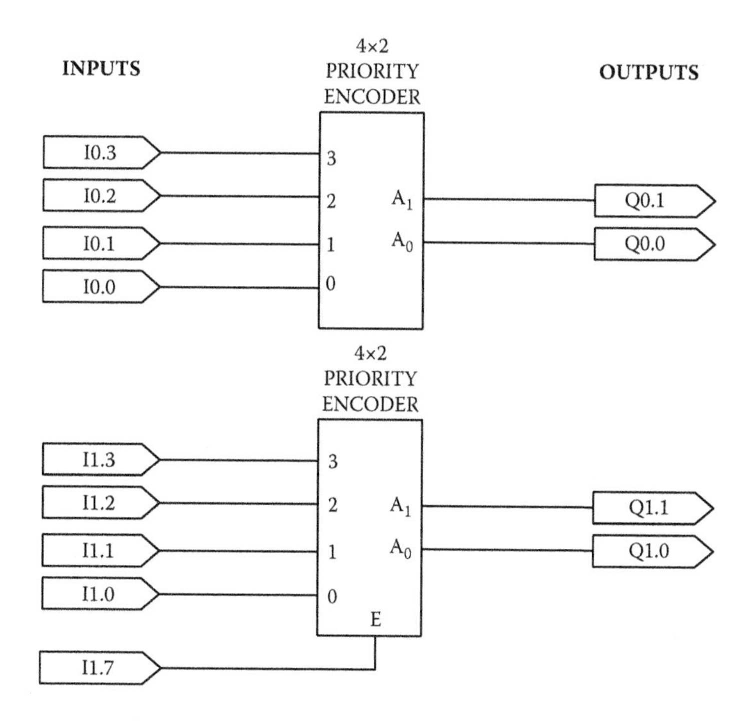

FIGURE 14.13
The schematic diagram of the user program of UZAM_plc_16i16o_ex29.asm.

```
.
;--------------- user program starts here ------------------------
encod_8_3_p   I0.7,I0.6,I0.5,I0.4,I0.3,I0.2,I0.1,I0.0,Q0.2,Q0.1,Q0.0    ;rung 1
;--------------- user program ends here --------------------------
.
```

FIGURE 14.14
The user program of UZAM_plc_16i16o_ex30.asm.

In the first rung, the priority encoder macro `encod_4_2_p` (4 × 2 priority encoder) is used. In this priority encoder, four input lines, 3, 2, 1, and 0, are defined as I0.3, I0.2, I0.1, and I0.0 respectively, while the output lines A_1 and A_0 are defined as Q0.1 and Q0.0, respectively.

In the second rung, the priority encoder macro `encod_4_2_p_E` (4 × 2 priority encoder with enable input) is used. In this priority encoder, four input lines, 3, 2, 1, and 0, are defined as I1.3, I1.2, I1.1, and I1.0, respectively, while the output lines A_1 and A_0 are defined as Q1.1 and Q1.0, respectively. In addition, the active high enable input E is defined to be E = I1.7.

The second example program, UZAM_plc_16i16o_ex30.asm, is shown in Figure 14.14. It shows the usage of the priority encoder macro `encod_8_3_p` (8 × 3 priority encoder). The schematic diagram of the user program of UZAM_plc_16i16o_ex30.asm, shown in Figure 14.14, is depicted in Figure 14.15. In this priority encoder, eight input lines, 7, 6, 5, 4, 3, 2, 1, and 0, are defined as I0.7, I0.6, I0.5, I0.4, I0.3, I0.2, I0.1, and I0.0, respectively, while the output lines A_2, A_1, and A_0 are defined as Q0.2, Q0.1, and Q0.0, respectively.

The third example program, UZAM_plc_16i16o_ex31.asm, is shown in Figure 14.16. It shows the usage of the priority encoder macro `encod_8_3_p_E` (8 × 3 priority encoder with enable input). The schematic diagram of the user program of UZAM_plc_16i16o_ex31.asm, shown in

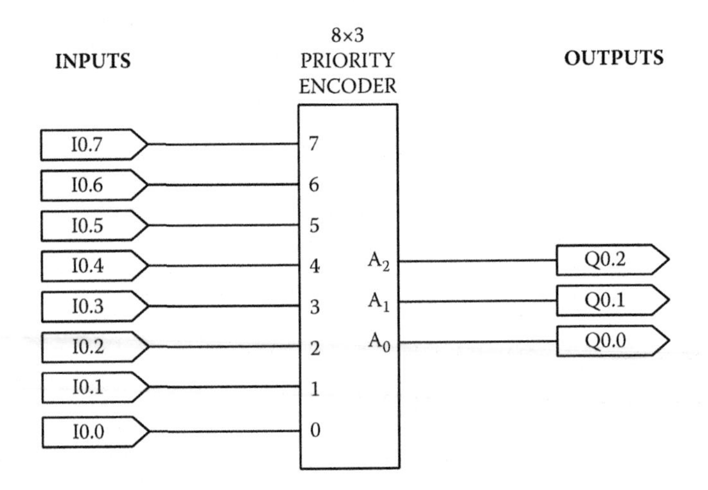

FIGURE 14.15
The schematic diagram of the user program of UZAM_plc_16i16o_ex30.asm.

```
.
;--------------- user program starts here ------------------------
ld              I1.7                                                  ;rung 1
encod_8_3_p_E   I0.7,I0.6,I0.5,I0.4,I0.3,I0.2,I0.1,I0.0,Q0.2,Q0.1,Q0.0
;--------------- user program ends here --------------------------
.
```

FIGURE 14.16
The user program of UZAM_plc_16i16o_ex31.asm.

Figure 14.16, is depicted in Figure 14.17. In this priority encoder, eight input lines, 7, 6, 5, 4, 3, 2, 1, and 0, are defined as I0.7, I0.6, I0.5, I0.4, I0.3, I0.2, I0.1, and I0.0, respectively, while the output lines A_2, A_1, and A_0 are defined as Q0.2, Q0.1, and Q0.0, respectively. In addition, the active high enable input E is defined to be E = I1.7.

The fourth example program, UZAM_plc_16i16o_ex32.asm, is shown in Figure 14.18. It shows the usage of the priority encoder macro `encod_dec_bcd_p` (decimal to BCD priority encoder). The schematic diagram of the user program of UZAM_plc_16i16o_ex32.asm, shown in Figure 14.18, is depicted in Figure 14.19. In this priority encoder, 10 input lines, 9, 8, 7, 6, 5, 4, 3, 2, 1, and 0, are defined as I1.1, I1.0, I0.7, I0.6, I0.5, I0.4, I0.3, I0.2, I0.1, and I0.0, respectively, while the output lines A_3, A_2, A_1, and A_0 are defined as Q0.3, Q0.2, Q0.1, and Q0.0, respectively.

The fifth and last example program, UZAM_plc_16i16o_ex33.asm, is shown in Figure 14.20. It shows the usage of the priority encoder macro encod_dec_bcd_p_E (decimal to BCD priority encoder with enable input).

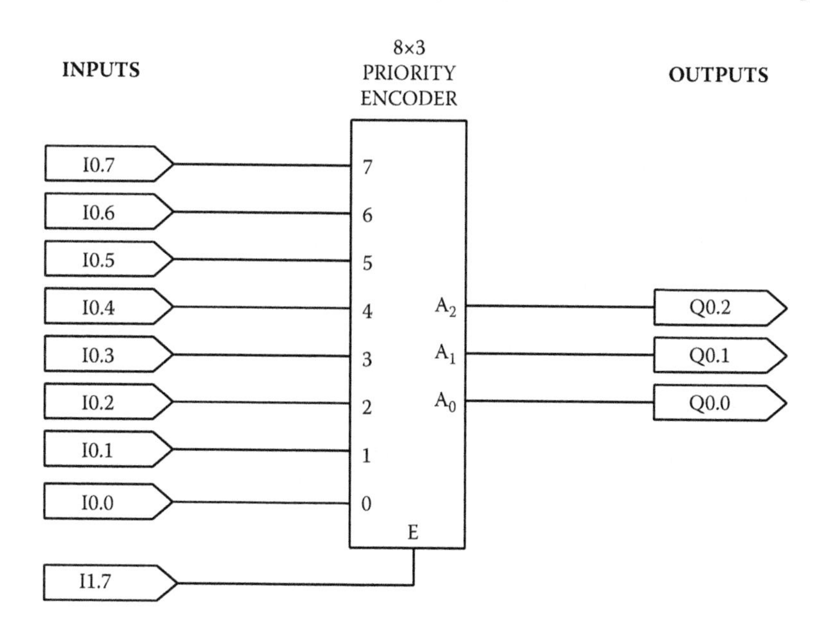

FIGURE 14.17
The schematic diagram of the user program of UZAM_plc_16i16o_ex31.asm.

```
.
;--------------- user program starts here -----------------------
encod_dec_bcd_p  I1.1,I1.0,I0.7,I0.6,I0.5,I0.4,I0.3,I0.2,I0.1,I0.0,Q0.3,Q0.2,Q0.1,Q0.0 ;rung 1
;--------------- user program ends here -------------------------
.
```

FIGURE 14.18
The user program of UZAM_plc_16i16o_ex32.asm.

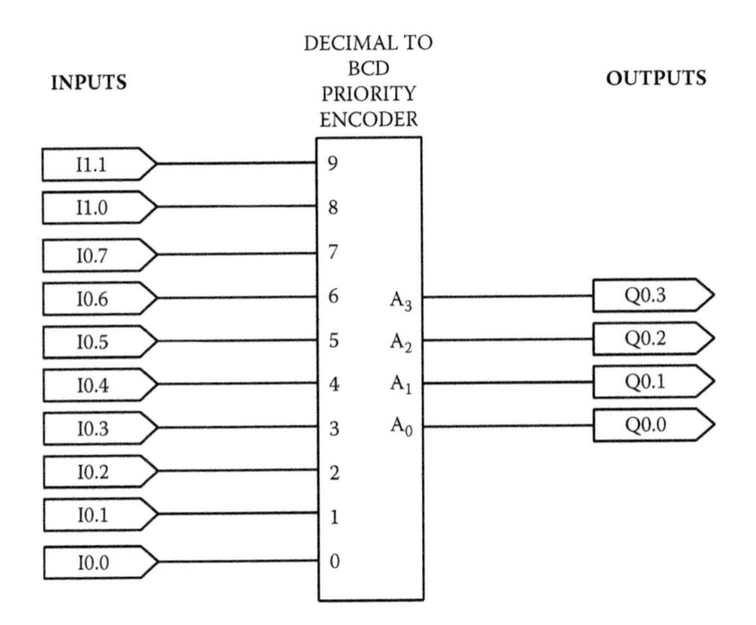

FIGURE 14.19
The schematic diagram of the user program of UZAM_plc_16i16o_ex32.asm.

```
.
;--------------- user program starts here -----------------------
ld                  I1.7                                                           ;rung 1
encod_dec_bcd_p_E   I1.1,I1.0,I0.7,I0.6,I0.5,I0.4,I0.3,I0.2,I0.1,I0.0,Q0.3,Q0.2,Q0.1,Q0.0
;--------------- user program ends here -------------------------
.
```

FIGURE 14.20
The user program of UZAM_plc_16i16o_ex33.asm.

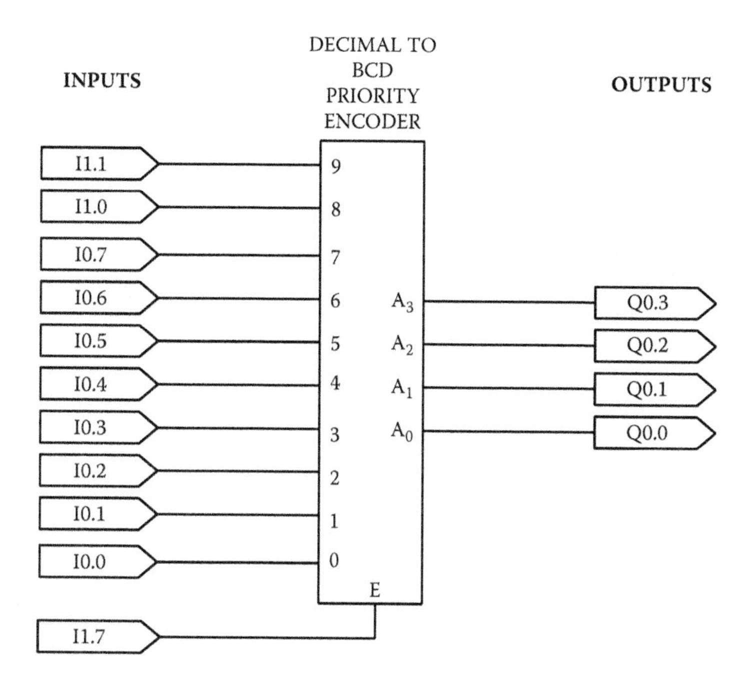

FIGURE 14.21
The schematic diagram of the user program of UZAM_plc_16i16o_ex33.asm.

The schematic diagram of the user program of UZAM_plc_16i16o_ex33.asm, shown in Figure 14.20, is depicted in Figure 14.21. In this priority encoder, 10 input lines, 9, 8, 7, 6, 5, 4, 3, 2, 1, and 0, are defined as I1.1, I1.0, I0.7, I0.6, I0.5, I0.4, I0.3, I0.2, I0.1, and I0.0, respectively, while the output lines A_3, A_2, A_1, and A_0 are defined as Q0.3, Q0.2, Q0.1, and Q0.0, respectively. In addition, the active high enable input E is defined to be E = I1.7.

15

Application Example

This chapter describes an example remotely controlled model gate system and makes use of the PIC16F648A-based PLC to control it for different control scenarios.

15.1 Remotely Controlled Model Gate System

Figure 15.1 shows the remotely controlled model gate system, used in this chapter as an example to show how the PIC16F648A-based PLC can be utilized in the control of real systems. In this system, when the DC motor turns backward (respectively forward) the gate is opened (respectively closed). To control the DC motor in backward and forward directions, PLC outputs Q0.0 and Q0.1 are used, respectively. In the system, there are two buttons, B0 and B1, and they both have only one normally open (NO) contact. When pressed, the button B0 (respectively, B1) is used to give the control system the following order: "open the gate" (respectively, "close the gate"). PLC inputs I0.0 and I0.1 are used for identifying the ON or OFF states of the buttons B0 and B1, respectively. When the gate is completely open, it applies the F1 force, shown in Figure 15.1, to the limit switch 1 (LS1). In this case, the NO contact of LS1 is closed. To detect whether or not the gate is completely open, the input I0.2 is utilized. When the gate is completely closed, it applies the F2 force, shown in Figure 15.1, to the limit switch 2 (LS2). In this case, the NO contact of LS2 is closed. To detect whether or not the gate is completely closed, the PLC input I0.3 is utilized. An infrared (IR) transmitter/receiver sensor is used to detect if there is any obstacle in the gate's path. This is very important because when the gate is closing, there should not be any obstacle in its path in order not to cause any damage to anybody or anything. When the light emitted from the IR transmitter is received from the IR receiver, the NO contact of the sensor is closed. In this case, we conclude that there is no obstacle in the path. When the light emitted from the IR transmitter is not received from the IR receiver, the NO contact of the sensor is open, i.e., in its normal condition. This means that there is an obstacle in the path. To detect whether or not there is an obstacle in the path, the PLC input I0.4 is utilized. In addition, there is also a radio frequency (RF) transmitter/receiver used as a remote control mechanism within the system. In the RF transmitter, there

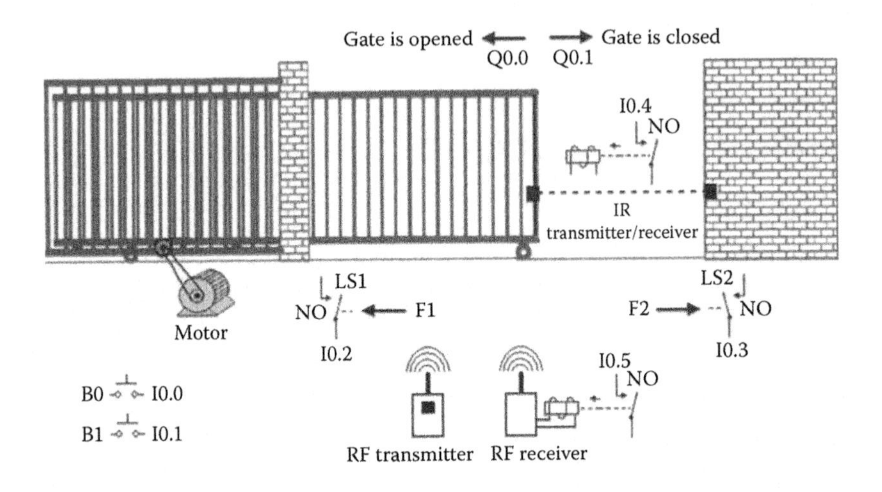

FIGURE 15.1
The remotely controlled model gate system.

is a button. When this button is pressed, the RF waves are emitted from the transmitter, and they are received from the RF receiver. In this case, NO contact at the RF receiver is closed, signaling the button press from the RF transmitter counterpart. To detect whether or not the RF transmitter button is pressed, the PLC input I0.5 is utilized.

The DC motor control circuit embedded within the model gate system is depicted in Figure 15.2, where there are two relays, Relay 1 and Relay 2,

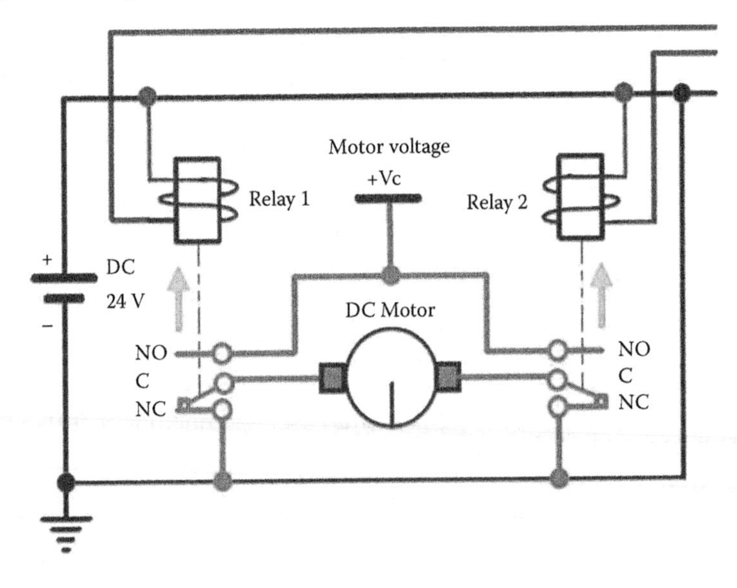

FIGURE 15.2
The DC motor control circuit embedded within the model gate system.

TABLE 15.1

State of the DC Motor Based on the Two Relays

Relay 1	Relay 2	DC Motor
OFF (Q0.1 = 0)	OFF (Q0.0 = 0)	OFF (not working)
OFF (Q0.1 = 0)	ON (Q0.0 = 1)	Turns backward (the gate is opened)
ON (Q0.1 = 1)	OFF (Q0.0 = 0)	Turns forward (the gate is closed)
ON (Q0.1 = 1)	ON (Q0.0 = 1)	OFF (not working)

operating on 24 V DC. These relays both have a single-pole double-throw (SPDT) contact, with the terminals named normally open (NO), common (C), and normally closed (NC). As can be seen, terminal C is shared between the other two contacts. The normal states of the contacts are shown in Figure 15.2. In this case, the C and NC terminals of both relays are closed, while C and NO terminals are open. If any of these relays' coils are energized, then the contacts are actuated, and thus the C and NC terminals of the relay are open, while C and NO terminals are closed. With this setup, by means of the two relays we can have the DC motor turning forward or backward, as shown in Table 15.1. It is important to note that if both relays are ON, then the DC motor will not be working. One terminal of each relay coil is connected to 24 V DC, while the other one is left unconnected. To operate any relay it is necessary to connect its open terminal to the ground of the 24 V DC. The control of the DC motor is achieved by means of the Q0.0 and Q0.1 outputs of the PLC. As can be seen from Figure 15.2, when Q0.0 is ON (and Q0.1 is OFF), the NO contact of Q0.0 will switch on Relay 2, in which case the motor turns backward and the gate is opened. Similarly, when Q0.1 is ON (and Q0.0 is OFF), the NO contact of Q0.1 will switch on Relay 1, in which case the motor turns forward and the gate is closed. Figure 15.3 shows the wiring of the PIC16F648A-based PLC with the remotely controlled model gate system. In this setup, when any of the NO contact of the model gate system is closed or a button is pressed, 5 V DC is applied to related PLC input.

15.2 Control Scenarios for the Model Gate System

In this section we will declare eight different control scenarios for the remotely controlled model gate system as follows:

1. When B0 is being pressed, the gate shall open.
2. Once B0 is pressed, the gate shall open.
3. Once B0 is pressed, the gate shall open. The motor shall stop when the gate is completely open.

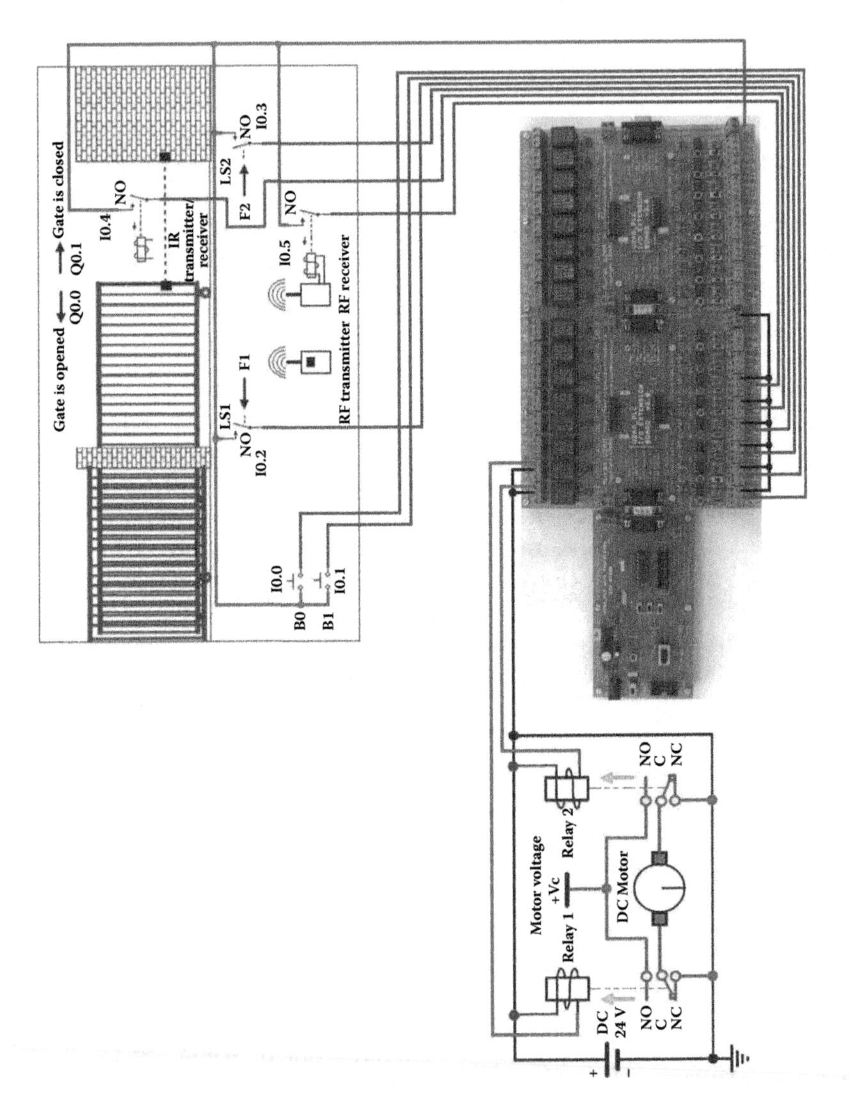

FIGURE 15.3
Wiring of the PIC16F648A-based PLC with the model gate system.

4. Once B0 is pressed, the gate shall open. The motor shall stop when the gate is completely open. Once B1 is pressed, the gate shall close. The motor shall stop when the gate is completely closed.
5. If the gate is not closing, then once B0 is pressed, the gate shall open. The motor shall stop when the gate is completely open. If the gate is not opening, then once B1 is pressed, the gate shall close. The motor shall stop when the gate is completely closed.
6. If the gate is not closing, then once B0 or the RF transmitter button is pressed, the gate shall open. The motor shall stop when the gate is completely open. When the gate is completely open, it shall wait 5 s before automatically closing. The motor shall stop when the gate is completely closed.
7. If the gate is not closing, then once B0 or the RF transmitter button is pressed, the gate shall open. The motor shall stop when the gate is completely open. When the gate is completely open, it shall wait 5 s before automatically closing. The motor shall stop when the gate is completely closed. When the gate is closing, if there is an obstacle in the gate's path, the gate shall open. In this case it shall wait 5 s before automatically closing as defined above.
8. Combine the previous seven control scenarios in a single program. By using three inputs, I1.2, I1.1, and I1.0, only one of the scenarios will be selected and will work at any time.

15.3 Solutions for the Control Scenarios

In this section, we will consider the solutions to the above-declared eight control scenarios for the remotely controlled model gate system, namely, UZAM_plc_16i16o_ex*X*.asm (*X* = 34, 35, 36, 37, 38, 39, 40, 41). In order to test one of these examples, please take the related file UZAM_plc_16i16o_ex*X*.asm (*X* = 34, 35, 36, 37, 38, 39, 40, 41) from the CD-ROM attached to this book, and then open the program by MPLAB IDE and compile it. After that, by using the PIC programmer software, take the compiled file UZAM_plc_16i16o_ex*X*.hex (*X* = 34, 35, 36, 37, 38, 39, 40, 41), and by your PIC programmer hardware send it to the program memory of PIC16F648A microcontroller within the PIC16F648A-based PLC. To do this, switch the 4PDT in PROG position and the power switch in OFF position. After loading the file UZAM_plc_16i16o_ex*X*.hex (*X* = 34, 35, 36, 37, 38, 39, 40, 41), switch the 4PDT in RUN and the power switch in ON position. Finally, you are ready to test the respective example program.

Let us now consider the example programs in the following sections.

```
.
;--------------- user program starts here -
        ld          I0.0         ;rung  1
        out         Q0.0
;--------------- user program ends here ---
.
```

FIGURE 15.4
The user program of UZAM_plc_16i16o_ex34.asm.

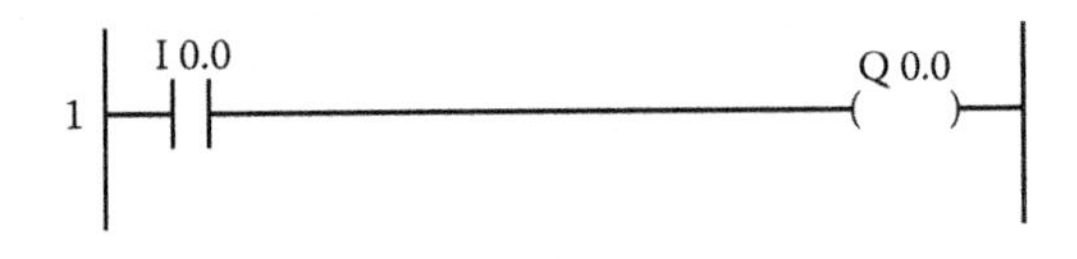

FIGURE 15.5
The ladder diagram of the user program of UZAM_plc_16i16o_ex34.asm.

15.3.1 Solution for the First Scenario

The user program of UZAM_plc_16i16o_ex34.asm, shown in Figure 15.4, is provided as a solution for the first scenario. The ladder diagram of the user program of UZAM_plc_16i16o_ex34.asm is depicted in Figure 15.5. In this example, when B0 (I0.0) is being pressed, the gate will open (Q0.0 will be ON). However, in this case, if B0 is released, then the gate will stop. This means that the program does not remember whether or not B0 was pressed.

15.3.2 Solution for the Second Scenario

The user program of UZAM_plc_16i16o_ex35.asm, shown in Figure 15.6, is provided as a solution for the second scenario. The ladder diagram of the user program of UZAM_plc_16i16o_ex35.asm is depicted in Figure 15.7. In this example, once B0 (I0.0) is pressed, with the help of NO contact Q0.0 connected parallel to NO contact I0.0, the gate will open (Q0.0 will be ON). Here,

```
.
;--------------- user program starts here --
        ld          I0.0         ;rung 1
        or          Q0.0
        out         Q0.0
;--------------- user program ends here ----
.
```

FIGURE 15.6
The user program of UZAM_plc_16i16o_ex35.asm.

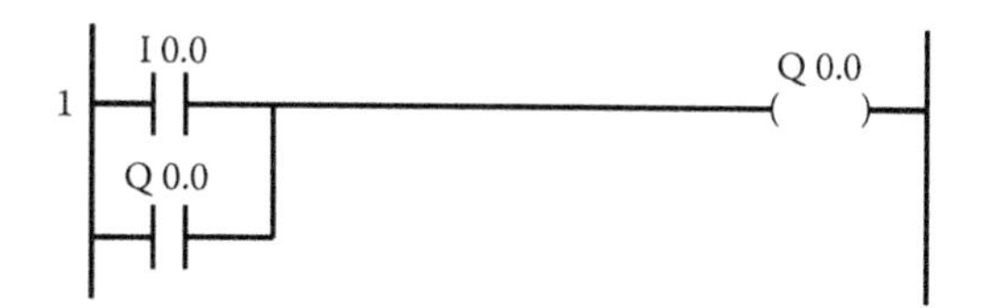

FIGURE 15.7
The ladder diagram of the user program of UZAM_plc_16i16o_ex35.asm.

the NO contact Q0.0 is a "sealing contact," and helps the program to remember whether B0 was pressed. The problem is that when the gate is completely opened, the motor will not stop.

15.3.3 Solution for the Third Scenario

The user program of UZAM_plc_16i16o_ex36.asm, shown in Figure 15.8, is provided as a solution for the third scenario. The ladder diagram of the user program of UZAM_plc_16i16o_ex36.asm is depicted in Figure 15.9. In this example, once B0 (I0.0) is pressed, with the help of NO contact Q0.0 connected parallel to NO contact I0.0, the gate will open (Q0.0 will be ON). Here, when the gate is opened completely, the motor will stop with the help of the NC contact of I0.2 inserted before the output Q0.0.

```
.
;--------------- user program starts here --
        ld          I0.0        ;rung 1
        or          Q0.0
        and_not     I0.2
        out         Q0.0
;--------------- user program ends here ----
.
```

FIGURE 15.8
The user program of UZAM_plc_16i16o_ex36.asm.

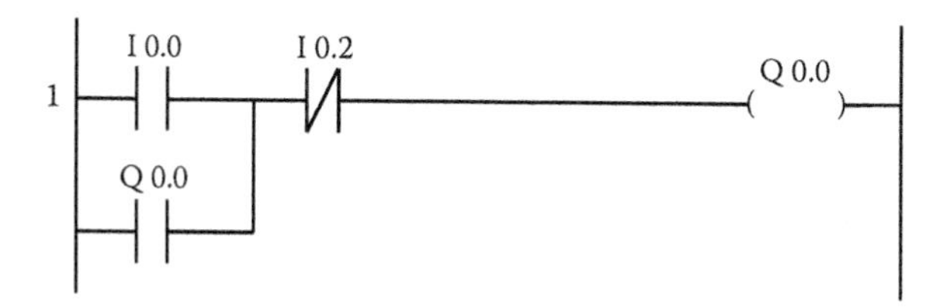

FIGURE 15.9
The ladder diagram of the user program of UZAM_plc_16i16o_ex36.asm.

```
.
;--------------- user program starts here --
        ld          I0.0        ;rung 1
        or          Q0.0
        and_not     I0.2
        out         Q0.0

        ld          I0.1        ;rung 2
        or          Q0.1
        and_not     I0.3
        out         Q0.1
;--------------- user program ends here ----
.
```

FIGURE 15.10
The user program of UZAM_plc_16i16o_ex37.asm.

15.3.4 Solution for the Fourth Scenario

The user program of UZAM_plc_16i16o_ex37.asm, shown in Figure 15.10, is provided as a solution for the fourth scenario. The ladder diagram of the user program of UZAM_plc_16i16o_ex37.asm is depicted in Figure 15.11. In this example, once B0 (I0.0) is pressed, with the help of NO contact Q0.0 connected parallel to NO contact I0.0, the gate will open (Q0.0 will be ON). Here, when the gate is opened completely, the motor will stop with the help of the NC contact of I0.2 inserted before the output Q0.0. Similarly, once B1 (I0.1) is pressed, with the help of the NO contact of Q0.1 connected parallel to NO contact I0.1, the gate will close (Q0.1 will be ON). Here, when the gate is closed completely, the motor will stop with the help of the NC contact of I0.3 inserted before the output Q0.1. The problem with this example is that if both B0 and B1 are pressed at the same time, then both outputs will be ON. This is not a desired situation. The solution to this problem is given in the next example.

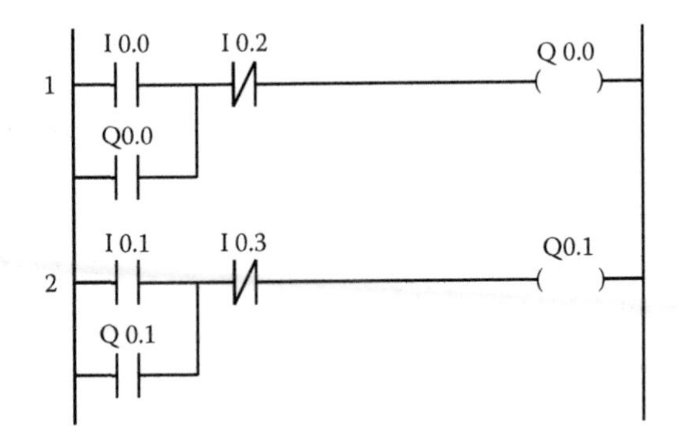

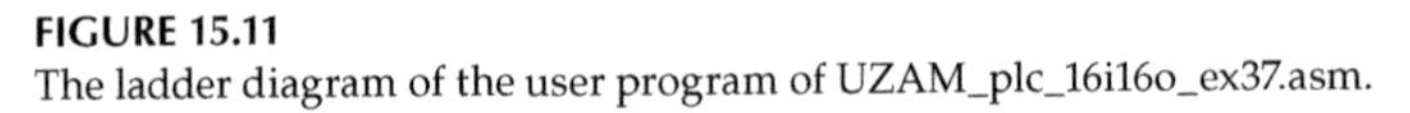

FIGURE 15.11
The ladder diagram of the user program of UZAM_plc_16i16o_ex37.asm.

```
.
;-------------- user program starts here --
        ld          I0.0        ;rung 1
        or          Q0.0
        and_not     I0.2
        and_not     Q0.1
        out         Q0.0

        ld          I0.1        ;rung 2
        or          Q0.1
        and_not     I0.3
        and_not     Q0.0
        out         Q0.1
;-------------- user program ends here ----
.
```

FIGURE 15.12
The user program of UZAM_plc_16i16o_ex38.asm.

15.3.5 Solution for the Fifth Scenario

The user program of UZAM_plc_16i16o_ex38.asm, shown in Figure 15.12, is provided as a solution for the fifth scenario. The ladder diagram of the user program of UZAM_plc_16i16o_ex38.asm is depicted in Figure 15.13. In this example, if the gate is not closing (Q0.1 = 0), once B0 (I0.0) is pressed, then the gate will open (Q0.0 will be ON) with the help of the NO contact of Q0.0 connected parallel to NO contact I0.0. In this case, when the gate is opened completely (I0.2 = 1, and therefore the NC contact of I0.2 will open), the motor will stop with the help of the NC contact of I0.2 inserted before the output Q0.0. Similarly, if the gate is not opening (Q0.0 = 0), once B1 (I0.1) is pressed, then the gate will close (Q0.1 will be ON) with the help of NO contact Q0.1 connected parallel to the NO contact of I0.1. Here, when the gate is closed completely (I0.3 = 1, and therefore the NC contact of I0.3 will open), the motor

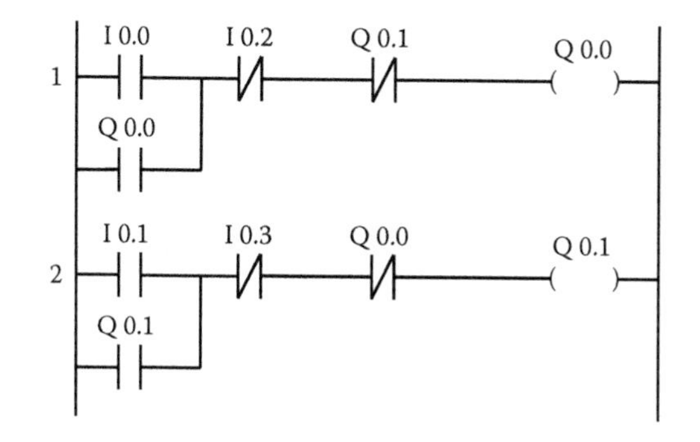

FIGURE 15.13
The ladder diagram of the user program of UZAM_plc_16i16o_ex38.asm.

will stop with the help of the NC contact of I0.3 inserted before the output Q0.1. Therefore, once the gate is being opened, we cannot force it to close, and vice versa.

15.3.6 Solution for the Sixth Scenario

The user program of UZAM_plc_16i16o_ex39.asm, shown in Figure 15.14, is provided as a solution for the sixth scenario. The ladder diagram of the user program of UZAM_plc_16i16o_ex39.asm is depicted in Figure 15.15. In this example, if the gate is not closing (Q0.1 = 0), once B0 (I0.0) or the RF transmitter button (I0.5) is pressed, then the gate will open (Q0.0 will be ON) with the help of the NO contact of Q0.0 connected parallel to the NO contact of I0.0. In this case, when the gate is opened completely (I0.2 = 1, and therefore the NC contact of I0.2 will open), the motor will stop with the help of the NC contact of I0.2 inserted before the output Q0.0. When the gate is completely open (I0.2 = 1), an on-delay timer (TON_8) is used to obtain a (100 × 52.4288 ms) 5.24 s time delay. After waiting 5.24 s, the status bit TON8_Q0 of the on-delay timer becomes true. If the gate is not opening (Q0.0 = 0), and if the NO contact of TON_8Q0 is closed (i.e., 5.24 s time delay has elapsed), then the gate will close (Q0.1 will be ON) with the help of the NO contact of Q0.1 connected parallel to the NO contact of TON8_Q0. Here, when the gate is

```
;
;--------------- user program starts here --
        ld          I0.0            ;rung 1
        or          I0.5
        or          Q0.0
        and_not     I0.2
        and_not     Q0.1
        out         Q0.0

        ld          I0.2            ;rung 2
        TON_8       0,T1.1,.100

        ld          TON8_Q0         ;rung 3
        or          Q0.1
        and_not     I0.3
        and_not     Q0.0
        out         Q0.1
;--------------- user program ends here ----
;
```

FIGURE 15.14
The user program of UZAM_plc_16i16o_ex39.asm.

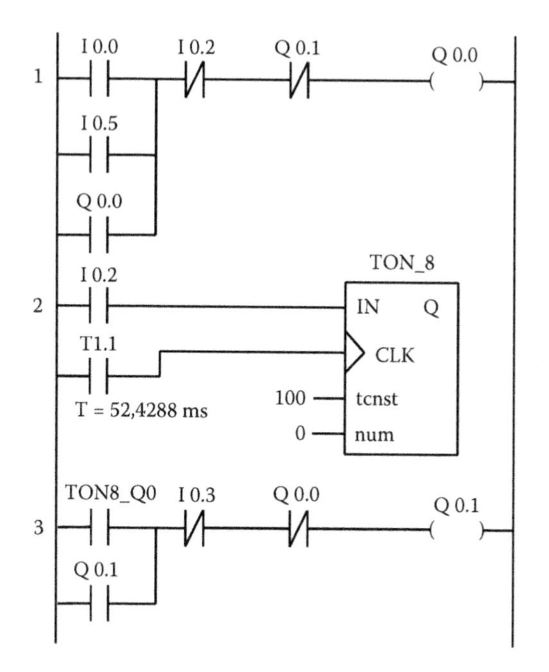

FIGURE 15.15

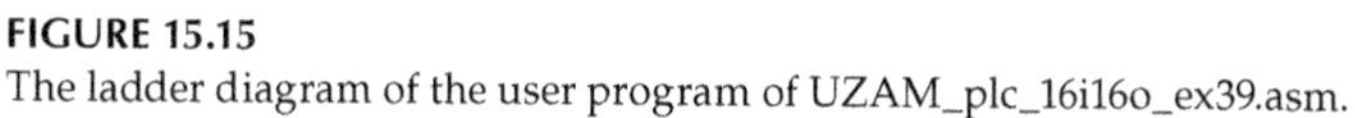
The ladder diagram of the user program of UZAM_plc_16i16o_ex39.asm.

closed completely (I0.3 = 1, and therefore the NC contact of I0.3 will open), the motor will stop with the help of the NC contact of I0.3 inserted before the output Q0.1.

15.3.7 Solution for the Seventh Scenario

The user program of UZAM_plc_16i16o_ex40.asm, shown in Figure 15.16, is provided as a solution for the seventh scenario. The ladder diagram of the user program of UZAM_plc_16i16o_ex40.asm is depicted in Figure 15.17. In this example, if the gate is not closing (Q0.1 = 0), once B0 (I0.0) or the RF transmitter button (I0.5) is pressed, then the gate will open (Q0.0 will be ON) with the help of NO contact Q0.0 connected parallel to NO contact I0.0. In this case, when the gate is opened completely (I0.2 = 1, and therefore the NC contact of I0.2 will open), the motor will stop with the help of the NC contact of I0.2 inserted before the output Q0.0. If the gate is closing (Q0.1 = 1) and the presence of an obstacle is detected in the gate's path (I0.4 = 0), then the gate will open (Q0.0 will be ON). When the gate is completely open (I0.2 = 1), an on-delay timer (TON_8) is used to obtain a

```
.
;--------------- user program starts here --
        ld          Q0.1                ;rung 1
        and_not     I0.4
        out         M0.0

        ld          I0.0                ;rung 2
        or          I0.5
        or          Q0.0
        and_not     I0.2
        and_not     Q0.1
        or          M0.0
        out         Q0.0

        ld          I0.2                ;rung 3
        TON_8       0,T1.1,.100

        ld          TON8_Q0             ;rung 4
        or          Q0.1
        and_not     I0.3
        and_not     Q0.0
        and         I0.4
        out         Q0.1
;--------------- user program ends here ----
.
```

FIGURE 15.16
The user program of UZAM_plc_16i16o_ex40.asm.

(10 × 52.4288 ms) 5.24 s time delay. After waiting 5.24 s, the status bit TON8_Q0 of the on-delay timer becomes true. If the gate is not opening (Q0.0 = 0), and if the NO contact of TON8_Q0 is closed (i.e., the 5.24 s time delay has elapsed), then the gate will close (Q0.1 will be ON) with the help of NO contact Q0.1 connected parallel to the NO contact of TON8_Q0. Here, when the gate is closed completely (I0.3 = 1, and therefore the NC contact of I0.3 will open), the motor will stop with the help of the NC contact of I0.3 inserted before the output Q0.1. If the gate is closing (Q0.1 = 1) and the presence of an obstacle is detected in the gate's path (I0.4 = 0), then the output Q0.1 will be switched OFF by means of the NO contact of I0.4 inserted before the output Q0.1.

15.3.8 Solution for the Eighth Scenario

In this last solution, the previous seven solutions are all combined in a single program. In order to choose one of the previous solutions, three inputs, I1.2, I1.1, and I1.0, are used. Table 15.2 shows the selected scenarios based on the logic signals applied to these three inputs.

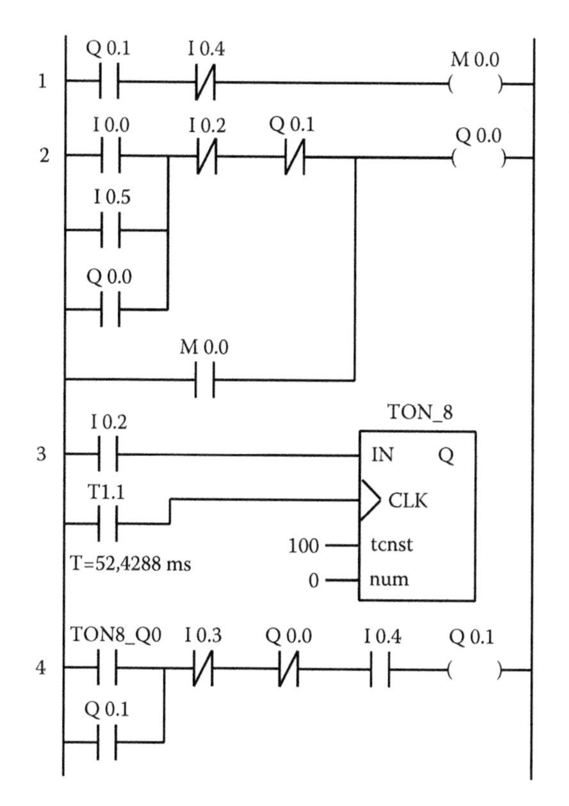

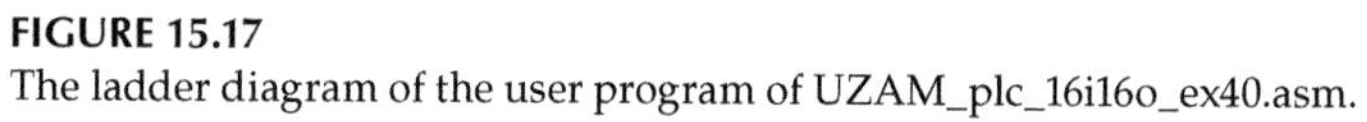

FIGURE 15.17
The ladder diagram of the user program of UZAM_plc_16i16o_ex40.asm.

TABLE 15.2
Scenarios Chosen Based on the Input Signals

Input Signals			Selected Memory Bit	Chosen Scenario
I1.2	**I1.1**	**I1.0**		
0	0	0	M0.0	—
0	0	1	M0.1	1
0	1	0	M0.2	2
0	1	1	M0.3	3
1	0	0	M0.4	4
1	0	1	M0.5	5
1	1	0	M0.6	6
1	1	1	M0.7	7

The user program of UZAM_plc_16i16o_ex41.asm, shown in Figure 15.18, is provided as a solution for the eighth scenario. The ladder diagram of the user program of UZAM_plc_16i16o_ex41.asm is depicted in Figure 15.19. In the first rung, a 3 × 8 decoder is implemented, whose inputs are I1.2, I1.1, and I1.0, and whose outputs are markers M0.0, M0.1, M0.2, M0.3, M0.4, M0.5, M0.6, and M0.7. The Boolean signals applied to the inputs

```
;--------------- user program starts here -----------------------
;--------------- code block for 3x8 decoder ------------
   decod_3_8 I1.2,I1.1,I1.0,M0.7,M0.6,M0.5,M0.4,M0.3,M0.2,M0.1,M0.0;rung 1

;----------------------------------------------------------
;--------------- code block for the 1st scenario -----------
   ld         I0.0                                           ;rung 2
   and        M0.1
   out        M1.1
;----------------------------------------------------------
;--------------- code block for the 2nd scenario ---------------
   ld         I0.0                                           ;rung 3
   or         Q0.0
   and        M0.2
   out        M1.2
;----------------------------------------------------------
;--------------- code block for the 3rd scenario -----------
   ld         I0.0                                           ;rung 4
   or         Q0.0
   and_not    I0.2
   and        M0.3
   out        M1.3
;----------------------------------------------------------
;--------------- code block for the 4th scenario -----------
   ld         I0.0                                           ;rung 5
   or         Q0.0
   and_not    I0.2
   and        M0.4
   out        M1.4

   ld         I0.1                                           ;rung 6
   or         Q0.1
   and_not    I0.3
   and        M0.4
   out        M2.4
;----------------------------------------------------------
;--------------- code block for the 5th scenario -----------
   ld         I0.0                                           ;rung 7
   or         Q0.0
   and_not    I0.2
   and_not    Q0.1
   and        M0.5
   out        M1.5

   ld         I0.1                                           ;rung 8
   or         Q0.1
   and_not    I0.3
   and_not    Q0.0
   and        M0.5
   out        M2.5
;----------------------------------------------------------
```

FIGURE 15.18
The user program of UZAM_plc_16i16o_ex41.asm. (*Continued*)

```
;--------------- code block for the 6th scenario -----------
    ld        I0.0                                          ;rung 9
    or        I0.5
    or        M1.6
    and_not   I0.2
    and_not   Q0.1
    and       M0.6
    out       M1.6

    ld        I0.2                                          ;rung 10
    and       M0.6
    TON_8     0,T1.1,.100

    ld        TON8_Q0                                       ;rung 11
    or        M2.6
    and_not   I0.3
    and_not   M1.6
    and       M0.6
    out       M2.6
;----------------------------------------------------------
;--------------- code block for the 7th scenario -----------
    ld        Q0.1                                          ;rung 12
    and_not   I0.4
    and       M0.7
    out       M3.0

    ld        I0.0                                          ;rung 13
    or        I0.5
    or        M1.7
    and_not   I0.2
    and_not   M2.7
    or        M3.0
    and       M0.7
    out       M1.7

    ld        I0.2                                          ;rung 14
    and       M0.7
    TON_8     1,T1.1,.100

    ld        TON8_Q1                                       ;rung 15
    or        M2.7
    and_not   I0.3
    and_not   M1.7
    and       I0.4
    and       M0.7
    out       M2.7
;----------------------------------------------------------
;--------------- code block for outputs --------------------
    ld        M1.1                                          ;rung 16
    or        M1.2
    or        M1.3
    or        M1.4
    or        M1.5
    or        M1.6
    or        M1.7
    out       Q0.0

    ld        M2.4                                          ;rung 17
    or        M2.5
    or        M2.6
    or        M2.7
    out       Q0.1
;----------------------------------------------------------
;--------------- user program ends here -------------------------
.
```

FIGURE 15.18 (*Continued*)
The user program of UZAM_plc_16i16o_ex41.asm.

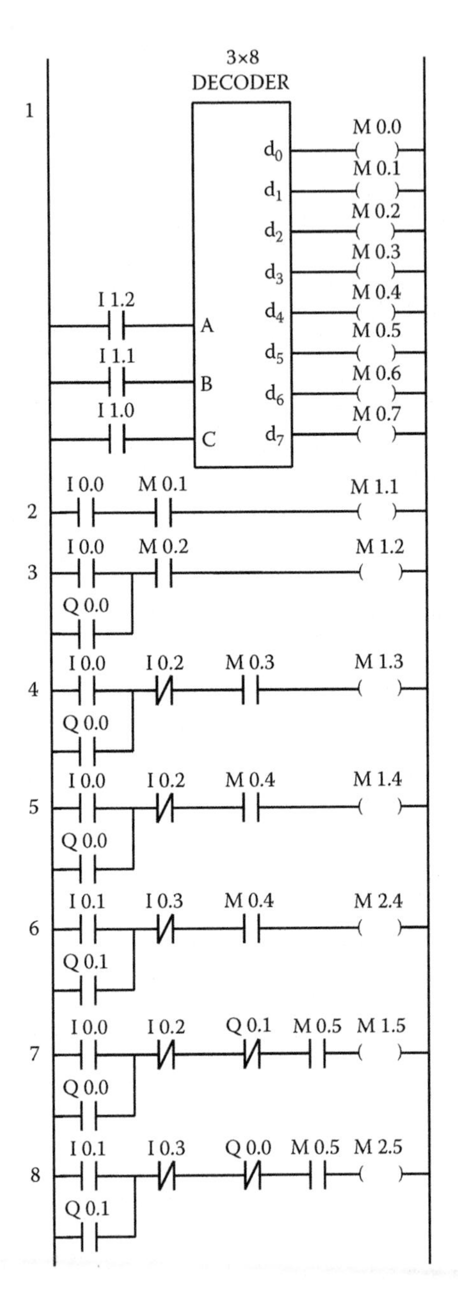

FIGURE 15.19
The ladder diagram of the user program of UZAM_plc_16i16o_ex41.asm. (*Continued*)

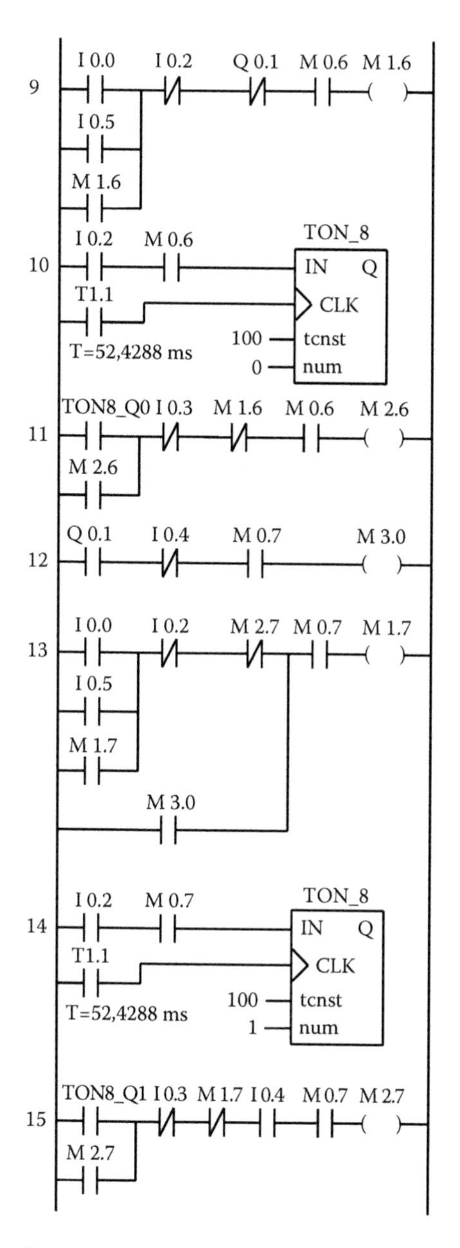

FIGURE 15.19 (*Continued*)
The ladder diagram of the user program of UZAM_plc_16i16o_ex41.asm. (*Continued*)

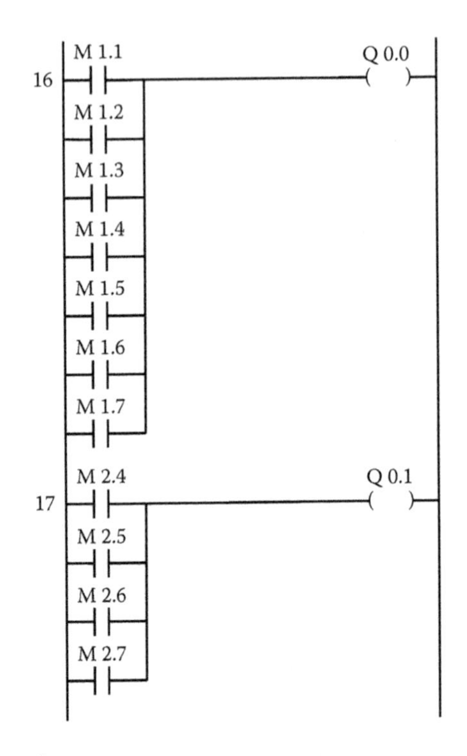

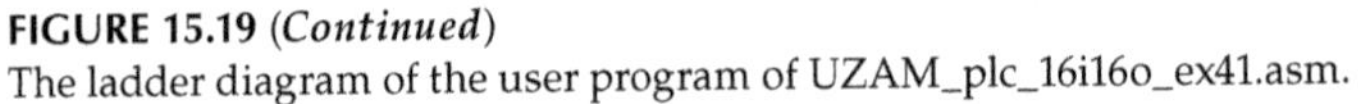
FIGURE 15.19 (*Continued*)
The ladder diagram of the user program of UZAM_plc_16i16o_ex41.asm.

I1.2, I1.1, and I1.0 select one of the outputs, and that particular output represents one of the scenarios as shown in Table 15.2. If I1.2,I1.1,I1.0 = 000 (respectively, 001, 010, 011, 100, 101, 110, and 111), then M0.0 (respectively, M0.1, M0.2, M0.3, M0.4, M0.5, M0.6, and M0.7) is set to 1. When M0.0 = 1, none of the scenarios are selected. When M0.1 (respectively, M0.2, M0.3, M0.4, M0.5, M0.6, and M0.7) is set, the code block for the first (respectively, second, third, fourth, fifth, sixth, seventh) scenario is activated, shown in rung 2 (respectively, 3; 4; 5 and 6; 7 and 8; 9, 10, and 11; 12, 13, 14, and 15). In this example, in order to operate the motor backward and forward, PLC outputs Q0.0 and Q0.1 are used as shown in rungs 16 and 17, respectively.

About the CD-ROM

The CD-ROM accompanying this book contains source files (.ASM) and object files (.HEX) of all the examples in the book. In addition, printed circuit board (PCB) (gerber and .pdf) files are also provided in order for the reader to obtain both the CPU board and I/O extension boards produced by a PCB manufacturer. A skilled reader may produce his or her own boards by using the provided .pdf files.

The files on the CD-ROM are organized in the following folders:

EXAMPLES
- PLC definitions (definitions.inc)
- Example source files (.ASM)
- Example object files (.HEX)

PIC16F648A_Based_PLC_16I_16O
- Web-based explanation of the PIC16F648A-based PLC project including
  - The schematic diagram of the CPU board
  - Photographs of the CPU board
  - The schematic diagram of the I/O extension board
  - Photographs of the I/O extension board
  - PCB design files for the CPU board (gerber files and .pdf files)
- PCB design files for the I/O extension board (gerber files and .pdf files)

References

M. Uzam. PLC with PIC16F648A Microcontroller—Part 1. *Electronics World*, 114(1871), 21–25, 2008.

M. Uzam. PLC with PIC16F648A Microcontroller—Part 2. *Electronics World*, 114(1872), 29–35, 2008.

M. Uzam. PLC with PIC16F648A Microcontroller—Part 3. *Electronics World*, 115(1873), 30–34, 2009.

M. Uzam. PLC with PIC16F648A Microcontroller—Part 4. *Electronics World*, 115(1874), 34–40, 2009.

M. Uzam. PLC with PIC16F648A Microcontroller—Part 5. *Electronics World*, 115(1875), 30–33, 2009.

M. Uzam. PLC with PIC16F648A Microcontroller—Part 6. *Electronics World*, 115(1876), 26–30, 2009.

M. Uzam. PLC with PIC16F648A Microcontroller—Part 7. *Electronics World*, 115(1877), 30–32, 2009.

M. Uzam. PLC with PIC16F648A Microcontroller—Part 8. *Electronics World*, 115(1878), 30–32, 2009.

M. Uzam. PLC with PIC16F648A Microcontroller—Part 9. *Electronics World*, 115(1879), 29–34, 2009.

M. Uzam. PLC with PIC16F648A Microcontroller—Part 10. *Electronics World*, 115(1880), 29–34, 2009.

M. Uzam. PLC with PIC16F648A Microcontroller—Part 11. *Electronics World*, 115(1881), 38–42, 2009.

M. Uzam. PLC with PIC16F648A Microcontroller—Part 12. *Electronics World*, 115(1882), 36–41, 2009.

M. Uzam. PLC with PIC16F648A Microcontroller—Part 13. *Electronics World*, 115(1883), 42–44, 2009.

M. Uzam. PLC with PIC16F648A Microcontroller—Part 14. *Electronics World*, 115(1884), 40–42, 2009.

M. Uzam. PLC with PIC16F648A Microcontroller—Part 15. *Electronics World*, 116(1885), 35–39, 2010.

M. Uzam. PLC with PIC16F648A Microcontroller—Part 16. *Electronics World*, 116(1886), 41–42, 2010.

M. Uzam. PLC with PIC16F648A Microcontroller—Part 17. *Electronics World*, 116(1887), 41–43, 2010.

M. Uzam. PLC with PIC16F648A Microcontroller—Part 18. *Electronics World*, 116(1888), 41–43, 2010.

M. Uzam. PLC with PIC16F648A Microcontroller—Part 19. *Electronics World*, 116(1889), 39–43, 2010.

M. Uzam. PLC with PIC16F648A Microcontroller—Part 20. *Electronics World*, 116(1890), 38–40, 2010.

M. Uzam. PLC with PIC16F648A Microcontroller—Part 21. *Electronics World*, 116(1891), 40–41, 2010.

M. Uzam. PLC with PIC16F648A Microcontroller—Part 22. *Electronics World,* 116 (1892), 40–42, 2010.

M. Uzam. The earlier version of the PIC16F648A based PLC project as published in *Electronics World* magazine is available from http//www.meliksah.edu.tr/muzam/UZAM_PLC_with_PIC16F648A.htm.

PIC16F627A/628A/648A Data Sheet. DS40044F. Microchip Technology, Inc., 2007. http://ww1.microchip.com/downloads/en/devicedoc/40044f.pdf.

MPASM™ Assembler, MPLINK™ Object Linker, MPLIB™ Object Librarian User's Guide. DS33014J. Microchip Technology, Inc., 2005. http://ww1.microchip. com/downloads/en/devicedoc/33014j.pdf.

Index